Sustainable Land Management in Greater Central Asia

Central Asia encompasses a vast area that includes deserts, natural grasslands, steppes, shrub lands and alpine regions. Many of these land types are degraded and productivity is falling at a time when human populations and livestock inventories are on the rise. Ecosystem stability and biodiversity are under threat and there is an urgent need to develop more sustainable land management regimes. This book uses an integrated regional approach to provide a comprehensive exploration of sustainable land development in Central Asia. An interdisciplinary team of experts analyses the economic, ecological, sociological, technological and political factors surrounding sustainable land and water management in the region, sharing potential problems and solutions. As international concern about desertification grows, the book concludes by asking how the region is likely to develop in the future. This book will be of value to scholars, students, policy makers and NGOs with an interest in sustainable development in Central Asia.

Victor R. Squires is a Guest Professor in the Institute of Desertification Studies, Beijing and an Adjunct Professor at the University of Arizona, Tucson, USA.

Lu Qi is Director and Chief Researcher of the Institute of Desertification Studies, Chinese Academy of Forestry, China.

Routledge Studies in Asia and the Environment

The role of Asia will be crucial in tackling the world's environmental problems. The primary aim of this series is to publish original, high-quality, research-level work by scholars in both the East and the West on all aspects of Asia and the environment. The series aims to cover all aspects of environmental issues, including how these relate to economic development, sustainability, technology, society, and government policies; and to include all regions of Asia.

Sustainable Land Management in Greater Central Asia

An Integrated and Regional Perspective

**Edited by Victor R. Squires
and Lu Qi**

Routledge
Taylor & Francis Group

LONDON AND NEW YORK

First published 2018 by Routledge

2 Park Square, Milton Park, Abingdon, Oxfordshire OX14 4RN
52 Vanderbilt Avenue, New York, NY 10017

Routledge is an imprint of the Taylor & Francis Group, an informa business

First issued in paperback 2019

British Library Cataloguing in Publication Data
A catalogue record for this book is available from the British Library

Library of Congress Cataloging in Publication Data
A catalog record for this book has been requested

ISBN: 978-1-138-93216-6 (hbk)
ISBN: 978-0-367-87266-3 (pbk)

Typeset in Times New Roman
by Swales & Willis Ltd, Exeter, Devon, UK

Contents

About the editors

Dr. Victor Squires is a Guest Professor in the Institute of Desertification Studies, Beijing, China. He has a PhD in Rangeland Science from Utah State University, USA. He is a former Foundation Dean of the Faculty of Natural Resource Management at the University of Adelaide, where he worked for 15 years after a 22-year career in Australia's CSIRO. He is author/editor of 12 books, including *River Basin Management in the Twenty-first Century: Understanding People and Place* (2014) and *Rangeland Ecology, Management and Conservation Benefits* (2015). Since his retirement from the University of Adelaide, Dr. Squires was a Visiting Fellow in the East West Center, Hawaii, and an Adjunct Professor in the University of Arizona, Tucson and at the Gansu Agricultural University, Lanzhou, China. He has been a consultant to the World Bank, Asian Development Bank and various UN agencies in Africa, China, Central Asia and the Middle East.

He was awarded the International Award and Gold Medal for International Science and Technology Cooperation in 2008 by the Government of China and in 2011 was awarded the Friendship Award by the Government of China. The Gold Medal is the highest award for foreigners. In 2015, Dr. Squires was honored by the Society for Range Management (USA) with an Outstanding Achievement Award.

Professor Lu Qi is Chief Scientist and Director of the Institute of Desertification Studies (IDS) in Beijing, China. The IDS is one of two such institutes in the world. Dr. Lu has a PhD in Ecology from the Chinese Academy of Sciences, Beijing and is a specialist in dryland sustainable development, integrated ecosystem management/desert ecology, and combating desertification and land degradation. He is the recipient of several prestigious awards. Dr. Lu is author/editor of 20 books, notably *A Study of Kumtagh Desert* (in Chinese; 2012), *Desert Plants in China* (illustrated handbook, in Chinese; 2012) and *Rangeland Degradation and Recovery in China's Pastoral Lands* (2009).

Foreword

Sustainable Land Management in Greater Central Asia: An Integrated and Regional Perspective is a timely book, coming as it does at the start of major global shifts to strengthen sustainable development and combat climate change. The book takes a critical look at the nine countries in Greater Central Asia, tackles knotty policy questions and stakes out a strong position on the significance of land degradation neutrality for the future of the region. Few regions have done that or carried out regional analyses of this kind, which are vital for effective, pragmatic action in combatting land degradation.

The challenges the region faces bear strong similarities to those faced by populations in many other parts of the world – large aging populations or a growing bubble of young people, the threats of forced migration and of ecosystem change as well as the external pressures to exploit the untapped land resources. These are complex policy dilemmas but they offer possibilities for transformative change.

The post-2015 dialogue calls for a bold vision that is flexible, innovative and grounded in practical action. The prioritization of sustainable land and water management in the Greater Central Asia (GCA) countries in the context of achieving land degradation neutrality provides a road map for change. It offers employment possibilities for young people. It promises to secure ecosystems and resources that are threatened by climate change. And the land degradation neutrality target ensures governments and all land users, including speculators, are held accountable for restoring an equal amount of the land they use and degrade so that the amount of productive land available today remains stable going forward.

Future generations have the right to the wealth of resources that have served us so well. We have a duty to do the right thing and the sense of urgency to address land degradation on a global level is evident.

A month after the adoption of the Sustainable Development Goals in 2015, the Parties to the UN Convention to Combat Desertification agreed to achieve land degradation neutrality by 2030 and to set up effective drought monitoring and mitigation policies. At the Paris Climate Conference more than 100 countries signed up to address climate change mitigation through land-based approaches. More than 100 countries will also focus on land as an entry point for adaptation activities. Within the first six months of 2016, more than 90 countries have signed up to set their targets for land degradation neutrality.

But the task ahead cannot be tackled by one agency alone. Consistent and concerted action by all, from the individual level right up to partnerships such as the GCA, are vital. This publication serves an important policy role. Its insights should stimulate activism and mobilize political will to advance this momentum within GCA, and beyond.

Monique Barbut
Executive Secretary, UNCCD
Bonn, June 2016

Scope and purpose of the book

This book represents, in part, a response to the *Rio+20* document *The Future We Want*, the UN Sustainability Conference in New York, and the 2015 Paris Climate Conference, in that it focuses on the issues around sustainable land management (SLM) and use in the post-2015 context.

This book has been prepared as a contribution to promoting the management of Greater Central Asia's landscapes in a way that is more environmentally sustainable, economically profitable and socially equitable. The primary audience/readership are the people who shape, plan, move and implement policies that affect the sustainable use of natural resources (including humans) in the vast region designated as Greater Central Asia (GCA). This region (as defined by UNESCO) includes Mongolia, Western China and Afghanistan as well as the five 'stans' that were, until the early 1990s, part of the now defunct Soviet Union.

The principal aims of the book are to assess and, as far as possible, quantify the effects of current practices and policies on the development and sustained use of this region that is destined to become a new development frontier because of its strategic location and proximity to giants such as India, China and the Russia Federation and its underdeveloped resources such as arable land, water and energy (especially petroleum) and strategic minerals. In particular, we want to identify ecological, social and economic constraints to development; and to identify the nature of policy changes necessary to enhance prospects for economic growth and development throughout the GCA region.

The book focuses on a number of 'issues' that are addressed as constraints to implementing SLM, including demographic pressures, land tenure arrangements (including use rights) and the effects of international trade arrangements and the land management policies of other countries both in the region and elsewhere. Special emphasis is given to China's initiative 'Belt and Road', which will bring enormous infrastructure development and open a new era of international cooperation and development to GCA.

After a brief introduction to GCA landscapes and societies, the book considers determinants of and constraints to resource utilization. Current issues such as population, political instability, and misplaced government optimism, desertification and land degradation, lost cultural identity are considered. We provide an overview of policy challenges especially those arising from institutional and administrative issues. Resource use rights are given consideration, especially the legal frameworks for land use (including land privatization), land reforms currently being considered, measures to empower minority groups, and their relationship to common property systems. We also consider the opportunities to provide incentives for sustainable land management with special reference to recognition of the natural potential of GCA's vast land and water resources.

The book is in five parts.

Part I, of three chapters, is a set of background chapters that introduce the vastness of the region, its geography, climatic features, and cultural/ethnic diversity, and traces the history of the GCA region. This part allows the reader to appreciate the complexity of the problems faced by decision makers when trying to deal with constraints imposed by geography, climate, history, economics and politics.

Part II, of three chapters, focuses on sustainable land management beginning with a consideration of what is being sustained (resource base, biodiversity, culture, lifestyles, livelihoods). In particular, we consider the feasibility of making zero net land degradation operational. ZNLD (also called LDNW) was a major commitment for the post-2015 agenda. We also consider barriers to the adoption of sustainable land management, drawing on examples from the five former Soviet republics.

Part III, of three chapters, explains the nature and extent of land degradation illustrated via studies that rely on remote sensing, GIS and other technologies. The indicators of land degradation and recovery are elaborated; barriers to implementation of SLM are analyzed. The role of inter-regional cooperation and bilateral and multilateral donors in arresting and reversing land degradation is outlined.

Part IV, of two chapters, takes up thematic issues related to SLM in GCA. Water is a cross-cutting issue of major importance. GCA has many high mountains and these are sensitive to climate change and biodiversity loss. We explore the issues of SLM in mountainous areas of GCA (including shrinking glaciers and changes to precipitation and temperature regimes) under conditions of global climate change.

Part V, of three chapters, considers geopolitical factors in the light of the assertion that GCA is the new frontier in the twenty-first century. The roles of China, Russia and other actors are explored and the final chapter attempts a synthesis and provides unifying perspectives on land, water, people and development plans for GCA and an agenda for future social-ecological research. Policy opportunities include: strengthening social technology at the local level; formally documenting and registering traditional rights; adopting administrative arrangements which are ecologically sensitive to the episodic, variable nature of land use systems that include alpine regions that are cold and arid as well as deserts that are hot and dry; devolving responsibilities to local communities; and emphasizing community needs for regional rather than sectoral development.

We want to stress the timeliness of this book in that it is a post-2015 evaluation of the large and rapidly developing GCA region that many believe is the new development frontier in the twenty-first century. It is also a critical examination of the UN's 2015 agenda as applied to a huge region that is of major strategic importance and a test bed for development models after the post-Soviet reconstruction in the five 'stans' and Mongolia and Afghanistan and the emergence of China as a great power.

The contributors have profound knowledge of current and historical resource management issues in this vast region. Because much of the relevant information is in Russian, Chinese or Mongolian, the book represents a distillation of information that is unavailable to many of the English-speaking readership.

Editors' preface

This book arose out of discussions engendered by the publicity surrounding the notion of the 'New Silk Road' and in particular the concept of 'Belt and Road' put forward by China. There had, of course, been earlier suggestions about trying to revive the ancient overland trading route and the geopolitical implications of that idea, but none had encapsulated the breadth of vision and mobilized the resources required to undertake such an immense and daring scheme.

During the course of our consultations with prospective contributors to our book, we witnessed the UN Sustainability Conference and the Paris Climate summit at the latter end of 2015. We recognized that there was a neat convergence of two potentially game-changing events and that these two aspects were likely to impact quite heavily on what came to be called 'Greater Central Asia'. We recognized that the post-2015 era was a tipping point and that the multifaceted and multidimensional set of problems, challenges and opportunities presented by the emergence of Greater Central Asia (GCA) as the new frontier in the twenty-first century deserved closer analysis and explanation. The focus of UN agencies, particularly through the UN Conventions UNCCD and CBD on ecological restoration, has directed more attention to GCA. Attention is given here to international (inter-regional) efforts to arrest and reverse land and water degradation. There is potential to use the accumulated experience of Chinese researchers and land management practitioners in restoring degraded ecosystems in GCA.

What we have attempted through our multinational (and multidisciplinary) team of writers and reviewers is to present a digest of issues in ecology, sociology, economics and geopolitics that impinge on and intersect in this important region. We want to stress the timeliness of this book in that it is a post-2015 evaluation of the large and rapidly developing GCA region that many believe is the new development frontier in the twenty-first century. It is also a critical examination of the UN's Rio+20 outcome, *The Future We Want*, as applied to a huge region that is of major strategic importance and a test bed for development models after the post-Soviet reconstruction in the five 'stans', the transition to a market economy in Mongolia and Afghanistan and the emergence of China as a great power.

Greater Central Asia, as defined by UNESCO, includes the five former Soviet republics (the 'stans'), western China, Afghanistan and Mongolia. This encompasses a vast area. Many of the land and water resources are degraded and productivity is falling. Biodiversity is under threat in the (predominantly) upland areas that comprise much of the area of GCA. GCA encompasses a vast mosaic of diverse and contrasting landscapes, plant and animal species, and human populations leading very different and unique lifestyles. However, when we look at the entire region, we tend to be struck by one significant characteristic: much of it is dry. GCA has some of the world's driest deserts and highest mountains. Much of it is covered by barren deserts, savannah, grassland, scrubland, woodlands and dry forests. Indeed, drylands comprise 62% of the GCA region, and drylands are home to a rapidly growing population that currently stands at about 95 million people. In spite of their environmental sensitivity and perceived fragility, and despite the prevailing negative perceptions of drylands in terms of economic and livelihood potentials, these ecosystems have supported human populations for centuries.

Drylands livelihoods represent a complex form of natural resource management (NRM), involving a continuous ecological balance between pastures, livestock, crops and people. The people living in the drylands of GCA are heavily dependent on ecosystem services, directly or indirectly, for their livelihoods. But those services – from nutrient cycling, flood regulation and biodiversity to water, food and fiber – are under threat from a variety sources such as urban expansion, mining and unsustainable land uses. As a result, these fragile soils are becoming increasingly degraded and unproductive. Climate change is now aggravating these challenges. However, combating climate change and adapting communities to its impacts represents an opportunity for new and more sustainable investments and management choices that can also contribute to improved livelihoods and fighting poverty among dryland communities. An ecosystem approach which includes restoration and renovation is one important direction that needs to be urgently undertaken.

Outside the cities, many dryland inhabitants are either pastoralists – sedentary or nomadic – or agro-pastoralists, combining livestock rearing and crop production where conditions allow. Intensive irrigated agriculture is an important contributor to GDP and to people's welfare. Over millennia people have lived with variable rainfall and frequent droughts using a range of coping strategies, but changing circumstances (including the transition to the market economy) mean that these traditional methods must be capitalized upon and enhanced. Unsustainable practices are contributing to significant land degradation, and it is predicted that climate change will further compound the already tenuous situation, especially in irrigated areas. Without significant efforts to address the impact of climate change and land degradation, the livelihoods of the GCA drylands populations will be in jeopardy.

To exacerbate the problems that are local and ecological there is also the rapidly changing economic and policy environment – and the threat of accelerated climate change. For this reason we decided to expand our coverage to include some chapters on geopolitical matters. These include consideration of exploitation of mineral resources, the impact of large-scale infrastructure developments and foreign influence on politics, security and trade. The emergence of China, India and Pakistan as dominant regional players and major water and energy consumers is also altering the political, economic and environmental landscape of the GCA region.

Victor R. Squires
Lu Qi
Beijing, November 2016

Figures

Tables

Contributors

Akramkhanov, Akmal

International Center for Agricultural Research in the Dry Areas (ICARDA), Tashkent, Uzbekistan

Aralova, Dildora

Institute of Photogrammetry and Remote Sensing, Technical University, Dresden, Germany and Samarkand State University, Uzbekistan

Bekchanov, Maksud

International Water Management Institute (IWMI), Colombo, Sri Lanka and Center for Development Research (ZEF), Bonn, Germany

Birner, Regina

Division of Social and Institutional Change in Agricultural Development, Institute of Agricultural Economics and Social Sciences in the Tropics and Subtropics, University of Hohenheim, Germany

Cao, Xiaoming

Institute of Desertification Studies, Chinese Academy of Forestry, Beijing, China

Djanibekov, Nodir

Leibniz Institute of Agricultural Development in Transition Economies (IAMO), Halle, Germany

Djanibekov, Utkur

Institute for Food and Resource Economics, University of Bonn, Germany

Feng, Haiying

Institute of Administration and Management, Xining, Qinghai, China

Feng, Yiming

Institute of Desertification Studies, Chinese Academy of Forestry, Beijing, China

Gofurov, Dilshod

Scientific Research Institute for Cotton Breeding, Seeding and Cultivation Agro-technologies, Kibray, Uzbekistan

Hagg, Wilfried

Dipartimento di Scienze Geologiche e Geotecnologie, Università dagli Studi di Milano-Bicocca, Milan, Italy, Geography Department, Ludwig-Maximilians-University, Munich, Germany, and Commission for Geodesy and Glaciology, Bavarian Academy of Sciences and Humanities, Munich, Germany

Halik, Umut

Faculty of Mathematics and Geography, Catholic University of Eichstaett-Ingolstadt, Eichstaett, Germany and Key Laboratory of Oasis Ecology, College of Resources and Environmental Science, Xinjiang University, Urumqi, China

Jia, Xiaoxia

National Bureau to Combat Desertification, State Forestry Administration, Beijing, China

Kariyeva, Jahan

Office of Arid Lands Studies, University of Arizona, Tucson, Arizona, USA and Alberta Biodiversity Monitoring Institute, Alberta, Canada

Kassam, Shinan

International Center for Agricultural Research in the Dry Areas (ICARDA), Cairo, Egypt

Khujanazarov, Timur

Water Resources Research Center, Disaster Prevention Research Institute, Kyoto University, Japan

Lamers, J.P.A.

Center for Development Research (ZEF), Bonn, Germany

Low, Pak Sum

Adjunct Professor, Institute of Sustainable Development and Architecture, Bond University, Robina, Australia

Lu, Qi

Professor and Research Leader, Institute of Desertification Studies, Chinese Academy of Forestry, Beijing, China

Menzel, Lucas

Institute of Geography, Heidelberg University, Heidelberg, Germany

Nishanov, Nariman

International Center for Agricultural Research in the Dry Areas (ICARDA), Tashkent, Uzbekistan

Orlovsky, Leah

Jacob Blaustein Institutes for Desert Research, Ben-Gurion University, Sede Boqer, Israel

Orlovsky, Nikolai

Jacob Blaustein Institutes for Desert Research, Ben-Gurion University, Sede Boqer, Israel

Shaumarov, Makhmud

Division of Social and Institutional Change in Agricultural Development, Institute of Agricultural Economics and Social Sciences in the Tropics and Subtropics, University of Hohenheim, Germany

Squires, Victor R.

Guest Professor, Institute of Desertification Studies, Chinese Academy of Forestry, Beijing, China

Swanström, Niklas

Silk Road Institute, Norway

Toderich, Kristina

International Center of Biosaline Agriculture, ICBA-Dubai, Tashkent, Uzbekistan and Samarkand State University, Uzbekistan

Wang, Feng

Institute of Desertification Studies, Chinese Academy of Forestry, Beijing, China

Yan, Feng

Institute of Desertification Studies, Chinese Academy of Forestry, Beijing, China

Yang, Liu

Institute for Desertification Studies, Chinese Academy of Forestry, Beijing, China

Yang, Youlin

Asia-Pacific Regional Coordination Unit of UNCCD, Bangkok/Beijing

Acknowledgements

Some of the research reported here and this entire publication received support from both the International S&T Cooperation Program of China (No. 2015DFR31130) and China National Key Technology R&D Program (No. 2012BAD16B0).

Part I

Introduction

The three chapters here provide a comprehensive and in-depth analysis of the lands and their peoples, and the biogeographic and strategic importance of the Greater Central Asia region.

Squires and Lu define Greater Central Asia for the purposes of this book and analyze the demography and ethnicity, as well as providing a brief history of the region.

Orlovsky and Orlovsky lead a detailed discussion of biogeography, climate and the role of the former Soviet Union's scientific endeavors, particularly in the five former Soviet republics.

Shaumarov and Birner analyze the impact of the Soviet era and discuss the issues around managing the commons in the post-Soviet transition. Special attention is given to the challenges of institutional change in pastoral systems in Uzbekistan.

1   Greater Central Asia

Its peoples and their history and geography

Victor R. Squires and Lu Qi

INSTITUTE OF DESERTIFICATION STUDIES, BEIJING

Greater Central Asia defined

Of late, there has been much reference to Greater Central Asia (GCA), which is taken to mean a much broader definition than 'Central Asia', and usually refers to the five former Soviet republics of Kyrgyzstan, Tajikistan, Turkmenistan, Uzbekistan and Kazakhstan plus adjacent places, sometimes including Mongolia, parts of western China and Afghanistan (see Orlovsky & Orlovsky, 2018 for an elaboration). In the Russian-speaking scientific community, 'Middle Asia' unambiguously refers to the regions stretching from the Caspian Sea to the Chinese Dzungaria (Junggar Basin in Xinjiang) and from the southern Ustyurt Plateau and the Aral Sea to northern Iran and Afghanistan (Rachkovskaya et al., 2003). Biogeographic studies habitually use 'Central Asia' for the territory covering Mongolia (Jigjidsuren & Johnson, 2003) and the Gobi Desert in Chinese Inner Mongolia (Petrov, 1966, 1973). In this book, we will use the broader definition as proposed by UNESCO. This definition includes the five former Soviet republics (the five 'stans'), plus Afghanistan, Mongolia, and most of western China (Xinjiang and the Tibet Plateau).

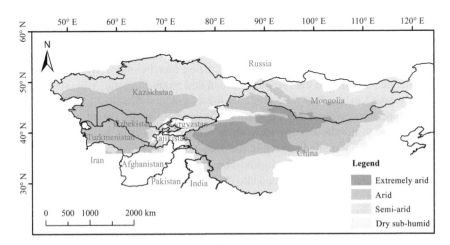

Figure 1.1   Map of Greater Central Asia, which comprises the five former Soviet
republics – Kyrgyzstan, Tajikistan, Turkmenistan, Uzbekistan and
Kazakhstan – plus adjacent places including Mongolia, parts of western
China and Afghanistan

The GCA region is located at a similar latitude range (35⁰–46⁰N) as the steppe-land of northern Syria–Turkey, Jordan, Iraq–Iran, Spain, the Maghreb in the Mediterranean, the US western Great Basin and Pacific Coast (California, Oregon, Washington, Idaho, Nevada, West Utah, West Arizona), and Patagonia, south-west and southern Africa, part of south and south-west Australia in the 35⁰–45⁰S range. The central Asian steppes of north-west China and Mongolia, albeit physiognomically similar, have a distinct flora and are in a summer-rain climate belt.

The lands and their peoples

The history of Central Asia has been determined primarily by the area's climate and geography. This vast region is an important link between East and West, but it would be to underestimate its importance if it were reduced to a corridor of traverse and link between seats of major powers. Central Asia should not be defined by the fact that people crossed it on particular occasions, but rather by the significant historical developments that occurred there. Two points need emphasis. One, there is a need to pin down crucial and genuinely transformative shifts between periods, beyond what might be termed 'motion', the regular changes that all societies experience. Two, there must be some possibility of grand comparison of processes such as secularization or state-formation and collapse. On the one hand, the GCA region is characterized by its relationship to outside influences and imperial forces that have shaped the boundaries, fate and destiny of principalities, kingdoms, states and regions until the present day. The aridity of the region makes agriculture difficult and distance from the sea cuts it off from much trade. Thus, few major cities developed in the region. Pastoralists have played an important role in shaping relationships, connecting regions, and exchanging goods and valuable information. The combination of highly productive and spatially concentrated oases in a wide-ranging environment with extensively used rangeland in deserts and steppe regions is modified by the third dimension represented by the verticality of GCA's high mountains and the Tibetan Plateau.

Nomadic horse peoples of the steppe dominated the area for millennia and used the high-altitude land for summer grazing (see below). Relations between the steppe nomads and the settled people in and around Central Asia (*sensu lato*) were marked by conflict. The nomadic lifestyle was well suited to warfare, and the steppe horse riders became some of the most militarily potent people in the world, due to the devastating techniques and ability of their horse archers. The dominance of the nomads ended in the 16th century, as firearms allowed settled people to gain control of the region. The Russian Empire, the Qing Dynasty of China and other powers expanded into the area and seized the bulk of Central Asia by the end of the 19th century.

After the Russian Revolution of 1917, the Soviet Union incorporated most of Central Asia; only Mongolia and Afghanistan remained nominally independent, although Mongolia existed as a Soviet satellite state and Soviet troops invaded Afghanistan in the late 20th century. The Soviet areas of Central Asia saw much industrialization and construction of infrastructure, but also the suppression of local cultures and a lasting legacy of ethnic tensions and environmental problems.

It was during the Sui and Tang Dynasties (581–907 AD) that China expanded into eastern Central Asia. Chinese foreign policy to the north and west now had to deal with Turkic nomads, who were becoming the most dominant ethnic group in Central Asia. In the 8th century, Islam began to penetrate the region, the desert nomads of Arabia could militarily match the nomads of the steppe, and the early Arab Empire gained control over parts of Central Asia. The Arab invasion also saw Chinese influence expelled from western Central Asia.

The most spectacular power to rise out of Central Asia developed when Genghis Khan united the tribes of Mongolia. Using superior military techniques, the Mongol Empire spread to comprise all of Central Asia and China, as well as large parts of Russia and the Middle East. After Genghis Khan died in 1227, most of Central Asia continued to be dominated by his successor until 1369, when Timur (Tamerlane), a Turkic leader in the Mongol military tradition, conquered most of the region. Even harder than keeping a steppe empire together was governing conquered lands outside the region. While the steppe peoples of Central Asia found the conquest of these areas easy, they found governing almost impossible. The diffuse political structure of the steppe confederacies was maladapted to the complex states of the settled peoples. Moreover, the armies of the nomads were based upon large numbers of horses, generally three or four for each warrior. Maintaining these forces required large stretches of grazing land, not present outside the steppe. Any extended time away from the homeland would thus cause the steppe armies to gradually disintegrate. To govern settled peoples, the steppe peoples were forced to rely on the local bureaucracy, a factor that would lead to the rapid assimilation of the nomads into the culture of those they had conquered.

Central Asian land trade bypassed by nautical advances in the 1500s

The lifestyle that had existed largely unchanged since 500 BCE began to disappear after 1500. An important change in the world economy in the 14th and 15th centuries was brought about by the development of nautical technology. Ocean trade routes were pioneered by the Europeans, who were cut off from the Silk Road by the Muslim states that controlled its western termini. The trade between East Asia, India, Europe and the Middle East began to move over the seas and not through Central Asia. The disunity of the region after the end of the Mongol Empire also made trade and travel far more difficult and the Silk Road went into steep decline. An even more important development was the introduction of gunpowder-based weapons. The gunpowder revolution allowed settled peoples to defeat the steppe horsemen in open battle for the first time. Construction of these weapons required the infrastructure and economies of large societies, and they were thus impractical for nomadic peoples to produce. The domain of the nomads began to shrink as, beginning in the 15th century, the settled powers gradually began to conquer Central Asia.

Russian expansion into Central Asia in the 19th century

The Russians also expanded south, first with the transformation of the Ukrainian steppe into an agricultural heartland, and subsequently onto the fringe of the Kazakh steppes. The slow Russian conquest of the heart of Central Asia began in the early 19th century. By the 1800s, the locals could do little to resist the Russian advance. Until the 1870s, for the most part, Russian interference was minimal, leaving native ways of life intact and local government structures in place. With the conquest of Turkestan after 1865 and the consequent securing of the frontier, the Russians gradually expropriated large parts of the steppe and gave these lands to Russian farmers, who began to arrive in large numbers. This process was initially limited to the northern fringes of the steppe and it was only in the 1890s that significant numbers of Russians began to settle farther south, especially in Zhetysu (Semirechye).

After being conquered by Bolshevik forces, Soviet Central Asia experienced a flurry of administrative reorganization. Under the Soviets, the local languages and cultures were systematized and codified, and their differences clearly demarcated and encouraged. New Cyrillic writing systems were introduced, to break links with Turkey and Iran. Under the Soviets, the southern border was almost completely closed and all travel and trade was directed north through Russia.

During the period of forced collectivization under Joseph Stalin, at least a million persons died, mostly in the Kazakh Soviet Socialist Republic (SSR). Islam, as well as other religions, were also attacked. In the Second World War, several million refugees and hundreds of factories were moved to the relative security of Central Asia; and the region permanently became an important part of the Soviet industrial complex. Several important military facilities were also located in the region, including nuclear testing facilities and the Baikonur Cosmodrome. The Virgin Lands Campaign, starting in 1954, was a massive Soviet agricultural resettlement program that brought more than 300,000 individuals, mostly from the Ukraine, to the northern Kazakh SSR and the Altai region of the Russian Soviet Federated Socialist Republic (SFSR). This was a major change in the ethnicity of the region. The Second World War sparked the widespread migration of Soviet citizens to Soviet Central Asia.

Similar processes occurred in Xinjiang and the rest of Western China, where the People's Republic of China (PRC) quickly established control. The area was subject to a number of development schemes and, like Soviet Central Asia, one focus was on the growing of the cotton cash crop. These efforts were overseen by the Xinjiang Production and Construction Corps (XPCC). The XPCC also encouraged Han Chinese to return to Xinjiang after many had migrated out during the Muslim revolts against the Qing Dynasty.

Dissolution of the USSR and the transition from central planning

After 1991, agriculture in the five 'stans' deteriorated due to disorganization and the ensuing slow reform, generating a breakdown in farming practices, fertilizer use and crop yield until 1995–1997, when it then picked up again. This revival was achieved by reassigning cotton- and rice-irrigated land to irrigated wheat cropping, and also to rainfed cereal cropping competing for the best rangelands, except in Kazakhstan where large areas of potential cropland was available.

The five former Soviet republics in Central Asia (the five 'stans') had no history as nation states before 1992, and during the Soviet era economic policy and development strategies were determined in Moscow. No one had anticipated the dissolution of the Soviet Union before its final months, and all were unprepared for the severing of Soviet ties. In these newly independent republics of the 5 'stans', the agricultural production systems of the former Soviet Union became instantly irrelevant. The Soviet system had integrated all aspects of food and industrial production into centrally controlled operations that included research and support services such as veterinary care and extension advice, supply of inputs, and marketing of end products. Each 'stan' was essentially a commodity-producing component of a larger system importing agricultural inputs from elsewhere and exporting its produce within the USSR. After independence, each republic had to face the challenge of developing a 'stand-alone' economy, a process that requires enormous efforts to diversify agricultural production in a sustainable manner.

The 'stans', with the exception of Kazakhstan, face three major challenges: (i) ensuring food security; (ii) alleviating poverty; and (iii) protecting the environment. During transition immediately after 1991, food output decreased by 15–45%, whereas per capita food need increased due to the increasing population, with the exception of Kazakhstan, where the population has shown a declining trend. There is a general decline in living standards (gross national product (GNP) is US$776) and 25–40% of the population is estimated to be living below the poverty line (Pomfret, 2006). Over 50% of these people live in rural areas, where farming is a primary source of livelihood for them. Fortunately, GCA countries have pockets of valuable minerals.

Mineral wealth in GCA

The Central Asia subsoil is rich in minerals as well as petroleum (oil and gas) (see Swanström, 2018, for more on oil and gas). Among the post-Soviet republics in Central Asia (the 5 'stans'), Kyrgyzstan has the smallest mineral reserves per capita, even though gold mining accounts for about 30% of the country's exports. Kazakhstan and Turkmenistan possess the greatest per capita reserves of mineral resources. The southern parts of Kazakhstan, Uzbekistan, Kyrgyzstan and Tajikistan all include rocky formations of the Tian Shan Mountains and the Pamirs, whereas Turkmenistan is part of another geological zone. Kazakhstan's central regions around Karaganda and the eastern regions in the Altay have proven rich in minerals. Uzbekistan has significant development of gold, copper and natural gas production. After signing a long-term contract in early 2000 for the delivery of natural gas to Russia, Turkmenistan sharply increased utilization of its productive capacity in this basic branch. It therefore receives significant additional hard-currency revenues and can enjoy a faster rate of economic growth in the short term. In all probability this will slow both the liberalization of foreign trade and the introduction of convertibility of the national currency through current account transactions (Pomfret, 2006).

Working gold and silver has been a traditional livelihood for artisans in sedentary Turkestani societies and these metals have been used in the animal art of nomads. But it was not until the Russians came to the region (beginning in the late 19th century) that the first large-scale extraction industries took shape. Later, the Soviet Union's thirst for energy and the priority it gave to heavy industry hastened the development of the extraction sector. Minerals such as bauxite, chrome, zinc, manganese and copper were used in abundance by the Soviet industries. Central Asian uranium was used in the nuclear industry, including in armaments that were tested at the Semipalatinsk site in Kazakhstan, and its gold (from Kyrgyzstan and Uzbekistan in particular) was sold on the international market. In recent decades, rare minerals such as titanium, beryllium, tantalum, cobalt and cadmium have been increasingly sought, especially by China.

Kazakhstan is the regional leader in mineral production and processing, while Uzbekistan is the world's ninth largest gold producer, but most of their mining projects are located in remote desert areas. In the mountains, the development of the mining sector has been significant since 2009, particularly in Kyrgyzstan and Tajikistan. At the end of the Soviet era and into the 1990s, there was only marginal gold mining in either country, and little state or international interest. With gold prices reaching record levels since the early decades of the 21st century, however, both local and global investors have become interested in developing even low-grade deposits. These are sources of revenue for national budgets, contributing up to 50% of national export earnings in Tajikistan (from alumina and gold) and up to 30% in Kyrgyzstan (mainly gold from the Kumtor mine). Kyrgyzstan, which foresaw the mining and energy sectors as having significant development potential, moved to create conditions favorable to mining operators by enacting economic reforms and by allowing access to geological information. Currently, almost all of its territory is licensed for mining activities. Tajikistan, in contrast, continues to consider its geological information semi-confidential, as in the Soviet era, and its legislation and the ease of doing business currently lag behind Kyrgyzstan's. As a result, Tajikistan has attracted fewer investors, and where Kyrgyzstan's mining sector has advanced, Tajikistan's remains stagnant. The World Bank is assisting both countries in reducing barriers in the mining sector.

The influx of new mining technologies and the launch of new projects have given rise to both opportunities and difficulties for governments and local communities. A reluctance on the part of governments and mining companies to share gold-mining profits equitably and a lack of transparency in decisions have led to feelings of discontent among poor and vulnerable groups in the mountains. Indeed, the benefit-sharing arrangement between mining projects, central government and local communities remains a lingering cause of resentment. The conflict between the use of land for traditional pasture and grazing, for nature conservation and for mining activities is also a source of friction in Kyrgyzstan. The experience of the Kumtor gold mine in Issyk-Kul Province in eastern

Kyrgyzstan has influenced all the developments that followed. In 1997, with the support of Canadian investment, operations started at the Kumtor mine, which now produces 90% of Kyrgyzstan's gold, about 15–18 tonnes per year. Kumtor tax payments contribute substantially to the national budget, and the mine provides significant employment opportunities to communities throughout the area. In addition, Kumtor sponsors local social development programs such as schools, kindergartens and summer camps, and has introduced a local development fund that is increasingly considered as a model by other mining companies. Kumtor maintains high safety standards, but a transport accident resulted in a spill of cyanide into a local river. The toxic material dispersed quickly, causing some environmental damage, but the psychological perception was significant and long-lasting. The accident galvanized local resistance to mining, whether or not cyanide would be used in operations, especially in areas with no mining history. The abandoned Soviet mining legacies across the country stand as stark reminders of possible grim scenarios not to be repeated. Now mining operators often encounter local opposition wherever they go, and find that environmental impact statements and the necessary permissions do not easily overcome the hostility and distrust they face.

Tajikistan has been famous for silver mining from ancient times, and a recent geological audit suggests that it probably has one of the largest silver reserves in the world. In 2012, the government officially announced a request for international tenders for the development of these deposits. Chinese investments and technology will likely support Tajikistan's plans to develop its own alumina mining and to expand cement production capacities across the country.

Kyrgyzstan has taken the lead in promoting an international initiative on transparency in extractive industries, and is working to involve as many mining companies as possible. The transparency initiative requires financial disclosure that shows how mining activities benefit governments. The initiative does not, however, require disclosure of how the activities may or may not benefit local communities. In both Kyrgyzstan and Tajikistan, the environmental problems associated with the increase in mining and related activities are offset to some extent by declines in all other industrial sectors. While the expansion in mining increases potential threats to the environment, the reduction in industry reduces other threats. Finally, both mountain countries have experienced a boom in small-scale mining for placer gold, particularly in Kyrgyzstan. Artisanal miners are a heterogeneous group of men aged from 16 to 60 years and over, and their reasons for mining are varied. For some, mining was and still is the main source of cash income. Gold helped them to survive in the turbulent economic transition period of 1992–2000. For others, it is an income supplement in the winter months when agricultural activities are limited in the mountains. In any case, artisanal gold mining is beyond the control of central and local authorities, and the increasing degree of labor mechanization and the use of mercury for fine gold extraction are growing threats to the mountain environment. In Mongolia, there is a major problem created by unfettered mining activities in the rangelands, due not only to the excavations that destroy the vegetation but also to stream diversion, the

Figure 1.2 Artisanal mining is destroying large areas of steppe in Mongolia

illegal damming of streams, unregulated disposal of mine wastes, and pollution from the use of mercury and other heavy metals. Abandoned mines, hazardous industrial waste sites and mine tailings – mostly legacies of the Soviet period – continue to be a major environmental concern for the mountain areas of Central Asia, including Mongolia.

Although their reserves are not large in modern terms, Kyrgyzstan and Tajikistan were among the pioneers in developing the uranium mining sector. When the Soviets left, they simply abandoned the mines and tailings with no remediation. These hazardous sites remain obstacles to sustainable development, environmental protection and population security in the region. Abandoned and active mining sites and metallurgy industries cause environmental problems in the Altai Mountains of Kazakhstan and in the Irtysh River Basin, where the country's mining sector was born in the 18th century. The cost of remediation is prohibitive for the countries involved, and, in the absence of legislation or financial resources to undertake the task, Kyrgyzstan and Tajikistan have no remediation plans in place but are looking to international partners for assistance. Abandoned mining sites pose as much or more danger to the countries' neighbors in the event of a flood or other mine failure, and regional cooperation is one prospective solution. Russia is participating in negotiations and may commit to helping resolve the problems. In Kyrgyzstan, abandoned uranium tailings are a national priority both politically and environmentally, but because of the scale of the problem the resources needed are overwhelming and no progress has been made. Continued efforts at cooperation with Russia and the other Central Asian countries are a promising path, as is the prospect of private sector involvement. Private firms may be interested in reopening some mines or in re-working some tailings.

Resource base degradation: a looming crisis?

The entire GCA region faces a serious challenge to its natural resource base. Croplands, rangelands, deserts and mountains are being degraded. The reduced availability of agricultural inputs, feed and fodder is resulting in a decline in livestock numbers (except in Turkmenistan). Water scarcity and misuse are compounding the threat to food security, human health and ecosystems (see Bekchanov et al., 2018).

In Mongolia, by 1992 liberalization of the economy was under way, and virtually all state-owned livestock had been privatized, dismantling herding collectives. For herders, privatization resulted in the loss of the formal institutions that regulated pasture use, in reduced social services, in declining trade and access to markets, in increased numbers of herding households, and in greater poverty and differentiation in wealth. These changes in herders' livelihoods altered patterns of pastoral land use and led to high rates of out-of-season and year-round grazing of key resources, to trespassing on customary winter and spring reserve pastures, and to declines in the distance and frequency of seasonal nomadic moves (Fernandez-Gimenez, 1999). Over the centuries there have been many sweeping changes. With each overt shift in the political economy of Mongolia, the territories of nomadic groups shrank in size, controls over animal movements became more rigid, the allocation of pasture was more closely controlled, tenure became more individuated, and the gap between formal and informal regulation of resource use widened. Individual livestock owners now decide how many animals to keep, what to feed them and where to move them. New systems of animal husbandry are resulting in different patterns of rangeland use, degradation and recovery from the Soviet period.

For the GCA countries to succeed in developing a market-driven economy, they must transform their agriculture to ensure that they can feed their people, that it is sustainable in the long term, and that it responds to the changes in global priorities and trends, such as climate change. Identifying agricultural research needs and priorities is important (Kijne, 2005; World Bank, 2006). These include research on water-use efficiency in irrigated systems; control of soil salinity and soil erosion; and development of mountain agriculture (Shigaeva, Wolfgramm, & Dear, 2013). The effects of the transition from centralized to private farms – especially the ways in which changes in ownership and management affect soil and water management, seed production, and livestock and range management – continue to be analyzed as more social science input becomes available (Lerman, 2012; Kurbanova, 2012; Halimova, 2012). This has led to the identification of important research activities under two themes – *livestock and environment*, and *policy and economics* – and the engagement of multilateral and bilateral donor organizations, such as the International Center for Agricultural Research in the Dry Areas (ICARDA), and the Central Asian Countries Initiative for Land Management (CACILM) initiative (Cardesa-Salzmann, 2014).

There is broad agreement as to what the agenda should be:

1   improve productivity of agricultural systems;
2   ensure better natural resource conservation and management;
3   foster the conservation and evaluation of genetic resources and biodiversity;

4   develop socio-economic and public policy research; and
5   strengthen national programs under the three Rio conventions, CITES (the Convention on International Trade in Endangered Species of Wild Fauna and Flora) and the Ramsar Convention on Wetlands.

The future challenge lies in implementing this ambitious agenda through collaborative activities in strategically important research areas.

According to Gintzburger et al. (2005), the newly independent countries (i.e. Kazakhstan, Kyrgyzstan, Tajikistan, Turkmenistan and Uzbekistan) had to produce their own food and livestock feed within their national territory after the Soviet system breakup in 1991. This presented real challenges for all except Kazakhstan, because arable land varied from 5% to 20% of territory in the other 'stans', with an additional 1.0–2.9% in permanent crops (mostly irrigated tree crops or vineyards). Rangelands cover most of these countries (80–95% of total agricultural areas), including Mongolia and Afghanistan.

The unexpected challenges of nation building were superimposed on the transition from a centrally planned economy, which had begun in the late 1980s but had little influence on Central Asia before the Soviet economic system began to unravel in 1991. All five 'stans' suffered serious disruption from the dissolution of the USSR. Demand and supply networks based on undervalued transport inputs quickly collapsed in the early 1990s. The economies of the five 'stans' gradually became more differentiated as their governments introduced national strategies for transition to a market-based economy. By the early 21st century, all five 'stans' had essentially completed the process of nation building and the transition from central planning. However, the typology of market-based economies varied substantially from the comprehensive price and trade liberalization and extensive privatization introduced in the Kyrgyz Republic between 1993 and 1998 to the non-reform in Turkmenistan (Pomfret, 2006).

After becoming independent, in 1991, the five 'stans' pursued different transition paths from the defunct central planning; a striking feature is that they followed divergent economic strategies after becoming independent. Despite similarities in culture, history, geography and economic structure, their transitions from Soviet central planning ranged from the most rapidly liberalizing (the Kyrgyz Republic) to the least reforming (Turkmenistan) of all the former Soviet republics. By the turn of the 21st century, when the transition from central planning was essentially complete, the 'stans' had created vastly different economic systems. Despite the differences, some commonalities remain, in particular the establishment of super-presidential political systems under autocratic rulers, obstacles to trade posed by geography (being landlocked), and unwillingness to engage in serious regional cooperation.

Performance over the two decades since independence has been determined by resource endowments rather than by policy: e.g. two 'stans' are resource rich (Turkmenistan and Kazakhstan) and others are resource poor (see Kerven et al., 2011). According to Pomfret (2010), prospects for significant change in the near future are limited because by the end of the 1990s the window of opportunity for policy initiatives had shut and entrenched political regimes had little incentive to sponsor major reforms.

Livestock and environment

According to Behnke (2006), policy failures at national level in all countries within the GCA region had promoted desertification, but especially in the 5 'stans', where such failures included:

- environmentally destructive and wasteful agricultural technologies dating from the period of the Soviet Union;
- post-independence monopolization of natural resources by national elites, which created poverty and forced poorer households to engage in unsustainable agricultural practices;
- ineptly formulated, unenforced and conflicting national land laws on natural resource ownership and control; and
- scientifically uninformed national policy formulation, which failed to address existing and documented resource management problems.

These issues are challenges even now to regional environmental policy formulation, with respect to: (i) over-grazing by livestock; (ii) natural resource degradation; and (iii) desertification control. Some of these issues are more fully explored in books by Kreutzmann (2012) and Squires (2012).

All GCA countries (not just the 5 'stans') have limited supplies of three basic natural resources: fertile land, clean water and a healthy environment. At least one of these resources, often two, is in particularly short supply because they are very unevenly distributed within the region. The situation is exacerbated when high population growth produces densely populated areas (e.g. the Ferghana Valley) where poor rural populations depend for their livelihoods on the quality and availability of natural resources. Under these conditions, desertification poses a 'soft security challenge' that contributes to the economic marginalization of certain sectors of the rural population and intensifies crises caused by a complex array of factors (Behnke, 2006). As a contributing element in political, social and economic instability, desertification is closely associated with a number of regional security concerns. Desertification affects food and health security when it contributes to malnutrition, increased incidence of disease and child mortality. Desertification influences livelihood security because it may force people to leave their homes and migrate when degraded land can no longer support them. Desertification has an effect on social and national security when it encourages civil unrest. Finally, desertification affects international security through cross-border migration and increased interethnic tensions. Dust and sandstorms are common events in the arid and semi-arid regions of Central Asia. The region is characterized by strong winds, scarcity of vegetation cover, a continental climate with long and dry summers, and frequent soil and atmospheric droughts. GCA drylands, encompassing a great variety of desert types, represent a powerful source of mineral and salt aerosols. Uplift and transport occur and transboundary problems arise (Indoitu, Orlovsky & Orlovsky, 2012).

It is unclear to what extent local knowledge of land management was lost in the 5 'stans' during the Soviet period, or whether the economic difficulties experienced over the decades since 1990 have led to new ways of thinking about people's relationships with their environments. It is also unclear whether decentralized approaches such as those mandated by the United Nations Convention to Combat Desertification (UNCCD) and that underpin its 10-year strategic plan can be easily implemented in a society where public participation has been historically absent.

Alibekov and Alibekov (2006) argue that agriculture in the 5 'stans' in particular was mismanaged for decades under a centralized command economy. The continuing legacy of this period is severe land degradation in the form of soil salinization and erosion, elevated groundwater levels caused by poorly managed irrigation systems, the drying of the Aral Sea, and the chemical and nuclear pollution of water and soil. They provide an overview of how these failures occurred and describe some of their negative social and economic impacts. Desertification takes place within naturally delimited geographical and ecological systems, and so it must also be studied and prevented within the framework of these natural systems.

The search for ways to sustainably develop vulnerable ecosystems

Given that a significant part of Central Asia's natural resources have been exhausted and the ecological situation continues to deteriorate, urgent action is required. But it is important to take the *right* action. The replication and scaling up of successful demonstrations are not always easy, nor is there a guarantee of success (Squires, 2013).

Grazing systems and desertification

There have been several analyses of rangeland degradation in Turkmenistan and Kazakhstan, two countries that have pursued contrasting policies in relation to their pastoral sector in the post-Soviet period (Behnke, 2006; Kerven et al., 2008). In Kazakhstan, the privatization of land and livestock began in the early 1990s and was complete by the end of the decade. In Turkmenistan, on the other hand, rangelands and many livestock still remain state property, with pastoralists working within what amount to reformed and renamed Soviet farms. Changes in pastoral land use reflect these national policy differences. In Kazakhstan, the sudden and chaotic privatization of livestock caused the loss of about three-quarters of the national herd in the mid-1990s. With the collapse of rural livelihood systems, there were high levels of emigration from pastoral areas into towns or larger rural settlements, rural farmsteads and wells were abandoned and destroyed, and many remote seasonal pastures were unused. Around 1999 or

2000 these downward trends were halted and then reversed as flocks expanded for the first time in a decade and larger livestock owners began to re-colonize isolated farmsteads and wells.

Turkmenistan represents the antithesis to Kazakhstan's radical reforms. Independent Turkmenistan operates a centralized agricultural economy modeled on farm reforms that were being implemented in the Soviet Union in the late 1980s, just prior to Turkmenistan becoming independent. Households are allowed to lease livestock from the state and private ownership of livestock is permitted, but there is no private ownership or leasing of pastures or water points. The slow pace of agricultural reform in Turkmenistan did not precipitate the catastrophic livestock losses that accompanied radical reform in Kazakhstan. The proportion of the national flock that is private or state-owned is unclear, though official statistics state that well over half of all small ruminants are now in private hands and that livestock are now more numerous than at any other time in Turkmenistan's history.

Strong and Squires (2012) examine pasture-based livestock systems in Tajikistan. Herd or flock dispersal and seasonal mobility limit the environmental impact of livestock. One of the main findings of a multi-country study (Kerven et al., 2008) of pastoralism in Inner Asia, the region lying to the east of Central Asia and consisting of Mongolia, western China and southern Siberia, was as follows:

> The highest levels of degradation were reported in districts with the lowest livestock mobility; in general, mobility indices were a better guide to reported degradation levels than were densities of livestock. This pattern corresponded with the experience of local pastoralists. At six sites, locals explicitly associated pasture degradation with practices that limited the mobility of livestock. (Sneath, 1998)

In Tajikistan, Strong and Squires (2012) concluded that the importance of the livestock economy to the rangelands is clear. The current and potential significance of pasturelands and rangelands to the current and future Tajikistan economy cannot be overstated. Specifically, the rangelands, mainly through livestock, are a source of direct rural income support for over 4 million people as well as providing much of the nation's meat and milk requirements. It will be productive innovations within the livestock sector that will largely influence and determine whether sustainable development will be possible in Tajikistan and in several neighboring countries (Afghanistan, Kyrgyzstan, parts of Kazakhstan, Uzbekistan and north-west China). The assured future utilization of the GCA rangelands depends on a number of factors; these are discussed *inter alia* in Lerman (2012) and Sedik (2012).

Legislation in Tajikistan allows farmers to access heritable land shares for private use, but reform has been geographically uneven. In mountainous areas, reform has led to privatization of arable land, but farming households are often renters or shareholders in 'collectives' whose managers still have much control over the land. In the productive areas of the lowlands, there has been little distribution to households of land suitable for cultivation, and much farming is still conducted by laborers working on collective farms. Access to pasture is generally good but some remote pastures have been abandoned due to the risks and costs associated with travelling to them, increasing pressure on pastures near villages (Robinson et al., 2012; Halimova, 2012). In general, where distribution has occurred, some households have prospered, yet many have been left landless or with insecure tenure. Poorer households generally have the least secure tenure arrangements, the worst quality land and the smallest land areas (Halimova, 2012; Lerman, 2012). The unsustainable use of land and vegetation is most likely to occur among these groups and makes the shift to sustainable land management more of a dream than a reality (Shigaeva, Wolfgramm & Dear, 2013).

Poverty, inequality and unequal opportunity

An important factor to consider in post-soviet Central Asia is the demography of the population, since nearly 40% of the population is under 18 years of age. In spite of the many inadequacies of the Soviet system, health and education indicators for child development in Central Asia were high, especially when compared with those in other developing countries in the world. It is now proving exceedingly difficult to maintain the same sort of commitment to the needs of children in the uncertain economic and political predicaments of the Soviet republics of Central Asia. Furthermore, with the collapse of central planning, many of the new states are having difficulty maintaining basic infrastructure, such as roads and public transportation. In the rural areas (where over half of the population of the 5 'stans' resides) the repercussions of poverty are particularly devastating. Poverty destabilizes society and creates stress lines and fissures. As a result, many children have experienced an abrupt diminishment in the quality of their lives. They have also felt the trickle-down effect of economic crises, including such problems as the increasing rate of school dropouts, the spread of debilitating communicable diseases (tuberculosis, syphilis and hepatitis), malnutrition among younger children, unprecedented homelessness, and an increase in youth crimes, mental depression and suicide among teenagers.

Clearly, the problems of children today foreshadow the human development issues of tomorrow. What things should we be focused on? What is the impact of poverty on democratic reforms? Current poverty levels in Kyrgyzstan and Tajikistan are having an impact on national stability. The widespread poverty cannot be underestimated in terms of breeding political discontent or intensifying illegal activities. Under such circumstances it is difficult for governments at all levels to pay attention to more sustainable land management.

What are the implications of the increasing stratification, not only between the rich and the poor, but also between the urban and rural regions? We see a rapidly disappearing 'middle class' (at least by former Soviet levels). Using the consumption-based measurements of the World Bank (versus an income measure), 51% of the population of Kyrgyzstan and Tajikistan now lives below the poverty line. The Gini Inequality Coefficient, which calculates the degree of class inequality, indicates a value for Kyrgyzstan of 0.41, which is higher than that of Kazakhstan at 0.35, although still lower than Russia's 0.46.

The rural poverty rate in Kyrgyzstan (65%) is more than twice that of urban areas (29%), and extreme poverty rates in rural areas (21%) are about four times those of urban areas (5%). With the most onerous economic problems affecting those who live in rural areas, it is important to explore how rural poverty can further weaken borders that are already quite porous and potentially dangerous. For example, three of Kyrgyzstan's four borders are certainly troublesome, especially those with Tajikistan and Uzbekistan in the Ferghana Valley, and, of course, the border with western China, most notably with Xinjiang. In each instance, these borders are for the most part in remote, mountainous areas, which means that they are often poorly guarded. Where guard posts do exist on the roadways, it is a well-known fact that these guards are generally susceptible to bribes, thus affording easy access for the cross-border drugs, human trafficking or even arms trade.

Two important developments occurred simultaneously a few years ago with respect to drugs. First, a Russian market for drugs has opened up. Second, Iran, once on the drug-smuggling route, launched a strict antidrug campaign, virtually sealing its border with Afghanistan. This has redirected the drug trade northward through Tajikistan and Kyrgyzstan. It is essential to consider that as economic conditions worsen, or even if they remain the same, the only viable business for many of the rural poor may be drug trafficking. Extensive rural poverty makes this gainful, albeit illegal, opportunity appear reasonable and necessary to desperate people. Able-bodied men still seek work in Russia as migrant labor and their remittances are important to sustain families back home. For some countries, such as Tajikistan, remittances represent a significant part of the nation's gross domestic product (GDP) (Olimova & Bosc, 2003).

Economic disparity may significantly exacerbate pre-existing tensions between ethnic groups. This is especially true since agricultural resources are limited, and there are relatively few fertile valleys. For instance, the Kyrgyz retain a latent resentment towards successful Chinese and Uighur farmers. Tensions between Russians and ethnic Kyrgyz living around Lake Issyk-Kul have been serious in the past. And, of course, there were the now-infamous Osh riots, which erupted from time to time over land rights in the Ferghana Valley, home to both Uzbeks and Kyrgyz. Certainly, limited access to arable lands and resources may be a catalyst for conflicts of a regional or ethnic nature.

What is the impact of poverty on social networks? The weakening of rural social networks as a result of increasing isolation and poverty has only recently begun to be considered as a significant social development concern. Since many of the transactions among people living in the rural regions pertain to survival issues – securing food, obtaining health services, finding fuel and water, and so forth – social networks, as a type of informal institution, are a critical dimension of day-to-day rural survival. The Kyrgyz Republic Social Networks Study (Kuehnast & Dudwick, 2004) indicates that the ability of the social networks of the poor to insulate them from the mounting problems of rural life is diminishing rapidly. As a result, the rural poor are finding themselves in a patronage relationship in which they borrow goods and food from a wealthier neighbor and then become indentured to that neighbor as a means of paying their debt. High population growth and the lack of alternatives to agriculture hinder farm consolidation and leave many households in a subsistence trap, unable to accumulate capital to make long-term investments in their land.

Sustainable management of pastures is an important factor for the ecological and socio-economic stability of Central Asian countries, especially under changing climatic conditions. Tajikistan and Kyrgyzstan are currently undergoing reform of their pasture management systems. Legislation formalizing common property regimes on grazing lands has been adopted for the first time in post-Soviet Central Asia. The other three republics are currently developing pasture-specific legislation, with both individual and common management approaches under consideration (Kerven et al., 2011). Smaller-scale livestock keepers who graze their animals and keep them moving to avoid resource depletion are among the most sustainable producers of meat, milk and other animal products. By feeding their ruminant livestock grass or forage rather than crops, they increase global food security and reduce pressure on land. Pastoralists should be supported with targeted policies and investments. These include: estimating pastoralists' *total economic value* (e.g. food and ecosystem services); protecting their *mobility* (e.g. via access rights); providing custom *extension services* (e.g. herd or pasture management); offering tailored *health and social services* (e.g. mobile clinics and schools); helping them *integrate in local and regional markets* (e.g. meeting consumer safety standards, packaging and transport); and *certifying and labelling* their goods for consumers so as to qualify for the premium 'organic' label.

Vulnerability to climate change and other hazards

Climate change and other hazards constitute a critical set of interactions between society and the environment. As transitional economies emerging from the collapse of the Soviet Union, the republics of Central Asia (and Mongolia and Afghanistan) are particularly vulnerable due to: (i) their physical geography (which is dominated by temperate deserts and semi-deserts); (ii) relative underdevelopment resulting from an economic focus on monocultural agricultural exports (mainly cotton) before 1991; and (iii) traumatic social, economic and institutional upheavals following independence. Aridity is expected to increase across the entire Central Asian region, but especially in the western parts of Turkmenistan, Uzbekistan and Kazakhstan. Temperature increases are projected to be particularly high in summer and autumn, accompanied by decreases in precipitation (Lioubimtseva & Henebry, 2009). The GCA region is very rich in biological resources and diversity, with many biodiversity hotspots that are vulnerable to deforestation, land degradation and desertification. It is also a region that is most vulnerable to climate change, which exacerbates land degradation and desertification, and vice versa (see also Squires & Lu, 2018 for further discussion).

Summary and conclusions

The collapse of the Soviet Union raised numerous questions and problems in many GCA countries regarding the transformation of political, economic, social and cultural life. While the five former Soviet republics proclaimed the building of developed democratic states as an overall goal, they inherited a deeply politicized public administration system in which government bodies attempted to control almost all aspects of society. The process of creating new public administration institutions and mechanisms in GCA states occurred simultaneously with their economic and political transformations. Despite common social and cultural traditions and institutional frameworks, during the first years of independence the five 'stans' have pursued varying state reform models. As a result, they currently differ in the completeness of political and economic reforms, decision-making mechanisms, and the development of civil society and the private sector. These differences can hinder the rapid development of regional cooperation and integration.

The record of actions taken to care for the environment, particularly land, in GCA makes for sad reading. The use or misuse of the land by humans is at the center of a complex web of competing environmental, social and economic pressures. Decisions about the use of land are driven by powerful competing economic and social forces that frequently have little regard for the long-term care of the environment, or of land in particular. Consequently, throughout the history of GCA in the past 100 years, land degradation has been a feature accompanying economic development in most parts of GCA. Throughout GCA, land and soil have been exploited, often abused by agri-business (often government-owned) to obtain increased profits, or by impoverished, hungry subsistence farmers and herders struggling to feed their families. At both extremes, the pressures of today outweigh any concerns that the natural resources – especially land, water and biodiversity – may be required to sustain future generations. It is apparent that there is a need to adopt a new philosophy (or paradigm) concerning land – one that regards land in an ecological sense, and that considers both humans and the natural world as parts of the same ecosystem (Tongway & Ludwig, 2011).

References and further reading

Alibekov, L. & Alibekov, D. 2006. 'Causes and Socio-economic Consequences of Desertification in Central Asia' in R. Behnke (ed.) *The Socio-economic Causes and Consequences of Desertification in Central Asia. Proceedings of the NATO Advanced Research Workshop, Bishkek, Kyrgyzstan, June 2006*. Springer, Dordrecht.

Babaiev, A. (ed.). 1999. *Desert Problems and Desertification in Central Asia*. Springer-Verlag, Berlin Heidelberg.

Bekchanov, M., Djanibekov, N. and Lamers, J.P.A. 2018. 'Water in Central Asia: a cross-cutting management issue' in this volume, pp. 211–36.

Behnke, R. 2006. *The Socio-economic Causes and Consequences of Desertification in Central Asia. Proceedings of the NATO Advanced Research Workshop, Bishkek, Kyrgyzstan, June 2006*. Springer, Dordrecht.

Cardesa-Salzmann, A. 2014. 'Combating Desertification in Central Asia: Finding New Ways to Regional Stability through Environmental Sustainability?', *Chinese Journal of International Law* 13 (1): 203–31. Available online at: http://chinesejil.oxfordjournals. org/content/13/1/203.short

Ellis, J. and Lee, R.Y. 2002. 'Collapse of the Kazakhstan Livestock Sector: A Catastrophic Convergence of Ecological Degradation, Economic Transition and Climate Change'. DARCA Project, European Union. Macaulay Land Use Research Institute, Aberdeen. Available online at: http://www.macaulay.ac.uk/darca/index. htm, accessed November 2014.

Fernandez-Gimenez, M. 1999. 'Sustaining the Steppes: A Geographical History of Pastoral Land Use in Mongolia', *Geographical Review* 89 (3): 315–42.

Gintzburger, G., Le Houerou, H. and Toderich, K.N. 2005. 'The Steppes of Middle Asia: Post-1991 Agricultural and Rangeland Adjustment', *Arid Land Research and Management* 19: 215–39.

Gintzburger, G., Toderich, K.N., Mardonov, B.K. and Mahmudov, M.M. 2003. *Rangelands of the Arid and Semi-arid Zones in Uzbekistan*. Centre de Cooperation Internationale en Recherche Agronomique pour le Developpement (CIRAD), Montpellier.

Halimova, N. 2012. 'Land Tenure Reform in Tajikistan: Implications for Land Stewardship and Social Sustainability – A Case Study' in V.R. Squires (ed.) *Rangeland Stewardship in Central Asia: Balancing Improved Livelihoods, Biodiversity Conservation and Land Protection*. Springer, Dordrecht, pp. 305–29.

Indoitu, R., Orlovsky, L. and Orlovsky, N. 2012. 'Dust Storms in Central Asia: Spatial and Temporal Variations', *Journal of Arid Environments* 85: 62–70.

Jiaguo, Q. and Evered, K. 2008. *Environmental Problems of Central Asia and their Economic, Social and Security Impacts*. Springer Business & Economics, Dordrecht.

Jigjidsuren, S. and Johnson, D.A. 2003. *Forage Plants in Mongolia*. RIAH, Ulaanbaatar, Mongolia.

Kerven, C. (ed.). 2003. *Prospects for Pastoralism in Kazakhstan and Turkmenistan: From State Farm to Private Flocks*. Routledge Curzon, London.

Kerven, C., Shanbaev, K., Alimaev, I.I., Smailov, A. and Smailov, K. 2008. 'Livestock Mobility and Degradation in Kazakhstan's Semi-arid Rangelands' in R. Behnke (ed.) *The Socio-economic Causes and Consequences of Desertification in Central Asia*. Springer, Dordrecht, pp. 113–14.

Kerven, C., Steimann, B., Ashley, L., Dear, C. and Rahim, I. 2011. *Pastoralism and Farming in Central Asia's Mountains: A Research Review*. Mountain Societies Research Institute, University of Central Asia, Bishkek.

Kharin, N.G. 2002. *Vegetation Degradation in Central Asia under the Impact of Human Activities*. Springer Nature, Dordrecht.

Kijne, J.W. 2005. *Aral Sea Basin Initiative: Towards a Strategy for Sustainable Irrigated Agriculture with Feasible Investment in Drainage. Synthesis Report*. IPTRID/FAO, Rome.

Kreutzmann, H. 2012. *Pastoral Practices in High Asia: Agency of 'Development' Effected by Modernisation, Resettlement and Transformation*. Springer, Dordrecht.

Kuehnast, K. and Dudwick, N. 2004. *Better a Hundred Friends than a Hundred Rubles?: Social Networks in Transition – The Kyrgyz Republic*. World Bank, Washington DC.

Kurbanova, B. 2012. 'Constraints and Barriers to Better Land Stewardship: Analysis of PRAs in Tajikistan' in V.R. Squires (ed.) *Rangeland Stewardship in Central Asia: Balancing Improved Livelihoods, Biodiversity Conservation and Land Protection*. Springer, Dordrecht, pp. 129–61.

Lerman, Z. 2000. 'From Common Heritage to Divergence: Why the Transition Countries Are Drifting Apart by Measures of Agricultural Performance', *American Journal of Agricultural Economics* 82 (5): 1140–48.

Lerman, Z. 2012. 'Rural livelihoods in Tajikistan: what factors and policies influence the income and well-being of rural families?' in V.R. Squires (ed.) *Rangeland Stewardship in Central Asia: Balancing Improved Livelihoods, Biodiversity Conservation and Land Protection.* Springer, Dordrecht, pp. 165–87.

Li, P., Qian, H., Howard, K.W.F. and Wu, J. 2015. 'Building a New and Sustainable "Silk Road Economic Belt"', *Environmental Earth Sciences* 74: 7267–70.

Lioubimtseva, E.U. and Henebry, G.M. 2009. 'Climate and Environmental Change in Arid Central Asia: Impacts, Vulnerability, and Adaptations', *Journal of Arid Environments* 73 (11): 963–77.

Nechaeva, N.T. (ed.). 1985. *Improvement of Desert Ranges in Soviet Central Asia.* Harwood Academic Publishers, Amsterdam.

Olimova, S. and Bosc, I. 2003. *Labour Migration from Tajikistan.* International Organisation for Migration/Sharq Scientific Research Centre, Moscow.

Orlovsky, L. and Orlovsky, N. 2018. 'Biogeography and natural resources of Greater Central Asia: an overview' in this volume, pp. 23–47.

Petrov, M.P. 1966. *The Deserts of Central Asia. Vol. I: The Ordos, Alshan and Peishan; Vol. II: The Hoshi Corridor, Tsaidam and Tarim Basin.* Nauka Publishing House, Moscow (in Russian). Translation: US Department of Commerce, Joint Publication Research Service (ref. TT 66-35568 and 67-33399), Washington DC.

Petrov, M.P. 1973. *The Deserts of the Globe.* Nauka Publishing House, Leningrad (in Russian).

Pomfret, R. 2006. *The Central Asian Economies since Independence.* Princeton NJ: Princeton University Press.

Pomfret, R. 2010. 'Central Asia after Two Decades of Independence'. World Institute for Development Economics Research (UNU-WIDER) Working Paper 2010/53, UN University, Helsinki.

Rachkovskaya, E.I., Volkova, E.A. and Khramtsov, V.N. (eds). 2003. *Botanical Geography of Kazakhstan and Middle Asia (Desert Region).* Komarov Institute of Botany, Russian Academy of Sciences, St Petersburg.

Robinson, S., Wiedemann, C., Michel, S., Zhumabayev, Y. and Singh, N. 2012. 'Pastoral Tenure in Central Asia: Theme and Variation in the five Former Soviet Republics' in V. R. Squires (ed.) *Rangeland Stewardship in Central Asia: Balancing Improved Livelihoods, Biodiversity Conservation and Land Protection.* Springer, Dordrecht, pp. 239–74.

Schuette, S. 2012. 'Pastoralism, Power and Politics: Access to Pastures in Northern Afghanistan' in H. Kreutzmann (ed.) *Pastoral Practices in High Asia: Agency of 'Development' Effected by Modernisation, Resettlement and Transformation.* Springer, Dordrecht, pp. 53–69.

Sedik, D. 2012. 'The Feed-livestock Nexus: Livestock Development in Tajikistan' in V. R. Squires (ed.) *Rangeland Stewardship in Central Asia: Balancing Improved Livelihoods, Biodiversity Conservation and Land Protection.* Springer, Dordrecht, pp. 189–212.

Shigaeva, J., Wolfgramm, B. and Dear, C. 2013. *Sustainable Land Management in Kyrgyzstan and Tajikistan: A Research Review.* Mountain Societies Research Institute, University of Central Asia, Bishkek.

Sneath, D. 1998. 'State Policy and Pasture Degradation in Inner Asia', *Science* 281 (5380): 1147–48.

Squires, V.R. 2012. *Rangeland Stewardship in Central Asia: Balancing Improved Livelihoods, Biodiversity Conservation and Land Protection.* Springer, Dordrecht.

Squires, V.R. 2013. 'Replication and scaling up: where to from here?' in G.A. Heshmati and V.R. Squires (eds) *Combating Desertification in Asia, Africa and the Middle East: Proven Practices.* Springer, Dordrecht, pp. 445–59.

Squires, V.R. and Lu, Q. 2018. 'Unifying perspectives on land, water, people, national development and an agenda for future social-ecological research' in this volume, pp. 283–305.

Strong, P.J.H. and Squires, V.R. 2017. 'Rangeland-based Livestock: A Vital Subsector under threat in Tajikistan' in V.R. Squires (ed.) *Rangeland Stewardship in Central Asia: Balancing Improved livelihoods, Biodiversity Conservation and Land Protection.* Springer, Dordrecht, pp. 213–37.

Swanström, N. 2018. 'Greater Central Asia: China, Russia or multilateralism?' in this volume, pp. 273–82.

Tongway, D.T. and Ludwig, J.A. 2011. *Restoring Disturbed Landscapes: Putting Principles into Practice.* Island Press, Washington DC.

Walter, H. and Box, O. E. 1983. 'Middle Asia Deserts' in N. West (ed.) *Temperate Deserts and Semi-deserts. Ecosystems of the World Volume 5.* Elsevier Scientific Publishers, Amsterdam, pp. 71–103.

World Bank. 2006. *Priorities for Sustainable Growth: A Strategy for Agriculture Sector Development in Tajikistan.* World Bank, Washington DC.

2 Biogeography and natural resources of Greater Central Asia

An overview

Leah Orlovsky and Nikolai Orlovsky

BEN-GURION UNIVERSITY, ISRAEL

Introduction: background to terminology

What countries/regions are included in the concept of "Greater Central Asia"? About two hundred years ago, Alexander von Humboldt took 44.5°N as the median latitude to define the region, and all areas five degrees to the north and to the south were considered as Central Asia (von Humboldt, 1843). According to Humboldt, the western border of the region was the Caspian Sea, and the eastern limit was undefined. Since then, in the Russian (and later Soviet) literature, there has been a division between "Middle" and "Central" Asia. In the narrow sense, Middle Asia is an area between the Caspian Sea in the west and the Pamir Mountains in the east, the Aral-Irtysh watershed in the north, and the Kopetdag-Hindukush Mountains in the south. At present, administratively, "Middle" Asia includes the newly independent states of the former Soviet Union (FSU): Turkmenistan, Uzbekistan, Tajikistan, Kyrgyzstan and southern Kazakhstan (Alexeyeva and Ivanova, 2003). According to the same source, "Central" Asia is an inland part of Asia between the Big Khingan mountain range in the east, the upstream valleys of the Indus and Brahmaputra Rivers in the south, and the mountains of eastern Kazakhstan in the west and north.

The cultural and geographical space of Central Asia in a significantly broader interpretation was given, for the first time, by UNESCO in six volumes of *History of Civilizations of Central Asia* – a project started in 1981 and completed in 2005. According to the UNESCO version, Central Asia includes eastern Russia, the area south of the taiga belt in Siberia, Mongolia, western China, Tibet, northeastern Iran, Afghanistan, Pakistan, the Soviet Central Asian republics, and the Indian provinces of Uttar Pradesh, Jammu, Kashmir, Himachal Pradesh, Haryana, and Punjab.

American political scientists introduced the concept of "Greater Central Asia" in the 2000s to facilitate the realization of the Modern Silk Road project and structured regional cooperation for a safe and stable Afghanistan (Davis and Sweeney, 2004; Starr, 2005; Blank, 2007). According to this concept, Greater Central Asia includes five Central Asian republics of the FSU, the northern regions of Iran and Afghanistan, Xinjiang region in China, Mongolia, southern Siberia and the Volga regions of Russia.

Geographic features

In our understanding, Greater Central Asia is a vast landlocked region including the five newly independent states (Kazakhstan, Kyrgyzstan, Tajikistan, Turkmenistan, and Uzbekistan), western China (Xinjiang, and the Qinghai-Tibet Plateau), Mongolia and Afghanistan (Figure 2.1). The territory of Central Asia is located between 27^0 and 55^0 North and 64^0 and 119^0 East; it stretches 5300 km from west to east, and 3090 km from north to south, covering an area of more than 11.5 million square km. The population of Greater Central Asia (as defined) is about 185 million (Atlas of China, 2008; Maclean, 2010).

The region consists of two natural zones: the Plains and High Asia. The hypsometric marks range from −154 meters MSL (the Turpan Depression in China) to 8611 m (the Chogori peak in the Karakorum range). The western plain part of the region (Kazakhstan, Turkmenistan, and Uzbekistan) lies at the lowest level of 100–200 m MSL on average; the altitudes vary here from −132 m (the Karagiye Depression in the Mangyshalk Peninsula, Kazakhstan) to 930 m (Tamdytau, Uzbekistan).

Mountainous terrains are typical for the central (Kyrgyzstan, Tajikistan and Afghanistan) and eastern (western China, Mongolia) parts of the region, with distinctive differences between the lower and upper sub-regions. The lower sub-region consists of the Gobi, Alashan, Ordos, Dzungar (Junggar) and Tarim plains located at the altitudes of 500–1500 m. The upper sub-region includes

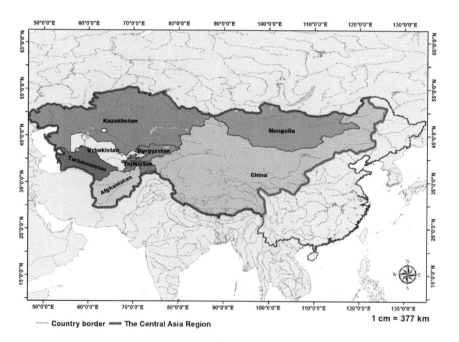

---- Country border ▬▬ The Central Asia Region

1 cm = 377 km

Figure 2.1 Central Asian Region

the Qinghai-Tibet Plateau with altitudes of 4000–4500 m MSL. Lowlands and uplands are separated by the high mountain systems of Tian Shan, Kunlun, Nan Shan, Karakorum, and Mongolian Altai. These mountain systems are the most important factor in the climate formation of GCA. They prevent the intrusion of the cold and dry northern winds to the south, and, at the same time, are impassable barriers for warm and moist air masses from the south. The glaciers of the Tian Shan and Pamir-Alai Mountains are places of origin for the large rivers flowing down to the plains, where irrigated agriculture and other economic activities are being developed. Thus, the mountains are the locations where water resources are being formed and accumulated, and the plains are the water consumption area.

The location of GCA in the middle of the vast Eurasian continent, its significant distance from the oceans, the presence of high mountain systems, high temperatures in summer and a long dry period predetermined by specific atmospheric circulation all cause the high aridity of the region. GCA can be divided into four bioclimatic zones:

1   Sub-humid zone with 400–800 mm of rainfall, represented by a chernozem steppe. Area of traditional rainfed agriculture; irrigation is needed for highly productive farming. The zone is characterized by a progressive increase in aridity mainly due to anthropogenic factors.
2   Semi-arid zone with precipitation of 200–300 mm. Dominated by shrub vegetation communities with patchy herbaceous cover. Area of rainfed crops and livestock breeding.
3   Arid zone with 100–200 mm of rainfall. Sparse vegetation represented by annual and perennial xerophytic species. Crop cultivation is impossible without irrigation. Zone of migratory livestock breeding.
4   Extra-arid zone with precipitation less than 100 mm, devoid of vegetation, except sparse vegetation consisting of shrubs and ephemerae along watercourses. Farming and livestock breeding is impossible (except in oases). This is so-called true desert.

Ecographic division of Central Asia

Central Asia can be divided into four physiographic regions/provinces (Petrov, 1966): the **Turanian, Junggar-Kazakhstan, Mongolian**, and **Qinghai-Tibetan provinces**. They differ significantly by climatic peculiarities and vegetation composition, while their litho-edaphic features are more homogeneous.

The *Turanian province* comprises the territories of Turkmenistan, Uzbekistan, Tajikistan and Afghanistan and has a continental climate of subtropical deserts. This area is located to the south of 43–44°N and includes the Karakum, Kyzylkum, Khash, and Dasht-e-Margo Deserts. Garmser, Reghistan and the mountain regions include western Tian-Shan and the Ferghana Valley, the Pamir-Alay, Kopet-Dag, and Parapamiz Mountains and the western Hindu Kush range.

This vast territory includes two climatic regions: a dry climate with very hot summers and temperate mild winters in the north, and a dry climate with very hot summers and mild winters in the south (the southern areas of Turkmenistan, Uzbekistan and Afghanistan).

The rainfall pattern is characterized by a winter–spring maximum and a summer minimum, which is distinctive for the Mediterranean type of climate. In the period of July–September, the rainfall amount is negligible in the southern zone of ephemeral deserts – less than 15 mm even at the northern edges of the region. In most of Afghanistan, the period of July–September is dry with zero precipitation (Titov, 1976). Precipitation during the cold season (November to March) amounts to more than 70% of the annual total; this amount varies from 50 mm in southern Afghanistan to 250 mm and more in the piedmonts. Snowfall is a rare phenomenon, and snow cover doesn't last long. Springs are wet and warm, summers are hot and cloudless. Soils are desert soils, light sierozems, strong-carbonate gray-brown soils, solonchaks and takyrs.

The vast massifs of the western Tian Shan, Pamir-Alay, Kopet Dag and western Hindu Kush mountains are related to the zone of the southern deserts. The landscapes of almost all the vertical zones of these mountain ranges, with a few exceptions, bear the marks of a desert in varying degrees (Babaev et al., 1986; Kharin, 2002). This zone (especially the windward slopes) is dominated by xerophytic vegetation, because of the high temperatures and lack of rainfall in summer.

All of the western slopes of Tian Shan are occupied by steppe vegetation, since they receive a higher amount of precipitation, though in summer they remain dry up to an altitude of 1000 m. The maximum precipitation occurs in spring.

The peculiarities of rainfall and temperature regimes affect vegetation development. The growing season is characterized by a marked spring peak of biomass and a dormant summer period. Ephemerals and ephemeroids are widely distributed. The vegetation is characterized by widespread shrubby psammophytic formations (*Calligonum, Haloxylon, Salsola*) on the sandy massifs, gypsophytes (*Salsola, Anabasis*) on the tertiary plateau and ephemeroid-ephemeral herbaceous communities on the piedmonts (*Carex, Poa, Gagea, Tulipa*).

The ***Junggar-Kazakhstan province*** comprises the entire Republic of Kazakhstan and the northern part of the Junggar Basin, located in the northwestern part of China. The province can be divided into three sub-regions: a) a zone of steppes and the adjacent mountain areas of western and southern Altai; b) a zone of semi-deserts and northern deserts; and c) the adjacent mountain areas of Alatau, northern and central Tian-Shan, the Issyk Kul Lake Valley, and Junggar itself.

The *steppe zone* of Kazakhstan occupies an area of 77 million hectares or 29% of the country (Valikhanova et al., 2005). It includes the north of the Turgay Plateau, almost all the area of the Mugodzhar Mountains, the northern edge of the Caspian lowland, and the northern Kazakh upland. The climate is continental with hot and dry summers and severe winters with a lasting snow cover up to 10–15 cm. Precipitation is about 300 mm, decreasing gradually from the north to the south. Rainfed agriculture is possible. A large-scale development project in the Virgin Lands, begun in the 1950s, evolved into agriculture production in more than 28 million hectares (Mha) of territory. The continuous misuse of this area has led to the development of soil erosion processes, resulting in dust storms, which have destroyed the fertile soil horizon.

Semi-deserts and northern deserts occupy 58% of Kazakhstan (Iskakov et al., 2006). In the west and northwest, this area is bounded by the Caspian Sea; in the north, it is limited by the southern border of the steppes along the southern slope of the most elevated part of the Kazakh upland. The southern part of the zone is occupied by the Turanian lowland with the sandy deserts of the Small and Big Barsuki, Priaral (Circum-Aral) Karakum, and the Moyunkum. On the southern edge of the Kazakh upland, there are Lakes Balkhash, Zaysan, Alakol and Sasykkol. The Aral Sea is situated in the southern part of the region.

The northern deserts border, to the south and east, the *mountainous region*, including the Junggar, the Alatau, the ridges of the northern and central Tian Shan, and the Issyk Kul Lake Basin. The western and northern slopes receive more precipitation. The average annual rainfall on the northern slopes is about 750 mm; the southwestern slopes of the Ferghana Ridge receive 950 mm of rainfall, and the northeastern slopes get 230 mm. The slopes of Tian-Shan are occupied by semi-desert and steppe vegetation, sub-alpine and alpine meadows; forests occupy an insignificant amount of areas.

The Junggar Basin, the circular intermountain depression, is situated between the Tian Shan and Altai Mountains at the altitude of 600–800 m – relatively low compared with the other plains of western China. The terrain includes lowlands, pebble-cobble plains, solonchak depressions and hillock sands. The Gurban Tonggut Desert – the largest stabilized and semi-stabilized sandy desert in China – is located in the center and southern part of the region. The surrounding mountains allow the passage of the humid westerly winds, and precipitation can reach up to 70–150 mm annually, which is higher than the southern Tarim Basin. Snow cover lasts for a long time in winter – up to 100–160 days, with a snow depth of more than 20 cm. Vegetation covers up to 40–50% of the fixed sand dunes, and 15–25% of the semi-fixed sand dunes. There are more than 100 species of plants, including both Chinese shrub species and Kazakh desert flora. The Chinese plant species are dominant in the eastern desert, and include *Haloxylon ammodendron, Ephedra distachya* and *Hedysarum scoparium*, while in the western and central parts of the desert, the Kazakh desert plants, such as *Haloxylon persicum, Artemisia santolina, Artemisia terrae-albae* and *Carex physodes*, predominate (Sun et al., 2010).

The **Mongolian Province** includes all the territory of Mongolia and the arid northwestern region of China. It includes the Gobi Desert located in northern and northwestern China and southern Mongolia. The desert is bounded by the Altai Mountains and the Mongolian grasslands in the north, the Taklimakan Desert in the west, the Hexi Corridor and Tibetan Plateau in the southwest, and the Northern China Plain in the southeast.

The Mongolian province has some distinctive differences from the Turanian and Junggar-Kazakhstan provinces, such as terrain (higher elevation in the Mongolian province), atmospheric circulation, precipitation seasons and amount of rainfall, soil and vegetation cover, and fauna. The eastern part of the Mongolian province up to the Beishan Desert is influenced by the eastern monsoons that cause the rains in summer. Summer precipitation also occurs in the high plains of the Qaidam Basin and the Beishan Desert, but the rainfall amount is less here, and is not associated with monsoon activity. There is some kind of climate dividing line in this area between its western part, influenced by Atlantic air masses, and its eastern part, which experiences the impact of Pacific air streams (Babaev et al., 1986).

The flora of the region, although unique, is relatively poor by the number of species. The ancient, typical desert plant genera were formed here, endemic or closely related to Australian and South African species (Babaev et al., 1986). The most abundant in the Mongolian flora are the *Chenopodiaceae* species, followed by *Asteraceae*, *Gramineae* and *Leguminosae*. Herbaceous vegetation is poorly developed, and ephemerae are almost absent. Ephemeroids are more common; in contrast to the ephemeroids of the Turanian Province, their biomass peak occurs in summer (Petrov, 1966).

The terrain of Mongolia is constituted by an elevated plateau with heights of 900–1500 m MSL. The plateau borders the Mongolian and Gobi Altai mountain ridges in the west and southwest; the Sayan and Khentii mountain systems are located in the north of the country; the Khangai mountain range stretches from northwest to southeast and occupies the central and northern central part of Mongolia. The Great Lakes Basin is in the northwestern part of the country. The eastern part of Mongolia is occupied by the plains, which merge with the Gobi Desert in the south.

The climate is continental with extremely cold and dry winters, dry, cold and windy spring seasons, and short and warm summers. The total amount of precipitation is rather low, with the maximum occurring in July–August. The rainfall amount decreases from 600 mm/year in the Hudsugul and Khentii Mountains to 25 mm/year or less in the Gobi Desert. The northern part of Mongolia is a territory with permafrost (Gunin et al., 1999).

There are six natural zones changing from the north to the south: high mountains, boreal forest, forest-steppe, steppe, semi-desert and desert (National Plan of Action to Combat Desertification in Mongolia, 1997). Mountains cover more than 40% of the total area of Mongolia, while steppes occupy most of the country.

The northwestern region of China occupies 55.1% of the country, approximately 10% of all arable lands, and 6.4% of the total Chinese population lives here. The region has an arid, semi-arid and pronounced continental climate; natural vegetation is mainly constituted by desert and desert-steppe species, which makes it the main grazing area of the country (Shen, 2010).

The Tarim Basin, with the sandy Taklimakan Desert, is located in the west of the region. Harsh climatic conditions and extreme aridity are typical for its landscape. The relief is dominated by sandy ridges, barkhan and hillock sands. In the east, the Lop Nor Lake Depression is situated. The vegetation of the Taklimakan Desert is sparser; some species of *Aristida pennata*, *Agriophyllum arenarium* and *Corispermum* can be seen. Scattered growth of *Haloxylon ammodendron* and *Tamarix laxa* occurs in the dry river valleys (Zhu Zhenda et al., 1988).

The stony Beishan and Gashu Gobi Deserts[1] are situated between the Tarim Basin and the Alashan Desert. Low mountain ridges alternating with inter-mountain depressions are typical for the Beishan Desert. These are extremely arid barren areas with rare species of bushy *Haloxylon ammodendron, Nitraria sphaerocarpa* and *Ephedra przewalskii*. Shrublands, formed by *Caragana leu-cophloea* and *Calligonum mongolicum*, grow in dry riverbeds.

The Hexi Corridor, located to the south of the Beishan Desert, forms a desert plain with mountain ridges and uplands. The western part of the Hexi Corridor is mostly dry and deserted. Rivers and streams cascading from the Nan Shan Mountains irrigate the oases of the Corridor (Babaev et al., 1986).

The Alashan and Ordos Deserts, divided by the Huanghe River, are the eastern prolongation of the Gobian Desert belt. The Alashan is a vast desert covered by barkhan sands, forming the large sandy massifs, such as Badain Jaran, Tengger and Ulan Buh. Besides the barkhan ridges, depressions with saline and fresh lakes, solonchaks, meadow and marsh vegetation communities, and *Haloxylon ammodendron* are widely distributed. Vegetation is represented by sparse thick-ets of psammophytic bushes: *Hedysarum mongolicum, Atraphaxis frutescens, Caragana microphylla* and *C. bungei* (Zhu Zhenda et al., 1988).

The Ordos is located in the northern bend of the Huanghe River. The terrain is constituted by depressions and uplands with numerous gullies. About half of the area is occupied by Aeolian sands, barkhans in the north and overgrown by vegetation in the south. The vegetation of the central part consists of *Caragana korshinskii* and *Caragana intermedia* with participation of *Artemisia ordosica* (Sun et al., 2010).

The Gobi Desert is located between the mountains of the Mongolian Altai and Khangai, and the eastern Tian Shan, Altyntag, Beishan and Ini Shan. It stretches from the west to the east for 1750 km, and from the south to the north for 600 km. The Gobi is divided into the Gashun, Junggar and Zaaltai Gobi in the west, and the Eastern Gobi in the central and eastern parts. The Eastern Gobi is a plain with average altitudes of about 1000 m.

The ***Qinghai-Tibet Plateau*** covers most of the Tibet Autonomous Region and the Qinghai Province in western China. It stretches approximately 1000 km from north to south and 2500 km from east to west. With an average elevation exceed-ing 4500 m, the Tibetan Plateau is sometimes called "the Roof of the World" and is the world's highest and largest plateau, with an area of 2.5 million km[2]. The pla-teau is framed by the towering mountain ranges of the Himalayas (to the south), the Kunlun, Arjin and Qilian Mountains (to the north), the Himalayan, Kunlun and Karakorum and Pamir Mountains (to the west) and the highlands of Gansu, Sichuan and Yunnan (to the east) (Brown et al., 2008).

The area contains major eco-regions, including the Central Tibetan Plateau Alpine Steppe, the Southeast Tibet Shrublands and Meadows, the North Tibetan Plateau-Kunlun Mountains Alpine Desert, the Tibetan Plateau Alpine Shrublands and Meadow, the Karakorum-West Tibetan Plateau Alpine Steppe, the Eastern Himalayas Alpine Shrub and Meadow, the Qilian Mountains Sub-alpine Meadows, the Western Himalayan Alpine Shrub and Meadows, the Yarlung Tsangpo Arid Steppe, and the Qaidam Basin Semi-desert.[2]

From the southeast to the northwest, the vegetation changes from alpine grasslands to alpine meadow steppe to alpine steppe to high altitude desert steppe to high altitude desert. About 50% of the plateau is alpine meadow, which supports 80% of the domestic livestock and 70% of the population (Brown et al., 2008).

The region is high, dry and cold. About 80% of the plateau is above 4000 m in altitude, with lake basins and river valleys making up the other 20%. The east is relatively humid and has about 600 mm average annual precipitation; the south is semi-arid; the west is arid with around 60 mm precipitation; the center is sub-frigid; and the northwest is frigid and arid. Most precipitation occurs as snowfall and hail in the summer with some heavy snowfall in winter. The harsh climate impacts all living things in Tibet (Leber et al., 1995). Sparseness of flora and the absence of closed canopies characterize most of the Tibetan region. The central and western areas of Tibet have especially poor vegetation cover. Cold alpine deserts with perennial low vegetation predominate. Herbage consisting of *Poa alpine*, *Avena sp.*, *Carex sp.* and *Kobresis pygmaea* grows along the small watercourses up to the altitude of 5000 m.

Zones of temperate and cold sagebrush deserts occur at the height of 3000–4000 m in the northwestern region and 3900–4200 m in the northern part. Among the dominant species are *Artemisia sacrorum*, *A. webbiana* and *A. salsoloides*, as well as *Stipa glacerosa* and *S. purpurea*, *Christolea*, *Ceratoides*, *Ajania*, and xerophytic herbs.

In the southern Tibetan landscapes of deserts and highlands, dry steppes occur only in the watersheds. In the river valleys, especially in the wide valley of Tsangpo, rich and diverse vegetation communities appear, such as grassy meadows, with the growth of *Salix sp.*, *Populus* and *Taxus*. Mountain meadow communities prevail at the height of 4000 m. Highland pastures can be used from April to June; in summer, flocks move to the lower areas, and towards October, they move down for their winter stay.

The **Qaidam Basin** lies in the northwestern area of the Qinghai Province. It is a large inland basin in the northwestern region of the Qinghai-Tibetan Plateau. The basin is about 2500–3000 m above sea level and is the highest desert region in China. The eastern part of the basin is desert steppe, and the west is desert. The basin comprises a mosaic landscape of sand dunes, salty lakes and solonchaks. The sand dunes are mainly covered with shrubs of 3–5 m in height. The most common plants are *Tamarix ramosissima* and *Haloxylon ammodendron* (Sun et al., 2010).

Water resources in GCA

Water in GCA is the most valuable resource, acquiring ever increasing economic and social significance (Bekchanov et al., 2018). GCA water resources consist of river flows formed by atmospheric precipitation, melting glacial waters and underground waters. They comprise renewable surface and groundwater of natural origin, as well as return flow of anthropogenic origin. A distinction must be made between internal renewable water resources (IRWR) and common renewable water resources (TRWR). IRWR is that part of a country's water resources generated by endogenous precipitation (produced in the country). Calculation of IRWR involves adding surface water flow and groundwater recharge and subtracting the overlap. TRWR is calculated by adding IRWR and external flow. This is a measure of the maximum theoretical amount of water available to a specific country without considering its technical, economic or environmental nature. Natural flow is the average annual amount of water that would flow at a given point in a river without any human influence, while actual flow takes into account the volumes of water reserved by treaties and agreements (Frenken, 2013; Bekchanov et al., 2018).

Central Asia is a landlocked region, and its water supply is dependent mostly on precipitation. The volume of annual precipitation in the Central Asian countries is estimated at 2540 km^3 (Table 2.1). This volume is equal to a regional average depth of 247 mm/year, but with significant disparities between and within countries. Average annual precipitation varies from less than 70 mm in the plains and deserts to more than 2400 mm in the mountains of Central Tajikistan. The lowest precipitation (less than 50 mm/year) occurs in the southern Xinjiang and the western part of the Qaidam Basin (Shen, 2010). At the country level, the driest country is Turkmenistan, with an average of 161 mm/year, and the wettest is Tajikistan with 691 mm/year.

Long-term average annual internal renewable water resources (IRWR) in Central Asia account for 455.7 km^3 (Table 2.1).

In absolute terms, the northwestern part of China accounts for the largest amount of IRWR: 181.3 km^3/year or 39.8% of the region's water resources. This refers to 39.5% of the region's total area, thus giving a depth of only 47 mm. Tajikistan follows with 63.5 km^3, or 13.9% of the region's water resources, taking into account that the country represents only 1.4% of the total area of the region, resulting in the greatest depth of 445 mm. Kazakhstan, with 64.4 km^3/year, accounts for 14.1% of the region's water resources, while its area covers 26.5% of the region, giving a depth of 24 mm (see also Table 3.1 in Shaumarov and Birner, 2018).

Kyrgyzstan and Afghanistan account for 48.9 km^3 and 47.2 km^3, respectively; each represents 10% of the water resources in the region. Kyrgyzstan accounts for only 1.9% of the region's total area, giving a depth of 245 mm, while Afghanistan accounts for 6.3% of the total area, giving a depth of 72 mm. Uzbekistan, with 16.3 km^3/year, accounts for 3.6% of the region's water resources, while its area covers 4.4% of the region, giving a depth of 37 mm.

Turkmenistan has the least water resources, with 1.4 km^3/year or less than 1% of the water resources in Central Asia. Its area represents 4.8% of the region, giving the least depth of 3 mm (Table 2.1).

The distribution of the total actual renewable water resources (TARWR) is different due to transboundary rivers. Thus, in Turkmenistan, IRWR are 275 m^3/inhabitant, while TARWR are 4851 m^3/inhabitant, and in Uzbekistan these figures are 589 m^3 and 1760 m^3, respectively. Conversely, in Kyrgyzstan and Tajikistan, IRWR are higher than TARWR, 9073 and 9096 m^3/inhabitant

Table 2.1 Long-term average annual renewable water resources

Country	Average annual precipitation		Annual renewable water resources			Actual total, taking into consideration agreements		Dependency ratio
			Internal (IRWR)					
	Depth	Volume	Volume	Depth	Per capita (2011)	Volume	Per capita (2011)	
	mm	km³	km³	mm	m³/capita	km³	m³/capita	%
Afghanistan	327	213	47.2	72	1457	65.3	2019	29
Kazakhstan	250	681	64.4	24	3971	107.5	6632	40
Kyrgyzstan	533	107	48.9	245	9073	23.6	4372	1
Mongolia	241	377	32.7	21	11758	34.8	12446	0
NW China*	230	892	181.3	47	2264	202.7	2531	0
Tajikistan	691	99	63.5	445	9096	21.9	3140	17
Turkmenistan	161	79	1.4	3	275	24.8	4851	97
Uzbekistan	206	92	16.3	37	589	48.9	1760	80
Central Asia	247	2540	455.7	44	2505	529.5	–	–

Sources: AQUASTAT;[3] Frenken, 2013; Shen, 2010.

* Not including Tibet.

compared with 4372 and 3140 m³/inhabitant, respectively, because of the water allocation agreements between the GCA countries.

The GCA states are connected by the following transboundary river systems: Syr-Darya and Amu-Darya (Afghanistan, Kazakhstan, Kyrgyzstan, Tajikistan, Turkmenistan and Uzbekistan), Chu and Talas (Kyrgyzstan, Kazakhstan), Tarim (Kyrgyzstan, Tajikistan, China), Ili (China, Kazakhstan), Irtysh (China, Kazakhstan, Russia), Murgab (Afghanistan, Turkmenistan), Tedjen (Afghanistan, Iran, Turkmenistan), and Selenga (Mongolia, Russia). The sources of most of the big and small rivers are in high mountains, which feed many of the rivers, lakes, water reservoirs and canals. Thus, the mountains are the areas of water accumulation, and the plains are the areas of water consumption, evaporation, and filtration. Most of the Central Asian territory comprises closed basins (Tarim, Junggar, Great Lakes Basin, Balkhash Lake Basin); only the edges of the territory have discharge to the ocean (Huanghe, Irtysh, Selenga).

Groundwater

Groundwater resources can be classified according to their recharge processes. Two main types can be distinguished: (1) groundwater formed under natural conditions in the mountain zone and catchment areas by infiltration of rainfall (autochthonous groundwater); and (2) groundwater formed by the infiltration losses from the irrigated areas.

The groundwater resources in the Chinese drylands are related to the first type; their total volume is 107.4 km³ (Shen, 2010).

In the Aral Sea Basin, renewable groundwater resources are located in 339 aquifers with a total volume of 43.51 km³, of which 20.58 km³ are in the Amu-Darya Basin and 22.93 km³ in the Syr Darya Basin. The actual water extraction from aquifers is 11.04 km³ per year, though in 1990 it exceeded 14.0 km³. The greatest groundwater deposits are found in Uzbekistan and Tajikistan (Alikhanova, 2008). They are considered to be a strategic resource for the future and are reserved for the essential human needs of future generations.

Return flow

Return flow forms a high proportion of water resources in the Aral Sea Basin. At present, it comprises drainage water from irrigated fields and wastewater from industry and municipalities. The total volume of return flow varies from 28.0 km³ to 33.5 km³ annually; 92% of it consists of drainage water and about 8% of untreated municipal and industrial wastewater. About 13.5–15.5 km³ are produced annually in the Syr Darya River Basin, and about 15–18 km³ are formed in the Amy Darya River Basin (International Fund for Saving the Aral Sea, 2011). The high portion of drainage water shows that only 45–50% of total agricultural withdrawals actually go to irrigation.[4]

Natural lakes

Most of the Central Asian lakes are endorheic (closed drainage basins). Published sources indicate that there are more than 56,600 lakes within the area greater than 1 km^2 (National Report on the State of Environment in Kyrgyzstan, 2012; Water Resources of Kazakhstan in the New Millennium, 2002; State of Environment in Turkmenistan, 2008; Nikitin, 1986; ADB, 2014; Shen, 2010). The total area covered by the lakes is about 67,000 km^2, with the total volume more than 3140 km^3. Significant water sources are concentrated in the Issyk Kul Lake (1738 km^3), the Khuvsgul (381 km^3), and the Balkhash (112 km^3). Most of the lake water resources are accumulated in the slightly saline lakes – more than 90% of the total volume.

There were no significant changes in the mountain lakes during the last few decades, while the area and amount of the lakes in the plains changed dramatically (Shen, 2010; International Fund for Saving the Aral Sea, 2011). Several lakes dried up, such as the Taitema Lake and the Lop Nor Lake downstream of the Tarim River, and the Juyanhai Lake downstream of the Heihe River (Shen, 2010). The deltaic lakes of Amy Darya and Syr Darya disappeared or decreased significantly. New lake-accumulators of drainage water have been formed in the periphery of irrigated massifs (Orlovsky et al., 2014).

Water reservoirs

The total reservoir capacity in the five post-Soviet republics and Afghanistan is 180.5 km^3, of which 53% is in Kazakhstan. Sixteen reservoirs each have a capacity greater than 1 km^3, of which six are in Uzbekistan, four in Kazakhstan, two in Turkmenistan, two in Tajikistan, one in Kyrgyzstan and one in Afghanistan. Most are multipurpose reservoirs for hydropower production, irrigation, water supply and flood control.

Glaciers and snowfields

Glaciers are the main sources of rivers in GCA, providing 25–30% of the annual flow and up to 50% of the flow during the vegetation growing season. The area of the glaciers and, consequently, the volume of melting water that runs off the glaciers can significantly impact the water supply and affect the balance of water resources. Within the five former Soviet republics and Afghanistan there are 51,427 glaciers, which cover an area of 58,040 km^2; the potential water capacity of these glaciers is 4230 km^3 (National Report on the State of Environment in Kyrgyzstan, 2012; Water Resources of Kazakhstan in the New Millennium, 2002; State of Environment in Turkmenistan, 2008; Nikitin, 1986; ADB, 2014; Shen, 2010). They are distributed unevenly throughout the region. Northwestern China has 24,061 glaciers with an area of 35,720 km^2, and a volume of 3078 km^3 (Shen, 2010). In Kyrgyzstan, there are 8,200 glaciers with a total surface area of 8169.4 km^2, occupying 4.2% of the country's territory. Kyrgyzstan's glacier water reserve is estimated at 650 km^3. The number of glaciers in Tajikistan is 14,509, with a total area of 11,146 km², or about 8% of the country's territory. The total ice reserve in glaciers is 845 km³. Mongolia has 1,400 glaciers, which cover 910 km², and the total water reserve is 63 km³ (Report of the State of Environment, 2011; ADB, 2014).

The glaciers are one of the striking indicators of climate change and, to a certain extent, an environmental reaction of a flow formation zone to global warming. At present, there is an intensive reduction in glacier area in Central Asia, which could be explained by an increase in the temperature and a change in the precipitation regime/amount (Seversky and Tokmagambetov, 2004, Lioubimtseva and Henebry, 2009).

Temporal and alternative water resources

These are represented by temporal surface runoff water and large sub-sand lenses. In the deserts of Central Asia, the sole source of freshwater is rainfall, which generates temporary surface runoff on mountain slopes and takyrs (clayey pans). The estimated volume of this water amounts to 0.43 km³ (Lezhinsky, 1974). The local population has a long history of using rainwater for livestock and domestic supply on rangelands and for micro-oasis agriculture. However, at present, these indigenous and simple techniques have been forgotten almost everywhere in Central Asia.

Large freshwater lenses, floating on saline confining layers, are being formed under sandy massifs. According to different estimations, the capacity of a single lens varies from 1 to 10 km³ or even more. The total volume of freshwater lenses is about 69 km³ in Turkmenistan (Zonn, 2014). Only two of them, Yaskhan (10 km³) and Chilmamedkum (4 km³), are being exploited at an industrial scale.

Land resources

Agricultural lands occupy about 665.57 Mha or 72% of Central Asia (Table 2.2), including dry-farming and irrigated lands, as well as perennial meadows and pastures. These agricultural lands constitute from 34% (Tajikistan) to 77% (Kazakhstan) of the total area of Central Asian countries.

Although the total area of Central Asia is enormous, the portion of arable land is very low, and changes from 0.4% in Mongolia to 11% in Uzbekistan. In total, the share of arable lands in the region is 9.5% of the agricultural lands. About 62.38 Mha are considered suitable for irrigation (11%). Currently, the total irrigated area is about 35.14 Mha or 5.4% of agricultural lands (Table 2.2). More than 63% of the irrigated lands are concentrated in the Aral Sea Basin. Currently, most of lands potentially suitable for dryland farming are used as pastures for nomadic livestock husbandry. Increasing the productivity of non-irrigated land is an important task. Some crops, which are cultivated under irrigation, could be moved to rainfed lands, which will substantially reduce the volume of water for irrigation.

Dryland farming is distributed mainly in the foothills and oasis margins in China, Afghanistan, south and southeastern Kazakhstan, and the Aral Sea Basin.

Table 2.2 Land resources in Central Asia (thousand ha)

Country	Land area	Agricultural land	Arable land	Permanent crops	Permanent meadow and pastures	Forest	Irrigated land	Other
Afghanistan	65,286	37,910	7,790	120	30,000	1,350	3,208	26,026
Kazakhstan	269,970	207,975	22,900	75	185,000	3,297	2,066	58,697
Kyrgyzstan	19,180	10,591	1,277	75	9,240	988	1,023	7,601
Mongolia	155,356	113,395	648	3	112,744	10,734	84	31,226
NW China*	294,536	220,296	21,390	415	198,906	10,672	21,805	63,154
Tajikistan	13,996	4,875	860	140	3,875	410	742	8,711
Turkmenistan	46,993	33,838	1,940	60	31,838	4,127	1,995	9,028
Uzbekistan	42,540	26,690	4,350	340	22,000	3,268	4,215	12,582
Total	907,857	655,571	61,155	1,228	593,603	34,845	35,138	217,025

Sources: FAOSTAT;[5] Shen, 2010.

* Not including Tibet Plateau.

The latter has 23.463 Mha of rainfed lands located in piedmont plains, mountain plateaus and slopes, intermountain troughs and valleys of the Aral Sea Basin.

Potentially arable and already cultivated dryland farming lands occupy 7.353 Mha, or 33% of all rainfed area. An area of 1.0 Mha of rainfed lands is located in the rainfall secured zone. A similar area, of 7.562 Mha, is located in the 50% probability zone, and 4.51 Mha in the unsecured by rainfall zone (Babaev et al., 1991).

The arable agricultural resources of Afghanistan comprise about 7.5 Mha of cultivable land, both rainfed and irrigated land. The rainfed area is estimated at about 4 Mha, but the area actually cultivated in a given year varies considerably depending upon climatic factors, mainly precipitation (Afghanistan's Environment, 2008). The recent succession of dry years has reduced the annually cultivated rainfed area to less than 0.5 Mha from a maximum of around 1.4 Mha (ADB, 2002; Brown and Blankenship, 2013).

Natural pastures constitute about 90.6% of the region's agricultural lands. The total area of pastures in Central Asia is about 593.6 Mha (Table 2.2), which is about 65.4% of the region's total lands. About 83.7% of pastoral lands are located in China, Kazakhstan and Mongolia. The other states hold about 16.3% of pastureland; this portion varies from 0.4% in Tajikistan to 5.4% in Uzbekistan. Pastures play a very important role in the economy of all the Central Asian countries. In addition to their direct function, pastures are also sources of fuel and medicinal plants, and serve as recreational sites. They ensure carbon absorption, playing a vitally important role in decreasing greenhouse gases.

Forests occupy about 35 Mha or 3.8% of the land area in Central Asia. Other lands (cities, settlements, roads, lands unsuitable for agriculture, and other uses) occupy about 217 Mha, or 24% of the total area.

Environmental issues

There are several factors causing land degradation processes in Central Asia, such as the significant distance of the region from the ocean, climate aridity, the uneven distribution of water resources and their scarcity, the vulnerability of drylands, underdeveloped infrastructure, high population growth rates and urbanization processes. Any medium- or long-term changes in temperature and precipitation regimes lead to disturbances of the equilibrium in the fragile ecosystems of the region. Droughts, although not causing desertification, nevertheless are catalysts of the vegetation and soil degradation processes under excessive pasture exploitation. All the territory of Central Asia is experiencing significant anthropogenic pressures, including overgrazing, irrational use of water and land resources, cutting of trees and shrubs for firewood, mineral surveys and exploitation, and road and urban construction. All of these processes, along with natural factors, led to the development of land degradation.

By our estimation, more than 7.5 million km^2 of land has undergone desertification processes (Table 2.3).

Table 2.3 Areas affected by desertification in Central Asia (%)

Type of land cover/type of degradation	Percentage of desertified area			
	Weak	Moderate	Severe and very severe	Total
Forests and shrublands (vegetation degradation)	0.00	15.12	63.1	78.3
Forests and shrublands/water erosion	1.48	7.88	3.07	12.43
Forests and shrublands/wind erosion	0.00	9.44	0.00	9.44
Total	**1.48**	**32.45**	**66.17**	**100**
Pastures and hayfields/vegetation degradation	20.59	33.14	14.98	68.72
Pastures and hayfields/water erosion	1.08	4.05	0.00	5.13
Pastures and hayfields/wind erosion	2.40	12.92	10.84	26.15
Pastures and hayfields/waterlogging	0.09	0.34	0.15	0.59
Total	**24.16**	**50.45**	**25.98**	**100**
Rainfed lands/water erosion	4.80	37.16	2.40	44.36
Rainfed lands/wind erosion	32.03	22.96	0.66	55.64
Total	**36.82**	**60.11**	**3.06**	**100**
Irrigated lands/soil salinization	27.22	47.90	24.88	100
Dried bottom of the sea/waterlogging	0.00	52.62	47.38	100
Overall total	**24.14**	**49.95**	**25.91**	**100**

Source: Kharin et al. (1999).

Vegetation degradation is the most widely distributed type of desertification in GCA. It results from the overuse of pastures and the cutting of trees and shrubs. At the same time, in some areas of the region (Turkmenistan, Kazakhstan and Uzbekistan) it is due to overgrazing. The formation of biogenic crusts consisting of mosses, lichens and cyanobacteria was commonly observed (Kaplan et al., 2014; Orlovsky et al., 2004b). This process, developing in the absence of grazing domestic and wild animals, leads to the degradation of native vegetation communities and the disappearance of valuable palatable species. Significant areas are occupied by sandy massifs and the scarcity of vegetation cover make the region susceptible to wind erosion processes. As a result, GCA is a region with a high frequency of dust storms – the average annual number of days with dust storms reaches 30–37 days in Mongolia and China, and 40–60 days in Turkmenistan. In the areas of irrigated agriculture, a substantial part of the lands experience secondary salinization processes – about 35% of all 35.1 Mha of the irrigated GCA massifs. The main reason for this lies in irrational water-use practices, and outdated irrigation techniques leading to losses during transportation to the fields.

One-third of all irrigated lands in Central Asia – about 10.0 Mha – are located in the Aral Sea Basin, which includes all the newly independent Central Asian states of the FSU and northern Afghanistan (Bekchanov et al., 2018). In order to achieve the ambitious goal of "cotton independence" in the 1950s, intensive water works construction projects were started, and the area of irrigated lands in the

Aral Sea Basin increased rapidly. In the period from 1913 to 1950, the irrigated area enlarged by 0.5 Mha. From 1950 to 1990, it increased by 3.7 Mha (Orlovsky and Orlovsky, 2002). At present, the total area under irrigation reaches 10.0 Mha. However, before 1960, the increase in water use in the region did not significantly affect either the water balance of the Aral Sea or the quality of the river water. The capacity of the natural environment favored the self-purification process of water. The total length of the irrigation network reached up to 380,000 km; only 13,000 km or 3.4% of the total were constructed with a waterproof cover. Losses during transportation to evaporation and filtration reached up to 40% every year. The undesirable side effect of these losses was the formation of filtration lakes along the main irrigation canals. Thus, in the construction zone of the Karakum Canal, with a total length of 1380 km, the area of filtration lakes was 62 km^2 along the first stage of the canal (between the Amu Darya and Murgab Rivers); the total area of hydromorphic landscapes in the upstream of the Karakum Canal reached 533 km^2 (Orlovsky, 1999). Later, these newly developed lakes and marshes dried up and turned to solonchaks. It is difficult to control the volume of water delivered to the fields because of the outdated irrigation techniques (furrow and flooding irrigation). The irrigation water consumption equals 12,000 m^3/ha in Kyrgyzstan, about 16,000 m^3/ha in Kazakhstan and Uzbekistan, and about 20,000 m^3/ha in Tajikistan and Turkmenistan. Each of these figures far exceeds the biologically required norms of irrigation, and, very soon, they will lead to an increase in the groundwater level and, consequently, to waterlogging and secondary soil salinization. The area of salinized cultivated lands in the Aral Sea Basin is 4.408 Mha.

To control secondary soil salinization, a 200,550 km-long system of collector and drainage network has been constructed, which annually drains up to 32 km^3 of water from irrigated fields. The utilization of this volume of water is unregulated (Orlovsky et al., 2004a).

A significant volume of drainage water – c. 18.1 km^3/year – returns to the rivers and, consequently, deteriorates the water quality downstream. Mineralization of the water varies from 0.58 g/l upstream to 2.7 g/l in the lower reaches of the Amu Darya; in the Syr Darya River's downstream, this value reaches 1.8–2.3 g/l (Qadir et al., 2009). The rest flows to the adjacent desert areas and natural depressions (9.3 km^3/year); 4.9 km^3 are utilized again for irrigation in the areas of drainage flow formation. The consequences of this practice are: waterlogging of the natural pastures; the degradation and loss of valuable forage vegetation species; the development of unpalatable hydrophilic vegetation; disorders in traditional migratory schemes in livestock husbandry; and the propagation of disease vectors, such as malaria and leishmaniasis. For example, collector-drainage water, aside from its high salt content, is also polluted by poisonous chemicals, defoliants, and chemical fertilizers. More than 2,350 artificial lakes, serving as accumulators of drainage water, have appeared with a total area of 7,066 km^2; the largest lake, Lake Sarykamish, has an area of more than 3,700 km^2 (Orlovsky et al., 2014). Nearly 330,000 ha of pastures are affected by drainage collectors, in which valuable fodder species in vegetative communities are being changed into unpalatable species.

In spite of all these efforts to support the productivity of irrigated lands, the drainage network did not improve the situation of soil salinization. Rather, it transferred it to the marginal lands. The poor quality of the irrigation systems, the return flow of saline drainage water to the rivers and the re-use of these waters for irrigation have led to the ecological disturbance of the entire natural system of the Aral Sea Basin. The large-scale development of irrigated agriculture and water intake from the main rivers feeding the Aral Sea, namely the Amu Darya and Syr Darya Rivers, resulted in the desiccation of the Aral Sea – once the fourth largest inland water body in the world (Saiko, 1998; Mainquet and Letolle, 1997).

The demise of the Aral Sea has been called one of the 20th century's worst environmental catastrophes and has been referred to as a "quiet Chernobyl" (Glantz and Figueroa, 1997). The drastic desiccation of the Aral Sea led to the intensification of desertification processes in the region and the development of a new desert, the Aralkum, on the dried sea bottom. In the last few decades, the exposed bottom has become the new "hotspot" of dust and salt storms in the region (Orlovsky and Orlovsky, 2000). In 1960, the water flow from the Amu Darya and Syr Darya Rivers, which represented the major input for the Aral Sea, varied from 56 km^3/year to 60 km^3/year (Micklin, 2014); by the mid-1980s, the water flow had almost ceased, and the water volume discharged was 3.5 km^3/year (Saiko and Zonn, 2000). By 2007, the water inflow to the Aral Sea was about 5 km^3/year to 10 km^3/year (Dukhovny and Stulina, 2011). The lake has progressively shrunk due to a lack of recharge by the rivers and high evaporation during summertime. By 1999, the lake level had dropped by 18 m, down to the mark of 33.8 m MSL (Orlovsky et al., 2001). Following the division of the Aral Sea in 2005, the salinity level in the Large Aral increased by as much as 90 g/l (western part, depth 21 m) and 160 g/l (eastern part, depth 28.3 m), while in the Small Aral, it decreased and reached 17 g/l (Aladin et al., 2009). By 2009, the Aral Sea was divided into three separate water bodies: the Small Aral, which has been restored after the construction of a dam in 2005; the western part of the Large Aral adjacent to the Ustyurt Plateau; and a little pond that appears during rainy seasons in the eastern part of the former Large Aral (Bekchanov et al., 2018).

In the last few decades, significant temperature increases during summertime and decreases in winter values have been recorded (Novikova, 2004). Increases in extreme air temperatures (Micklin, 2007), as well as in the amplitudes of mean daily, monthly and annual values, have been observed (Khan et al., 2004).

For millions of years, the Aral Sea received the salts of the Aral Basin, and for the last four to five decades, it has served as a repository for the fertilizers, pesticides, herbicides and other chemicals washed away from the irrigated massifs of the region (Dedova et al., 2006; Orlovsky and Orlovsky, 2000). Nowadays, the exposed bottom of the sea has become the "distributer" of salts and chemicals over the adjacent areas (Indoitu et al., 2015). Thus, the salt storms, also referred to as "white dust storms" (Orlovsky and Orlovsky, 2000), have caused the serious pollution of the air, soil, water and agricultural products, as well as causing a variety of diseases and the degradation of natural ecosystems. From 1960 to 1984, the amount of salts and other toxic particles carried annually from the dried bottom of the Aral Sea to distances of hundreds of kilometers varied, by different estimations, from 15 million metric tons to 75 million metric tons, and raised considerable concerns about the impact on human health (Glantz, 1999; Saiko and Zonn, 2000). Estimates of the total transported by wind material, which were made in the late 1980s, ranged from 13 million metric tons per year to as high as 231 million metric tons per year (Glazovskiy 1990). At any rate, over 57,500 km^2 of the former seabed has been exposed (Indoitu et al., 2012), and the dried bottom of the sea is being turned into a powerful source of dust and salt emissions (Wiggs et al., 2003). Spivak et al. (2012) found that the average annual number of days with dust storms that originated on the dried bottom of the Aral Sea reached 13 between 2000 and 2009. The salt-dust plumes can be up to 400 km long, and finer particles can travel up to 1,000 km away (Semenov, 2012).

The other manifestations of environmental deterioration in the areas adjacent to the Aral Sea include the following:

- a decrease in the total surface areas of lakes in the Amu Darya Delta from 4,000 km^2 in 1960 to 260 km^2 currently (by a different estimation, this area in 1989 was 836 km^2) (Kozhoridze et al., 2012); in the Syr Darya Delta, the area of water bodies decreased from 273 km^2 in 1989 to 205 km^2 at present (Kozhoridze et al., 2012);
- a drop in groundwater tables at rates depending on the distance from the former sea shoreline;
- the incision of river beds up to a depth of 10 m;
- the transformation of the topsoil: an area of hydromorphic soils decreased from 6,300 km^2 to a current level of 62 km^2 in the Amu Darya Delta, and to 162 km^2 in the Syr Darya Delta (Dukhovny and Sokolov, 2003; Kozhoridze et al., 2012);
- an increase in the solonchaks' area from 850 km^2 to 2,730 km^2 in the Amu Darya Delta; the area of solonchaks in the Syr Darya Delta is 1,220 km^2 at present;
- the reduction of areas that are overgrown with reeds from 6,000 km^2 to 300 km^2 in the Amu Darya Delta;
- the reduction of areas of tugai (riverain) forests from 13,000 km^2 to 500 km^2 in the Amu Darya Delta;
- a drop in fish production from 40,000 to 2,000 tons/year or 20 times.

Wind soil erosion in the Aral Sea Basin has been developing over 29.5 million ha of arable land, and desertification spots around water wells have been observed on the 25 million ha of natural pastures in sandy deserts. Water erosion has been observed on an area of 11.9 million ha of arable lands and 7.3 million ha of pastures. Water erosion affects some 1.8 million ha of irrigated lands. About 11.2 million ha are affected by soil dehumidification and 181,300 ha by technogenic desertification (irregular movement of vehicles, geological surveys, etc.). Soil pollution (chemical, petrochemical and radioactive) has been observed in the areas of oil and gas production (4 Mha) and the testing areas of nuclear missiles (Semipalatinsk, Kapustin Yar, Circum-Balkhash region) (Valikhanova et al., 2005).

Afghanistan suffers from desertification due to the interminable military operations over the last several decades. Still, military desertification isn't the only degradation process affecting Afghanistan. More than 75% of the total land area in the northern, western and southern regions, where widespread grazing and deforestation have reduced vegetation cover, suffers from accelerated land degradation (Ahmad and Wasiq, 2004; Brown and Blankenship, 2013). In spite of the lack of real monitoring data on the extent and impact of desertification in Afghanistan, there are many indicators showing that soil fertility is being degraded by poor agricultural practices. Forests are being cut down for fuel and construction; grazing patterns have changed as conflict, land claims and drought have affected traditional migratory schemes; and irrigation systems are being affected by siltation and flooding. Areas affected by desertification are increasing as rainfed wheat production expands, herd sizes grow, grazing patterns change and household energy deficits drive the over-exploitation of brush and fuelwood collection. Impoverished soils have a reduced carrying capacity for livestock, resulting in overstocking, severe flooding, the cultivation of unsuitable land for food cropping, reduced pasture quality, the decimation of wildlife populations, air pollution, a decrease in the quality and quantity of water for irrigation and drinking, deforestation, and the exposure of soils to wind and water erosion. Conflicts result from competing land uses, decreasing yields and reduced natural resource and water availability. These dynamics are all compounded by macro-level climatic changes, especially related to increasing global temperatures and changes in precipitation dynamics (Brown and Blankenship, 2013).

In China, the main types of desertification are wind erosion (160.7 Mha) and water erosion (20.5 Mha) processes. The other damaging manifestation of desertification is the degradation of pastoral vegetation (developed on 10,523 Mha). Specific for the Tibet region are the freezing and melting processes (36.3 Mha), which lead to the destruction of rocks and, as a result, to the development of soil erosion. Soil salinization (both natural and secondary) affects 33.3 Mha of land. More than 100,000 ha of natural forests are undergoing degradation processes. Arable lands have degraded in the area of 7.7 Mha.

Approximately 77.8% or 1.22 million km^2 of Mongolian territory has been affected by desertification at different rates. About 35.3% of Mongolia is weakly degraded, 25.9% moderately degraded, 6.7% suffers from severe desertification, and 9.9% experiences very severe desertification (Desertification Atlas of Mongolia, 2013). Desertification processes include wind and water erosion, pasture degradation, the drying of lakes and rivers, and technogenic desertification in the mining areas. Wind erosion has developed in an area of 1.4 million km^2. Nearly 106,000 km^2 suffers from water erosion. Areas with degraded vegetation amount to 103,000 km^2 (Kharin and Natsag, 1992).

Notes

1 The term "gobi" in the Mongolian and Manchu languages is also called Gashun Gobi. It is situated in Xinjiang. The Beishan Mountains are in the south.
2 WWF, https://www.worldwildlife.org/biomes (accessed April 2015).
3 http://www.fao.org/nr/water/aquastat/water_res/index.stm (accessed April 2015).
4 CAWaterinfo, 2011, http://www.cawater-info.net (accessed in April 2015).
5 FAOSTAT, 2012, http://faostat.fao.org (updated January 2015; accessed March 2015).

References and further reading

ADB. 2002. *Afghanistan Natural Resources and Agriculture Sector Comprehensive Needs Assessment Final Draft Report.* Asian Development Bank (ADB), Metro Manila, Philippines. http://www.trade.gov/static/afghanistan_needsassessment.pdf

ADB. 2014. *Demand in the Desert: Mongolia's Water-Energy-Mining Nexus.* Asian Development Bank (ADB), Metro Manila, Philippines. http://www.adb.org/publications/mongolia-water-energy-mining-nexus

Afghanistan's Environment. 2008. United Nations Environment Programme (UNEP), Nairobi, 32 pp.

Ahmad, M. and Wasiq, M. 2004. *Water Resources Development in Northern Afghanistan and its Implications for Amu Darya Basin.* Word Bank, Washington DC, 66 pp.

Aladin, N.V., Plotnikov, I.S. and Micklin, P. 2009. Aral Sea: Water level, salinity and long-term changes in biological communities of an endangered ecosystem-past, present and future. *Natural Resources and Environmental Issues* 15 (article 35).

Alexeyeva, N. and Ivanova, I. 2003. Central or Middle Asia? *Geography* 30: 13–17.

Alikhanova, B.B. (ed.). 2008. *National Report on the State of Environment in Uzbekistan (Retrospective Analysis for 1988–2007).* Government of Uzbekistan, Tashkent, 298 pp.

Atlas of China. 2008. Sinomaps Press, Beijing, 182 pp.

Babaev, A.G. (ed.). 1999. *Desert Problems and Desertification in Central Asia: The Researches of the Desert Institute.* Springer, Berlin, 293 pp.

Babaev, A.G., Nikolaev, V.N. and Orlovsky, N. 1991. The modern state and perspectives of usage of natural grazing and rainfed lands in the Aral Sea basin. *Problems of Desert Development* 6: 3–11.

Babaev, A.G., Zonn, I.S., Drozdov, N.N. and Freikin, Z.G. 1986. *Deserts.* Moscow, Mysl, 318 pp.

Behnke, R. 2006. *The Socio-Economic Causes and Consequences of Desertification in Central Asia.* Springer, Dordrecht, 259 pp.

Bekchanov, M., Djanibekov, N. and Lamers, J.P.A. 2018. Water in Central Asia: a cross-cutting management issue. In this volume, pp. 211–36.

Blank, S. 2007. *US Interests in Central Asia and the Challenges to Them.* Strategic Studies Institute, Carlisle, PA. http://ssi.armywarcollege.edu/pdffiles/pub758.pdf

Brown, C., Waldron, S. and Lorgworth, S. 2008. *Sustainable Development in Western China.* Edward Elgar Publishing Limited, Cheltenham, 275 pp.

Brown, O. and Blankenship, E. 2013. *UN Country Team Report: Natural Resource Management and Peace Building in Afghanistan.* United Nations Environment Programme (UNEP), Nairobi, 62 pp.

Chen, J.Q., Wan, S.Q., Henebry, G., Qi, J.Q., Gutman, G. Sun, G. and Kappas, M (eds). 2013. *Dryland East Asia: Land Dynamics amid Social and Climatic Change.* High Education Press, Beijing, 470 pp.

China Country Paper to Combat Desertification. 1997. Beijing, China Forestry Publishing House, 47 pp.

Davis, J. and Sweeney, M. 2004. *Central Asia in US Strategy and Operational Planning, Where Do We Go From Here?* Institute for Foreign Policy Analysis, Cambridge MA and Washington DC.

Dedova, T.V., Semenov, O.F. and Tuseeva, N.B. 2006. Division of Kazakhstan territory by the repetition of very strong dust storms, and based on meteorological observations, remote sensing images and GIS (in Russian). In: *Republic of Kazakhstan. Vol. 3. Environment and ecology.* Almaty, pp. 287–92.

Desertification Atlas of Mongolia. 2013. University of Bern, Bern.

Desertification, Rangelands and Water Resources. Final Thematic Report. 2008. UNEP, 55 pp.

Dukhovny, V. and Sokolov, V. 2003. *Lessons on Cooperation Building to Manage Water Conflicts in the Aral Sea Basin.* UNESCO, Paris, 56 pp.

Dukhovny, V. and Stulina, G. 2011. *Water and Food Security in Central Asia.* NATO Science for Peace and Security Series C: Environmental Security. Springer, Dordrecht.

Emadi, M.H. 2012. Better land stewardship to avert poverty and land degradation: A viewpoint from Afghanistan. In: V. Squires (ed.) *Rangeland Stewardship in Central Asia: Balancing Improved Livelihoods, Biodiversity Conservation and Land Protection.* Springer, Dordrecht, pp. 91–108.

Frenken, K. (ed.). 2013. *Irrigation in Central Asia in Figures.* Food and Agriculture Organization of the United Nations (FAO), Rome, 246 pp.

Glantz, M. (ed.). 1999. *Creeping Environmental Problems and Sustainable Development in the Aral Sea Basin.* Cambridge University Press, Cambridge.

Glantz, M. and Figueroa, R. 1997. Does the Aral Sea merit heritage status? *Global Environmental Change* 7 (4): 357–80.

Glazovskiy, N.F. 1990. Aral'skiy krizis: prichiny vozniknoveniya i put' vykhoda. (The Aral crisis: causative factors and means of solution.) *Nauka* 20–23.

Gunin, P.D., Vostokova, E.A., Dorofeyuk, N.I., Tarasov, P.E. and Black, C. (eds). 1999. *Vegetation Dynamics of Mongolia.* Kluwer Academic Publishers, Dordrecht, 238 pp.

Indoitu, R., Kozhoridze, G., Batyrbaeva, M., Vitkovskaya, I., Orlovsky, N., Blumberg, D. and Orlovsky, L. 2015. Dust emission and environmental changes in the dried bottom of the Aral Sea. *Aeolian Research* 17: 101–15. doi:10.1016/j.aeolia.2015.02.004

Indoitu, R., Orlovsky, L. and Orlovsky, N. 2012. Dust storms in Central Asia: spatial and temporal variations. *Journal of Arid Environments* 82: 62–70.

International Fund for Saving the Aral Sea. 2011. Tashkent, 44 pp.

Iskakov, N.A., Kunaev, M.S., Malkovsky, I.M. and Medeu, A.R. 2006. Ecological state of natural ecosystems. In: *Republic of Kazakhstan. Vol. 3. Environment and ecology.* Almaty, pp. 13–196.

Jacobs, M.J. and Schloeder, C.A. 2012. Extensive livestock production: Afghanistan's Kuchi herders: Risks to and strategies for their survival. In: V. Squires (ed.) *Rangeland Stewardship in Central Asia: Balancing improved livelihoods, biodiversity conservation and land protection.* Springer, Dordrecht, pp. 109–28.

Kaplan, S., Blumberg, D., Mamedov, E. and Orlovsky, L. 2014. Land use change and land degradation in Turkmenistan in the post-Soviet era. *Journal of Arid Environments* 103 (4): 96–106.

Khan, V.M., Vilfand, R.M. and Zavialov, P.O. 2004. Long-term variability of air temperature in the Aral Sea region. *Journal of Marine Systems* 47: 25–33.

Kharin, N. 2002. *Vegetation Degradation in Central Asia under the Impact of Human Activities*. Springer, Dordrecht, 182 pp.

Kharin, N.G. and Natsag, J. 1992. The results of desertification study of the Mongolian arid territories. *Problems of Desert Development* 5: 54–46.

Kharin, N.G., Tateishi, R. and Harahsheh, H. 1999. *Degradation of the Drylands of Asia*. Chiba University, Chiba, Japan, 82 pp.

Kozhoridze, G., Orlovsky, N. and Orlovsky, L. 2012. Monitoring land cover dynamics in the Aral Sea region by remote sensing. *Proc. SPIE* 8538.

Leber, D., Holwe, F. and Hausler, H. 1995. Climatic classification of the Tibet Autonomous Region using multivariable statistical methods. *GeoJournal* 37 (4): 451–73.

Lezhinsky, G.T. 1974. *Resources of the Temporal Surface Runoff in the Deserts of Central Asia and Western Kazakhstan*. Ashgabat, 187 pp.

Lioubimtseva, E. and Henebry, G.M. 2009. Climate and environmental change in arid Central Asia: Impacts, vulnerability, and adaptations. *Journal of Arid Environments* 73: 963–77.

Maclean, J. (ed.). 2010. *Central Asia Atlas of Nature Resources*. Asian Development Bank, Manila, 223 pp.

Mainquet, M. and Letolle, R. 1997. The ecological crisis of the Aral Sea basin in the frame of a new time scale: The 'Anthropo-Geological Scale'. *Journal Naturwissenschften* 84 (8): 331–9.

Micklin, P. 2007. Aral Sea Disaster. *Annual Review of Earth and Planetary Sciences* 35: 47–72. doi:10.1146/annurev.earth.35.031306.140120

Micklin, P. 2014. Introduction to the Aral Sea and its region. In: P. Micklin, A. Aladin and I. Plotnikov (eds). *The Aral Sea: The Devastation and Partial Rehabilitation of a Great Lake*. Springer, Heidelberg, pp. 15–40.

National Plan of Action to Combat Desertification in Mongolia. 1997. Government of Mongolia, Ulaanbaatar, 67 pp.

National Report on the State of Environment in Kyrgyzstan for 2006–2011. 2012. Government of Kyrgyzstan, Bishkek, 126 pp.

Nikitin, A.I. 1986. Water balance of the lakes of Central Asia. *SANIGMI* 59: 3–13.

Novikova, N. 2004. Priaralye ecosystems and creeping environmental changes in the Aral Sea'. In: M. Glantz (ed.) *Creeping Environmental Problems and Sustainable Development in the Aral Sea*. Cambridge University Press, Cambridge.

Orlovsky, N. 1999. Creeping environmental changes in the Karakum canal's zone of impact. In: M. Glantz (ed.) *Creeping Environment Problems and Sustainable Development in the Aral Sea Basin*. Cambridge University Press, Cambridge, pp. 225–49.

Orlovsky, N. and Orlovsky, L. 2000. White sandstorms in Central Asia. In: Yang Youlin, V.R. Squires and Lu Qi (eds) *Global Alarm: Desert and sandstorms from the world's drylands*. UN, Beijing, pp. 169–201.

Orlovsky, N. and Orlovsky, L. 2002. Water resources in Turkmenistan use and conservation. Paper presented at workshop on water, climate, and development issues in the Amu Darya basin. Philadelphia PA.

Orlovsky, N., Birnbaum, E. and Orlovsky, L. 2004a. Drainage and land degradation: Consequences of desert reclamation in the Aral Sea basin. *Journal of Arid Land Studies* 14: 123–6.

Orlovsky, L., Dourikov, M. and Babaev, A. 2004b. Temporal dynamics and productivity of biogenic soil crusts in the central Karakum desert, Turkmenistan. *Journal of Arid Environments* 56: 579–601.

Orlovsky, N., Glantz, M. and Orlovsky, L. 2001. Irrigation and land degradation in the Aral Sea basin. In: S.-W. Breckle, M. Veste and W. Wucherer (eds). Springer, Berlin, pp. 115–25.

Orlovsky, L., Matsrafi, O., Orlovsky, N. and Kouznetsov, M. 2014. Sarykamysh Lake: collector of drainage water – The past, the present, and the future. In: S.-W. Breckle, W. Wucherer, L.A. Dimeyeva and N.P. Ogar (eds) *Aralkum – A Man-Made Desert: The Desiccated Floor of the Aral Sea (Central Asia)*. Ecological Studies Vol. 218. Springer-Verlag, Berlin and Heidelberg, pp. 107–40.

Petrov, M.P. 1966. *Desert of Central Asia. Vol. 1*. Moscow, Nauka, 274 pp.

Qadir, M., Noble, A.D., Qureshi, A.S., Gupta, R.K., Yuldashev, T. and Karimov, A. 2009. Salt-induced land and water degradation in the Aral Sea basin: A challenge to sustainable agriculture in Central Asia. *Natural Resources Forum* 33: 134–49.

Report of the State of the Environment of Mongolia (2008–2010). 2011. Government of Mongolia, Ulaanbaatar, 80 pp.

Saiko, T.S. 1998. Geographical and socio-economic dimensions of the Aral Sea crisis and their impact on the potential for community action. *J Arid Environ* 39 (2): 225–38.

Saiko, T. 2001. *Environmental Crisis*. Prentice Hall/Pearson Education, Harlow, UK, 320 pp.

Saiko, T. and Zonn, I. 2000. Irrigation expansion and dynamics of desertification in the Circum-Aral region of Central Asia. *Applied Geography* 20: 349–67.

Semenov, O.E. 2012. Dust storms and sandstorms and aerosol long-distance transport. In: S.-W. Breckle, W. Wucherer, L.A. Dimeyeva and N.P. Ogar (eds) *Aralkum – A Man-Made Desert: The Desiccated Floor of the Aral Sea (Central Asia)*. Ecological Studies Vol. 218. Springer-Verlag, Berlin and Heidelberg, pp. 73–82.

Seversky, I.V. and Tokmagambetov, T.G. 2004. *Current Glaciation Degradation of Mountains of the Southeast Kazakhstan*. Almaty.

Shaumarov, M. and Birner, R. 2018. Managing the commons in the post-Soviet transition: what are the challenges of institutional change in pastoral systems in Uzbekistan? In this volume, pp. 48–70.

Shen, Y. 2010. Natural resources and their utilization in the dryland of China. In: L. Ci and X. Yang (eds) *Desertification and its Control in China*. Springer, Berlin and Heidelberg, pp. 101–76.

Spivak, L., Terechov, A., Vitkovskaya, I., Batyrbayeva, M. and Orlovsky, L. 2012. Dynamics of Dust Transfer from the Desiccated Aral Sea Bottom Analysed by Remote Sensing. In: S.-W. Breckle, W. Wucherer, L.A. Dimeyeva and N.P. Ogar (eds) *Aralkum – A Man-Made Desert: The Desiccated Floor of the Aral Sea (Central Asia)*. Ecological Studies Vol. 218. Springer-Verlag, Berlin and Heidelberg, pp. 97–106.

Spoor, M. 1998. The Aral Sea Basin Crisis: Transition and Environment in Former Soviet Central Asia. *Develop. Change* 29 (3): 409–35.

Starr, F. 2005. A partnership for Central Asia. *Foreign Affairs*, July/August. https://www.foreignaffairs.com/articles/asia/2005-07-01/partnership-central-asia

State of Environment in Turkmenistan (Состояние окружающей среды Туркменистана). 2008. UNEP, Ashgabat, 145 pp.

Strengthening Cooperation for Rational and Efficient Use of Water and Energy Resources in Central Asia. 2004. UN, New York, 110 pp.

Sun, J., Zou, X., Gao, Q., Jia, B. and Yang, X. 2010. Natural background of China's drylands. In: L. Ci and X. Yang (eds) *Desertification and its Control in China*. Springer, Berlin and Heidelberg, pp. 29–99.

Titov, V.I. 1976. *The Climate of Afghanistan*. Moscow, 258 pp.

UNESCO. 1992–2005. *History of Civilizations of Central Asia*. 6 volumes. http://www. unesco.org/new/en/social-and-human-sciences/themes/general-and-regional-histories/ history-of-civilizations-of-central-asia/

Valikhanova, A., Belikh, A., Pavlichenko, L., Ni, V., Dostay, Zh., Chigarkin, A. and Kuatbaeva, G. 2005. *Thematic Overview: Desertification/land degradation*. Astana, 88 pp.

von Humboldt, A. 1843. *Asie centrale: Recherches sur les chaines de montagnes et la climatologie comparee*. 3 volumes. Paris: Gide.

Water Resources of Kazakhstan in the New Millennium. 2002. UNDP, Almaty, 145 pp.

Wiggs, G.F., O'Hara, S.L., Wegerdt, J., Van Der Meer, J., Small, I. and Hubbard, R. 2003. The dynamics and characteristics of aeolian dust in dryland Central Asia: possible impacts on human exposure and respiratory health in the Aral Sea basin. *The Geographical Journal* 169 (2): 142–57.

Williams, M. 2004. Creeping Environmental Problems and Sustainable Development in the Aral Sea Basin. *Earth-Science Reviews* 66 (3): 389–90.

Yang, Y. Squires, V. and Lu Qi. 2001. *Global Alarm: Dust and sandstorms from the world's drylands*. UN, Beijing, 200 pp.

Zhu Zhenda, Liu Shu and Di Xinmin. 1988. *Desertification and Rehabilitation in China*. Desert Research Institute, Lanzhou, 222 pp.

Zonn, I. 2014. Water resources of Turkmenistan. In: I.S. Zonn and A. Kostianoy (eds) *The Turkmen Lake Altyn Asur and Water Resources in Turkmenistan*. Springer, Berlin and Heidelberg, pp. 59–68.

3 Managing the commons in the post-Soviet transition

What are the challenges of institutional change in pastoral systems in Uzbekistan?

Makhmud Shaumarov and Regina Birner

DIVISION OF SOCIAL AND INSTITUTIONAL CHANGE IN AGRICULTURAL
DEVELOPMENT, INSTITUTE OF AGRICULTURAL ECONOMICS
AND SOCIAL SCIENCES IN THE TROPICS AND SUBTROPICS,
UNIVERSITY OF HOHENHEIM, GERMANY

Introduction

Degradation of dryland pastures is known as a challenging factor in many developing and transition countries in the Greater Central Asian (GCA) region – home to one of the largest trans-boundary desert ecosystems. Within the 'central core' of the five former Soviet Republics (see Squires and Lu, 2018), the dryland pasture degradation was a somewhat neglected aspect during the post-Soviet transformation period reforms. In two decades, after the collapse of the former Soviet regime, the total area of degraded rangelands in the five 'stans' exceeded 71 million ha, or 61.04% of the total pastoral area (IFPRI/ ICARDA 2008; Nishanov 2013; Otakulov 2013). The major institutional challenges currently in rangelands are expressed by limited mobility of animal flocks in remote areas and by access (use rights) to pastoral land resources (Robinson et al. 2012). However, the historical reports of the former Soviet experience in rangelands indicate that political decisions over the dryland production system had resulted in a better management of pastoral resources with higher productivity than those we see today.

Hence, considering that pasture lands are a common property resource (CPR), which – following the current commons discourse – might be best used by local communities, the question arises as to why the highly centralized former Soviet system was rather successful in managing pasture lands fairly sustainably on a large scale. The next question arises as to why those established organizational and institutional mechanisms collapsed during the current transition period, and why alternative institutional arrangements that allow for a sustainable management of pasture resources have not been established, so far. To answer these questions, the authors combine historical analysis and political economy approaches. The empirical evidence presented here was collected from two case studies in pastoral areas in Uzbekistan and from a survey of published literature.

Literature review

In this section we review the current literature on the current transformations in the pasture management system in the five 'stans', with particular emphasis on Uzbekistan, within the past two decades of transition.

Rangeland systems in five Central Asian republics

Pastures are characterized by aridity, and are primarily used to graze sheep, goats, horses and camels, and to a lesser extent for crop cultivation. In Kazakhstan, Turkmenistan and Uzbekistan, pastures are mainly located in steppes and semi-deserts, whereas the majority of rangelands in Kyrgyzstan and Tajikistan are based in mountains (Kerven et al. 2011). More intensive utilization for animal husbandry and cropping in dryland pastures are limited by water availability.

Rangeland management systems in the five 'stans' have several common features:

–   Firstly, all five countries have a similar history of state- and collectively oper-ated farming systems in pastoral livestock production (Kreutzmann 2012). Under the former Soviet regime, the overall driving force for intensive range-land use was to feed large numbers of livestock and to meet ever-increasing state production plans (Undeland 2005), centrally set by *Gosplan*. Rural households had restrictions on ownership of animals per household, and they mostly kept small numbers for self-consumption. Regional rural councils (*Sel'sky sovet*) controlled annual fulfilment of the production plan by state and collective farms (Kreutzmann 2012).
–   Secondly, all decisions on pasture use rights and terms of allocation were jointly taken *de facto* by collective farm managers and rural councils. Even though national land resources belonged to the citizens, as the Soviet Constitution states, in fact all decisions over land administration and distribu-tion were taken on their behalf by state authorities (ibid.).
–   The third common feature was the annual provision of detailed pasture use parameters by the State Agency for Land Planning (*Giprozem*) to rural

Table 3.1 Pastureland territories in Central Asia

Country	Land area (Mha)	Pastures (Mha)	Pastures (%)
Kazakhstan	272.2	185.5	68.2
Kyrgyzstan	19.8	9.1	46.0
Tajikistan	14.5	3.5	24.1
Turkmenistan	49.5	39.6	80.0
Uzbekistan	45.2	23.4	51.8
Total	**401**	**261.1**	65.4

Source: Adapted from Gilmanov (1996).

authorities and collective farms in each country. These parameters included detailed mapping of rangeland quality and carrying capacity, watering sources and climate risk zones.

– Fourthly, there was an organized scheme of cross-border livestock herd mobility: from summer pastures in the mountainous regions of Kyrgyzstan and Tajikistan to winter pastures in the steppes and semi-deserts of Kazakhstan, Turkmenistan and Uzbekistan. The seasonal mobility was supported by institutionalized transportation systems – trucks, trains and even airplanes (Undeland 2005) – as well as by better organized wintering infrastructure both for mobile shepherds and animals (Robinson et al. 2012).

– Finally, the Soviets developed pastoral infrastructure facilities such as watering points, animal and herder shelters, veterinary services, winter fodder supply and social-service infrastructure, as well as farm subsidies and incentives for the remote allocation of specialists (Rahimov 2012).

– The regional forestry institutes provided additional phyto-melioration and desert afforestation services to avoid desertification and land degradation in rangelands. These efforts, compared with the former nomadic practices, allowed an economy of scale and organization of a relatively balanced intensive use of greater pastoral areas on over 250 million ha.

After the collapse of the Soviet economy in the early 1990s, all five republics experienced shocks in socio-economic, fiscal and political dimensions (Wehrheim, Schoeller-Schletter, and Martius 2008). It caused price shocks and the rapid decline of subsidies, inputs and public service provision and led to the collapse of livestock mobility to more remote rangelands (Kerven, Channon, and Behnke 1996). The lack of water infrastructure support was coupled with unregulated grazing practices (Lerman 2005).

As reported by the FAO, the recent degradation of rangelands and decreasing livestock productivity is caused, first of all, by poverty, and is being exacerbated by the following interlinked challenges (Steinfeld et al. 2006):

– concentration of animals in peri-urban areas;
– disruption of transhumance grazing practices;
– lack of market access and infrastructure in distant pastures;
– lack of sustainable pasture management technologies;
– fragmentation of livestock flock composition.

Today, these changes lead to different levels of pastureland degradation: 24 million ha (13.2% of the total) in Kazakhstan, 6.8 million ha (74%) in Kyrgyzstan, over 3 million ha (90%) of pastures in Tajikistan, 20.8 million ha (50%) in Turkmenistan, and about 16.4 million ha (78%) in Uzbekistan (IFPRI/ICARDA 2008; Otakulov 2013).

The process of transition towards leasehold, private or community-based rangeland use rights is still in the formative stage in most of these states (Kreutzmann 2012; Robinson et al. 2012). Kyrgyzstan and Kazakhstan recognized private land

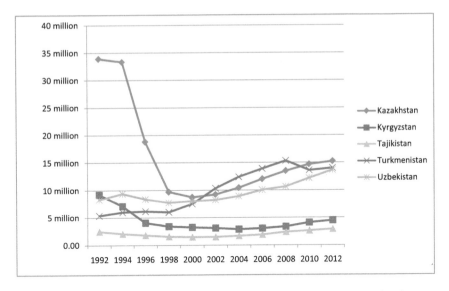

Figure 3.1 Transition-period dynamics of the sheep population in five Central Asian republics

Source: FAOSTAT (2014).[1]

ownership in 2000 and 2003 respectively, whereby land was distributed based on membership shares of collective farms (Lerman 2005). Leasehold land allocations were recognized in Tajikistan, Turkmenistan and Uzbekistan during farm restructuring reforms in the late 1990s. The major pastoral areas in Kazakhstan, Tajikistan, Turkmenistan and Uzbekistan are still run by cooperatives, the direct successors of the former *Sovhoz* farms. These farms underwent only minor institutional changes, and in practice production and management approaches still continued as before (Undeland 2005).

Community-based pasture management legislation in Central Asia was pioneered by Kyrgyzstan in 2009, whereby local pasture users associations are responsible for the allocation of grazing areas, for designing pasture rotation plans and for organizing respective infrastructure maintenance. Initial empirical studies on the implications of the pastoral system reforms in the Kyrgyz Republic, conducted by Crewett (2012), show that the issue of livestock mobility in remote pastures still persists due to insufficient services, inconvenient access, lack of sanctions and weak incentives in the new legal framework. Since 2012, Tajikistan has started developing a new legislation on community-driven rangeland use that entails decentralized responsibilities and pasture allocations for local user groups from communities (Kreutzmann 2012), but it is still in the early stages of implementation.

Overall, the current literature on rangeland management reveals two emerging obstacles to hamper further development of pastoral livestock production:

limited mobility to remote rangelands, and lack of access to productive pastures by communities (Robinson et al. 2012).

Pasture management reforms in Uzbekistan

Dryland pastures in Uzbekistan are characterized by aridity and low precipitation (100–150 mm), low soil humus content (SNC 2009), and remoteness from markets, infrastructure and services (Robinson et al. 2012). Natural rangelands cover 20.75 million ha, and two-thirds of the area is operated by large-scale cooperative (*Shirkat*) farms specialized in *Karakul*[2] sheep produce: mostly in pelt, wool and meat. Other rangeland users are rural households, private farms, and state enterprises such as forestry, nature protection areas, hunting zones and others. According to published estimates, the *Karakul* lamb population reached almost 6 million head in 2011 (see Table 3.2), whereas all types of sheep and goat reached 16.2 million head in 2013 (Teshaboev and Sharipov 2013). Rural households keep 94% of livestock, which are considered a major source of their livelihood and savings.

After independence in 1991, a number of state agencies were dissolved and their functions transferred to province-level administration: the former Pastoral Department within the Ministry of Agriculture is one example. The functions and inventories of '*Karakul-Trest*' were transferred to a newly formed state company '*Uzbek Karakuli*'. The former 'Republican Pasture Melioration and Construction Union', the major public agency for water-related services in drylands, was commercialized as the '*Obi-Hayot*' Association (Teshaboev and Sharipov 2013). Most of its territorial inventories and facilities were distributed to local municipalities and agro-producers (Gupta et al. 2009).

The former *Kolhoz* and *Sovhoz* farms were disbanded between 1992 and 2000, after adoption of the Law on Land (1990), whereby the Land Code of 1998 formally recognized three forms of market-oriented land users: households (*Dehkan*), private farms (*Fermer*) and agro-cooperatives (*Shirkat*) (UNDP 2010a). The 1998 Law on Agricultural Cooperatives stipulated a restructuring of agro-cooperative according to asset shareholding and family labor; thus, all former employees became their members. Despite the efforts of the state, the methods and management principles of managing *Shirkats* were similar to the approaches used in the former Soviet farms: e.g. internal employee–manager

Table 3.2 Distribution of *Karakul* sheep population (thousand head)

Year	Total		Private farms		Households		Shirkats	
	Karakul	*%*	*Karakul*	*%*	*Karakul*	*%*	*Karakul*	*%*
2007	4896.3	100	255	5.2	2704.7	55.2	1935.6	39.5
2011	5897.1	100	413	7.0	3638.5	61.7	1854.5	31.3

Source: State Statistics Committee, 2012, in Holmirzaev (2013).

relations, financial and production routines, as well as top-down decision making (Lerman 2008). In 2015, 106 *Shirkats* are left in the drylands and all specialize in *Karakul* sheep production (UNDP 2010a). The state retains these cooperatives for genetic conservation of the *Karakul* sheep, to maintain social-infrastructure services, and to ensure food security in rural areas (Holmirzaev 2013). *Shirkats* still try to meet state-sponsored production quotas for *Karakul* pelt and meat, and even for cotton or wheat that have to be supplied annually at fixed prices (Farmanov 2005; UNDP 2005).

Currently, the pastoral production system is administered by the Ministry of Agriculture, the state company '*Uzbek Karakuli*', the Association of Farmers and *Dehkans*, and by province- and district-level administration (*Hokims*). Despite the fact that *Hokims* develop territorial programs for rural development annually, including infrastructure supply, there is hardly any strategic program designed so far to improve rangeland productivity and to reduce its degradation (UNDP 2005; Robinson et al. 2012). In most cases, these plans are not realized due to the lack of funding, and in other cases all actions for pasture rehabilitation are simply delegated to *Shirkat* managers (ibid.). Meanwhile, all commands and decisions of *Hokims* are focused only on increasing the number of animals and on the fulfilment of the state procurement plans as a top priority that must be met by all means (UNDP 2005). Decisions on allocation of pasture land and animal stocking are also taken centrally by local *Hokims* and farm managers (Robinson et al. 2012).

Some targeted public services are available from state suppliers, such as logistics, equipment, seeds and fertilizers, irrigation water, and veterinary and mechanized services, but only for the production of quantities stipulated in state procurement contracts (Robinson et al. 2012). Most of the former public services, e.g. zoo-veterinary, machinery supply, pedigree insemination, field fertilization, bio-lab analysis, and phyto-melioration, were commercialized and the current pasture users cannot afford them due to budget constraints (UNDP 2005). Lack of quality service delivery, weak market infrastructure, increasing input prices, and the increasing cost of veterinary medicines in particular led to a productivity decline in cooperatives of two-thirds from 1991 to 1997 (Kreutzmann 2012).

In the early years of independence, the state farms considerably reduced the seasonal (summer–winter–spring) migratory system of grazing from 150–500 km to 10–50 km, due to abrupt cuts in public subsidies, provisions and wages that followed after the system's collapse (Kerven, Channon, and Behnke, 1996). Stall feeding is practiced only in snow seasons. At the same time, in communities, traditional systems of collective herding were quickly re-established in locally available pastures (Robinson et al. 2012). However, the legal status of access to pastures by household-owned animals remains unclear, and rural households mostly rely on verbal agreements with *Shirkat* managers or directly with shepherds. This resulted in open-access grazing on locally available rangelands followed by their overgrazing, on the one hand, and on the underuse of remote pastures and the gradual decay of water infrastructure and seasonal shelter facilities on the other. *Shirkat* managers instruct teams of shepherds via annual production contracts, but there is little incentive created for remote pasture rotation and seasonal animal mobility (UNDP 2005). Winter feeding stocks are also prepared by shepherd brigades for animals under their care and the shepherd brigade's privately owned animals have access to the same feedstocks to ensure their subsistence (Farmanov 2005).

The collapse of the former Soviet Union (FSU) also had impacts on agricultural research and development. Research institutes, previously highly subsidized and staffed by qualified scientists, have seen a significant decline of personnel due to low salaries, aging and shortages of funding (Gupta et al. 2009). In the past, agricultural scientists had strong linkages with former Soviet farms, but most of the formal communication channels were broken after the FSU collapse (Wall 2008). Agricultural research institutions had only 7000 scientists in 2007 (Paroda et al. 2007), and the technical capacities of institutions were run down and they no longer serviced remote pastures. Language barriers also prevented local scientists from accessing the benefits of international public goods and foreign scientific advancements. Coordination of the agricultural research system was weakly supported and scientists felt isolated (Gupta et al. 2009).

Methodology

This is qualitative research, in which the major part of the data come from secondary sources, and is complemented, to a lesser extent, with primary data sources. The secondary data were collected from peer-reviewed articles, archive materials, official state reports, project documents, administrative reports, and from farmers' papers.

We reviewed archive sources published from 1925 to 1977 by Soviet scientists such as Chayanov (1925), Arasimovich (1929), Sovetkina and Korovin (1941), Nechaeva, Mordvinov, and Mosolov (1943), Morozova (1946), Lobanov (1953), Panteleev (1954), Babaev (1977), Kayumov (1977), Fedorovich (1950), Sergeeva (1951), Khudayberdyev (1976), Gaevskaya and Krasnopolin (1957) and Gaevskaya and Salmanov (1975). The materials contain rich and unique historical data that represent a systematic study of Central Asian rangelands from the perspectives of natural and social sciences. Archive study was also complemented by a number of Soviet decrees, resolutions and normative reports.

We collected field data from two case studies and two control groups in the period July–October 2012 and July–August 2013. Each case study represented a district *Karakul* sheep production cooperative, where we observed pasture degradation mostly. Control groups were selected to contrast management practices with comparatively better biomass and livestock productivity within the same district pastures. The pastures in control groups are leased seasonally by the Uzbek Forestry Department to individual livestock owners. The first study area, with treatment and control groups, was located in *Karnabchul* steppe of *Navoi* province. The second study area was located in *Forish* district, *Jizzakh* province of Uzbekistan. The study areas were selected using criteria to represent semi-desert and desert ecosystems with the most commonly distributed pastoral vegetation and soil characteristics in drylands within the country. We selected these study areas with the purpose of contrasting two different outcomes (degraded and well-maintained categories) of land management practices both within the same socio-environmental conditions.

Field data were gathered from pasture land users, service providers, local- and national-level agencies, international development programs, and national and international research institutions. In total, 54 interviews were carried out, including 16 interviews with national- and district-level respondents, and 38 interviews with community-level actors from two case-study areas. We employed net mapping tools to better visualize institutional processes, organizations and actors in the rangeland production system. A purposive sampling method was used to select relevant respondents and data sources.

As a tool, we used individual in-depth interviews, formally and informally, to record changes in livestock production during the transition period. We discussed their opinions, perceptions and attitudes towards past and present service structures functioning in the system, as well as organizational and administrative conditions in pastoral animal production.

For data analysis we employed content analysis and a contrasting and constant comparison method. We applied a political economy lens to understand which political or economic incentives and challenges have played an important role in the past and in the present developments in dryland production systems. This approach helped to answer questions of how political and economic factors as well as institutions related to development affected reform decisions and political outcomes.

Results

In this section of the chapter we analyze historical perspectives of the evolution of the pastoral system in the former Soviet era, and during the current period of transition to a market economy. Based on this, we discuss the current political economy challenges to understand the broader context of the current agricultural policies in dryland systems.

Historical overview of dryland pastoral systems

The historical analysis reveals that rangeland degradation issues were tackled at regional scales more effectively during the former Soviet regime than they are now (Holland 2010). As reported by Kayumov (1977), according to a 1974 national inventory, 90% out of the total 29.6 million ha of pastures, although notable for its poorer yield, was suitable for *Karakul* sheep rearing. The rest of the territories needed special measures for pasture amelioration. The historical analysis presented in the next section shows that this was achieved by means of: (a) making intensive use of agricultural research on the one hand; and (b) setting up an effective institutional structure on the other. Science-based successful pasture utilization by the Soviets was also confirmed by the 1964 FAO study tour organized for international agricultural scientists to study rangeland research progress and pastoral livestock management in Kazakhstan, Turkmenistan and Uzbekistan. As reported by the group:

> the signs of plant cover, forage composition, soil stability and general good condition of pastures were clear to all of the Fellows, many of whom had never seen such expansive areas of good productive range. It was evident that these ranges had had the benefit of light to moderate stocking for many years. (FAO 1964, 308)

But the question of whether such a productive pastoral system was profitable for the Soviet economy or not requires closer attention. By the 1980s, rangeland animal production in the FSU was already in decline due to insufficient fodder, poor management and economic distortions led by state-mandated production plans (Kerven, Channon, and Behnke 1996). In the late 1980s, just before the collapse of the Soviet regime, a group of Uzbek economists, Kayumov, Kuvnakov and Farmanov (1993), estimated that the *Karakul* production sector of agriculture, including all public services, social infrastructure and subsidies, had become economically unattractive for public investments under Brezhnev's policy of rural subsidization of the 1970s. It was found that the profit margin of the *Karakul* sector was only 1.6%, whereas in intensive cattle production it was 8.5%, and in cotton production it was 31.2% (ibid., 63). They came to the conclusion that the central Planning Ministry (*Gosplan*) commanded too low procurement prices and inadequate profitability norms for the *Karakul* sector, and it failed to objectively estimate all the transaction costs of production, taking into account the geographic conditions and socio-economic and seasonal environmental challenges of animal rearing in remote dryland pastures (Kayumov, Kuvnakov, and Farmanov 1993, 62). Additionally, the production deficit for spare parts, petroleum, oil and lubricants as well as price inflations had shown that the Soviet centralized old business model was not profitable any longer in emerging market economy conditions. At the same time, the global demand for *Karakul* fur dropped significantly due to declining world market prices (Kreutzmann 2012).

Following the Soviet system collapse, animal husbandry became a major source of income for rural resource-poor households. Sheep and goat stocks rapidly increased from 10% in 1991 to a 70% share in 2010 (Lerman 2008). As a consequence, animal concentration near villages exceeded the pasture carrying capacity by 3.4-fold and overgrazing around these areas became widespread (Holmirzaev 2013). After lifting former government restrictions on the number of livestock per household, communities were further motivated to increase livestock numbers (Lerman 2005), although the grazing areas near villages belong to cooperative farms and pasture land use rights for individual households are insecure under the current Law on Land.

In contrast, remote pasture territories suffered from weed invasion and infrastructure decay due to underuse. Only 13.5 million ha of pastures were used from the total available 20.75 million ha in 2012 (Holmirzaev 2013). From 2008 to 2011, pastures with water supply declined by 568.2 thousand ha (Umarov,

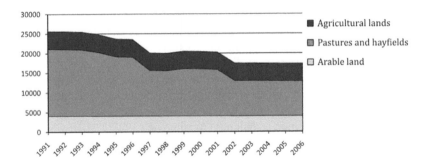

Figure 3.2 Dynamics of land distribution in Uzbekistan (1000 ha)

Source: State Committee for Geodesy, Cartography and Land Cadastre (in UNDP 2010a).

Shakirov, and Asilbekova 2013), mainly due to poor service delivery at remote pastures. As reported by the Ministry of Agriculture, 73% of pastures and grass-lands are affected by degradation processes, and productivity has declined by a third since 2010 (Holmirzaev 2013). Since 2005, 50,000–100,000 ha of rangelands were taken out of pastoral rotation annually due to overgrazing, desertification, wind/water erosion, obsolete infrastructure and remoteness (ibid.). On the other hand, the rapid growth rates of household animals led to a concentration of 1.2 livestock units per ha, which exceeded capacity norms of arid pastoral areas by 3.4 times (Holmirzaev 2013, 99). Hence, there is a constant deficit of pasturage and a high animal grazing intensity due to limited seasonal mobility and lack of services in remote rangelands.

Nonetheless, there are lessons to be learnt from the Soviet pastoral system that are worth documenting, and in the next two sub-sections we examine some of the key factors that allowed them to achieve such productivity results in rangelands.

Contribution of agricultural research to the development of rangeland systems

The review of archive materials shows that, in the early 1920s, the Soviet Union's central Politburo initiated extended land surveys in newly established colonies for their potential reclamation, resource extraction and expansion of agricul-ture production. For these purposes, groups of prominent soviet scientists from Moscow and Leningrad were sent (Nechaeva, Mordvinov, and Mosolov 1943) to establish large-scale land surveys and to find potential production and resource extraction schemes, mostly for gold, oil and gas, coal and gypsum resources, as well as for the expansion of cotton, fruit/vegetables and livestock produce. The initial assessment phase (1920–1925) in drylands included the geographical study of desert and semi-desert territories, a general inventory of rangeland areas, the exploration of water sources and traditional approaches of animal husbandry in drylands, soil geology and analysis of its physical features (Arasimovich 1929).

This then led to extended studies and conversion to cropland of rangelands in the period from 1925 to 1940 (Morozova 1946). This included: (a) botanical analysis and mapping of local plants; (b) groundwater mapping, including levels of mineralization, water-table analysis, carrying capacity of the groundwater lens; and (c) testing optimal utilization of distant rangelands (Gaevskaya and Salmanov 1975). Traditional nomadic practices were also studied and they were found useful in accessing distant pastures, in setting up seasonal rotation schemes, in organizing grazing techniques, in indigenous methods of water harvesting in deserts, and in natural desalinization techniques (Arasimovich 1929; UNDP-GM 2007).

The long-term stability of the fragile desert ecosystem and extensive animal production in drylands were found to be fundamental principles of dryland reclamation, as concluded by Soviet scientists (Morozova 1946), whereby fodder yields, grazing intensity and water quality in pastures were identified as the main criteria for setting up effective pasture rotation schemes (Nechaeva, Mordvinov, and Mosolov 1943). Science-based approaches, created with substantial efforts over the decades, had contributed to the development of productive rangeland systems in the FSU for over 250 million ha by the mid-1960s (Khudayberdyev 1976).

As the designers of pastoral production systems understood the fundamental problem of climate risks, a range of measures were introduced to reduce losses in cold seasons such as early frost detectors, shelters and storage facilities. Scientific departments were established to constantly assess agro- and zoo-climatic conditions and to estimate animal productivity effects (Babushkin, Sumochkina, and Sitnikova 2007). Meteorological stations and locators were installed in districts by the Hydro-meteorological Committee to predict and communicate potential frost days per season (ibid.). Plant phyto-melioration and afforestation were introduced by the Forestry Institute to rehabilitate degraded areas and to reduce desertification (Gaevskaya and Krasnopolin 1957). Genetic research experiments and artificial insemination practices were introduced by the Karakul Research Institute to enhance sheep's resistance to cold seasons (ibid.).

Materials from archives show that the results of in-depth research on distant pasture rotation and livestock mobility schemes were instrumental in providing evidence on the promising potential of the pastoral livestock production system (Lobanov 1953). Field experiments in drylands, conducted in Central Asian drylands from 1925 to 1940, demonstrated that experimental distant mobility methods of pasture grazing reduced the primary cost of pastoral livestock production by 50%, the labor costs were 30% lower, and animal maintenance was 40% less costly compared with stalled livestock keeping (Lobanov 1953; Nechaeva, Mordvinov, and Mosolov 1943; Babaev 1977). The application of plant phyto-melioration in semi-desert pastures increased fodder availability by two to three times (Fedorovich 1950). These study results then encouraged a massive expansion of political support and state investment programs in the drylands of Central Asia, the Caucasus, and Siberia (Nechaeva, Mordvinov, and Mosolov 1943; Fedorovich 1950). As estimated by Lobanov, Soviet *Karakul* farms' managed to generate huge profits in the late 1950s to the early 1970s, whereas the major share of earnings (up to 75–95%) was primarily gained from extended transhumant schemes (Lobanov 1953). This was a new approach in livestock production, whereas all-seasonal pasture rotation methods were coupled with social infrastructure services, whereby animal husbandry and feeding costs were 'close to gifted' (Lobanov 1953, 7). As a result, the number of *Karakul* sheep reached 6 million head by the early 1970s (Khudayberdyev 1976).

The role of institutional and organizational structures in the development of rangelands

Interview data and archive study indicate that, on the basis of the long-term empirical studies in drylands, extensive political support was provided to establish a number of public services, infrastructure facilities and production units in rural areas, as presented in Table 3.3, to scale up pastoral livestock production in the former Soviet Uzbekistan from the early 1930s to the late 1960s.

Large-scale *Kolhoz* and *Sovhoz* farms were administratively designed as rural towns with associated agro-production, social infrastructure and rural services attached to each territory (Swinnen and Rozelle 2006). Massive financial and political support for rural industrialization in the FSU led to high employment rates and livelihood improvements in rural areas (Razzakov 2009). For example, as archive materials indicate, the infrastructure construction investment programs of the early collectivization period (1930–1945) included provisions of the following to all state farms: warehouses, rural housing, wells, canals, reservoirs, pumping stations and communal water networks, roads, machinery, electricity and gas supply systems, etc. (Morozova 1946). Archive records also show that further development efforts in rural areas after 1945 established other social infrastructure and services, especially in remote rural settlements: health clinics, schools and professional colleges; transport and postal communication; pharmacy, bakery and grocery stores; veterinary offices and research stations (UNDP-GM 2007). Brezhnev's campaign program on 'Entire villages electricity supply' in Soviet Uzbekistan was fully accomplished in all rural areas by the end of the 1950s (Razzakov 2009).

Historical analysis indicates that communication between academia and Soviet farms was well organized. Every *Sovhoz* and *Kolhoz* had staff positions for agricultural scientists, engineers and specialists, who also served to monitor production processes and to regulate and report results to senior executives (Sovnarkom 1945). These staff played a key role in the organization and monitoring of animal rotation, grazing and watering in distant pastures. It is worth noting that the Soviets employed a range of incentive schemes to encourage the labor productivity of these staff in remote arid areas. Interviews and archive materials confirmed that financial rewards, promotions and recognitions were widely applied for the remoteness of service, productivity and years of experience (Lobanov 1953; Khudayberdyev 1976). For example, the most productive employees were granted public awards titled '*Stakhanovets*' and '*Udarnik*' (Khudayberdyev 1976) and had access to advanced education and recreation,

Table 3.3 Institutions and services established for pastoral system development

Organizational level	Type of institution	Functions in pastoral system
National agencies for sector coordination	*Karakul-Trest* (with status of the National Ministry for Pastures)	Planning and coordination of pastoral livestock production, mainly *Karakul* sheep
	Ministry of Forestry	Massive reclamation/afforestation in deserts
	Republican Corporation for Rangeland Melioration and Construction (*RPMSO*) with Mobile Mechanized Units (*PMK*)	Construction and maintenance of water facilities in pastures and villages
	State Committee for Nature Protection	Monitoring and maintaining ecosystems of drylands and to prevent their overuse
Research institutes and experimental stations	State Institute of Land Resources Assessment and Planning (*Uzgiprozem*)	Designing distant pasture rotation schemes and mapping, and scientific expeditions to conduct regular geobotanical assessments
	Soviet Research Institute of *Karakul* Production (1935)	Improving the quality of *Karakul* pelts through research in genetics, breeding, planting, water quality and desert melioration
	Research Institutes of Water Planning, Forestry, Veterinary Services, Livestock Breeding, Botany, and Plant Engineering	Wide range of public goods and services to improve pastoral system production
Additional services	Agro-meteorological and zoo-climatic assessments (based on national agency for hydrometeorology)	Monitoring and forecasting factors of animal productivity based on climatic changes, number of unfavorable days for grazing, animal productivity changes, pastoral vegetation yields, etc.
	State farms specializing in the artificial insemination of improved *Karakul* bloodlines	Distribution of semen of high-quality breeds
	Mobile veterinary brigades and zoo-technicians	Disease prevention and treatment services in remote grazing areas, as well as disinfection of water points and sheds
	Mobile water tanks, machinery services and tractor brigades	Supported remote watering, afforestation, phyto-melioration and construction of wells
Production, processing and construction	Units for primary processing of meat, pelts, wool and milk	
	Factories with brigades to construct furniture and mobile housing for shepherds	

Source: Modified from Sergeeva (1951), Khudayberdyev (1976) and Babaev (1977).

housing and automobile subsidies, a favorable pension, etc. (Lobanov 1953). Conducive policies and an enabling environment in the rural areas allowed better organization and the distribution of higher numbers of livestock herds across 23 million ha of pastures (Khudayberdyev 1976).

Not surprisingly, the former Soviet agricultural enterprises and centrally integrated supply chains also faced numerous organizational challenges that were inherent in the public sector, such as production inefficiency and corruption (Filtzer and Gregory 2006). In response to these challenges, as archive sources indicate, the Soviets developed a number of regulatory mechanisms for crime detection and sanctioning. In his notable Soviet archive study, William Clark (1993) identified a whole range of monitoring, conspiracy, investigation, prosecution and revisionary institutions created at all FSU levels to control political and organizational crime cases and to take radical measures. These included: State Party Control Committee, and Department to Fight Theft of Socialist Property for the state level; Committee of National Control and Criminal Investigation for the regional level; and a Soviet Whistleblowers' free-press section in the newspapers and anonymous complaint phonelines established at the local level. Thus, corruption in the FSU had a form of 'controlled corruption' and was a measurable expense, rather than the 'uncontrolled corruption' in the current transition period, which is unpredictable (Ferguson 2003). Obviously, the measures used to resolve management problems during the Soviet period are highly sensitive from a human rights and wellbeing perspective, and it is not our intention here to justify them in any way. The point is rather to highlight the role that sanctioning mechanisms played in achieving the observed outcomes.

Pastoral system challenges in transition period

As a result of the system collapse followed by inflation and employment constraints, animal husbandry became one of the main sources of income, and the number of animals increased to ensure savings for rural households (UNDP 2005). The recent economic assessment by UNDP revealed that revenues from pastoral meat production were, in fact, three times higher than they were from *Karakul* pelt sales (UNDP 2011). This explains why *Shirkats* have little interest in producing *Karakul* fur, and give preference to meat production by any means. As a result of this mismatch in production interests, the number of *Karakul* lamb owned by *Shirkats* is decreasing compared with increasing numbers of fat-tail sheep. Currently, 92.3% of *Karakul* sheep are produced in dryland pastures (Holmirzaev 2013), but their breed quality is deteriorating significantly (Robinson et al. 2012).

By early 2000, most of the formerly subsidized public agricultural services were commercialized (Gupta et al. 2009). As our interviews show, currently *Shirkats* experience hardships to cover the following commercial services: construction and maintenance of water wells, agro-meteorological assessments, provision of pastoral rotation schemes, and rangeland afforestation and phyto-melioration services. The poor service demand resulted in unstable rangeland service delivery and staff cuts at supply institutions. Institutional memory of rangeland service organizations is being lost too, mainly due to improper maintenance of knowledge database, poor funding for research and innovation, weak continuity of professionals, and aging among experts.

Poor demand for agricultural services in rangelands is also stimulated by other indirect expenses that *Shirkats* have to meet: sustaining district budgets through 'voluntary' fund allocations to primary schools, pension funds, cultural events, roads and other infrastructure maintenance. Additionally, they cover land taxes and social infrastructure bills, and have to meet state production quotas, and adapt to market uncertainties. Estimations from *Uzbek Karakuli* reports[3] show that these expenditures exceed a 45% share of *Shirkats'* revenue. Our archive sources indicate that social responsibility schemes were also practiced in the Soviet period, because *Sovhoz* were established as rural towns with corresponding administration, infrastructure and organizational schemes (Lobanov 1953). However, the former Soviet farms were well subsidized and they did not have to pay land taxes and other ad-hoc bills (Sovnarkom 1945). Therefore, it is not surprising that *Shirkats* are currently unable to afford additional services to fully utilize distant pastures, to maintain their productivity, and, consequently, to avoid land degradation.

Interview respondents confirmed that households normally graze animals communally on *Shirkat* farms' rangelands with prior verbal agreements. Kinship ties and social networks among farm management and local community families favorably contribute to informal land use regulations in agreed pastoral territories around villages and water sources. *Shirkats* have neither the capacity to monitor vast pastures nor any strong incentive to exclude households from pasture grazing (Farmanov 2005; UNDP 2005). However, in other cases the rangeland use competition between villages is increasing after the adoption of the Livestock Development Program (2006), which formally encourages increased animal numbers for food security (Lerman 2008). In practice, however, this and previous land-related legislation have neither regulated the current unsystematic pasture use by households nor introduced any clear property rights for pasture users and extended livestock mobility. For the time being, *Shirkats* accept this, but in the long term more adequate regulations will be necessary.

As reported by UNDP (2005), the administrative and municipal agencies at the provincial and district levels have not yet developed a clear plan of action or focused strategy for the sustainable management of livestock and pasture resources. In this respect, the local *Shirkat* farm managers keep working on routine production issues, while pasture rehabilitation is largely being neglected, as our interview data demonstrate. Hence, there is rising awareness in Uzbek state agencies about rangeland use competition between communities and *Shirkats*. Different models are under consideration: private versus collective leasehold of pastureland; remuneration of the *Shirkat* for pasture use versus leasehold directly from the state (Robinson et al. 2012).

A political economy perspective of transition period reforms

Based on the literature review, we now discuss some of the political economy determinants and challenges that considerably influenced political decisions on pastoral system development in Uzbekistan's transition period reforms.

Firstly, to answer the question of why the pastoral land was not given to the private sector or to community-based management, we go back to the historical legacy and traditions of land ownership in the region. Historically, in the pre-Soviet period of Uzbekistan, there was a feudal regime. All pastures were owned by two Khanates and the Bukhara Emirate (Valiev 1980). Arable land plots were leased to peasants, and grasslands were leased and used as common property resource (CPR) by wealthy landlords (ibid.). Apparently, there was neither any tradition nor a legacy of private land ownership, and no demand from the grassroots existed for privatization. Historically, a legacy and tradition of private land ownership existed in several former Soviet Central and East European nations, and one could observe strong demand for land privatization by households during the post-independence transition period (Rozelle and Swinnen 2009a).

Secondly, the characteristics of the geographic location have also contributed to political decisions on pastoral property rights. The region's drylands are characterized by relatively low rates of fertility and precipitation (SNC 2009), and the rangelands are often located at a greater distance from settled communities, which makes their use particularly problematic or costly. Input prices are expensive and services are often unavailable or costly in distant areas (Swinnen and Heinegg 2002), which often demotivates land users from potential livestock mobility efforts. Returns on investments are far lower and labor intensity is higher in drylands compared with intensive livestock rearing and arable farming. For example, cotton, being a strategic cash crop, has clearly higher economic incentives and the harvest season mobilizes the majority of the rural population in the country (Ferguson 2003). These contrasts clearly indicate why policies towards sustainable pasture management have been left behind in the development agenda.

Thirdly, changes in government structure and political regime have often induced changes in land reforms (Swinnen and Rozelle 2006). On the contrary, it is argued that even small changes in political leadership structure, level of participation of civil society and the private sector in political decision making affect the likelihood of reforms and the pace of liberalization in most of the Newly Independent States (NIS) (Swinnen and Heinegg 2002). We can observe this in the examples of Belarus, Kazakhstan, Tajikistan, Turkmenistan and Uzbekistan, which have been run by more or less the same political leadership since the early transition period (Swinnen and Rozelle 2006). In the case of Kyrgyzstan, the frequent land use conflicts between pastoral users have induced a need to address these issues with the help of grassroots NGOs, community leaders and municipalities (Crewett 2012). This has created strong pressure for parliamentarians, and they involved the World Bank to develop a new law on community-based pasture management in 2009 (Robinson et al. 2012). Thus, a positive correlation between political reforms and land reforms has been identified by Swinnen and Heinegg (2002) in most of the NIS, as shown in Figure 3.3.

Fourthly, one of the reasons why pastoral system institutions had deteriorated was the level of technological and capital integration into Soviet production systems. State farms in the FSU were organized as capital- and land-intensive, and were highly integrated into industrial systems; a complex network of exchange

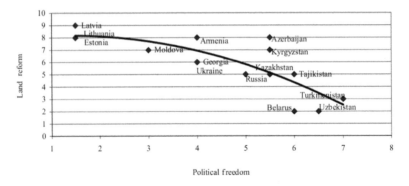

Figure 3.3 Positive correlation between political reforms and land reforms

Source: Adapted from Swinnen and Heinegg (2002, 19).

relations existed between input suppliers and processors (Rozelle and Swinnen 2009a) within and outside the Newly Independent States (NIS). Moreover, the increased outmigration of the Slavic population from the country after the Soviet collapse influenced the availability of highly qualified field specialists and service professionals significantly (Ferguson 2003). One can argue that the deterioration of the pastoral production system was inevitable after the disintegration of interdependent exchange mechanisms and the massive collapse of the centrally organized fiscal, economic and political structures.

Fifthly, the specificities of traditional institutions have also contributed internally to the collapse of the pastoral system. These institutions are characterized by the historical domination of ethnic, religious and clan networks. The central control, sanctioning and enforcement mechanisms of the Soviets were strong setback factors for them, but these networks survived and flourished after the system collapse due to their pervasiveness and extensiveness (Ferguson 2003). Clientelism and patron–client relationships are more extensive among political actors in the region than in the rest of the FSU (Swinnen and Heinegg 2002). Cronyism and kickbacks to officials have been at the heart of corruption (Ferguson 2003). Local elites have a strong influence on resource distribution, service providers and producers, both formally and informally. Therefore, unless a fairly accountable and transparent checks-and-balances mechanism is established, adopting a radical transition reform agenda could facilitate more tensions between these networks and interest groups than any positive development outcomes (Swinnen and Heinegg 2002).

Finally, the difference in pace of economic reforms towards market liberalization also affected liberal reforms in the pastoral production systems. Following China's successful example of transition reforms, Uzbekistan undertook gradual reforms of the agriculture sector towards a market economy (Rozelle and Swinnen 2009b). The agriculture sector has traditionally provided a high proportion of

Uzbekistan's GDP (over 45% in 1991) and supported about 65% of the country's population (UNDP 2010a). Uzbekistan's gradual reform agenda can also be explained by the mismatch of reform interests between top politicians and farmers (Swinnen and Heinegg 2002). The latter opposed reforms primarily due to the benefits derived from subsidies and high wages (ibid.).

Discussion

The current debates of CPR management largely neglect the state-managed regime of the commons. The CPR theory would suggest that the highly centralized state management of rangelands during the Soviet period should have been unsustainable and ineffective, whereas the political change should have created good conditions for the successful management of the CPRs by local pastoral communities. Nevertheless, in practice it was exactly the opposite: the Soviets managed to utilize pastoral systems fairly successfully under the central state regime, whereas the post-Soviet transition reforms have scarcely favored any community-based property rights approaches in order to better utilize such systems.

Our findings indicate that a strong political will, science-based approach and economic incentives are the key reasons for such contrasting outcomes. The historical analysis shows that the Communist leaders' industrial and ideological race against capitalistic economies was a major political driving force towards capital injections into all sectors of the national economy, including pastoral production systems. Sanctioning mechanisms and labor motivation schemes established by the Soviets also played a significant role in achieving the aimed-for results in drylands.

The development of the fragile vast rangelands required substantial innovations in pastureland and water management, which would not have been possible without major advances in agricultural research. While the economic theory of induced innovation has emphasized this factor (Hayami and Ruttan 1985), the literature on CPRs and studies on post-Soviet transition economies have both largely neglected this factor. This is rather surprising, as support for research and technological development was a key priority of both agricultural development and industrialization in the FSU (Gregory 2008). The historical analysis presented in this chapter demonstrates how important the results of agricultural research were, and how consistently they were utilized for decision making in the Soviet Politburo in the development of drylands.

The analysis of political economy incentives during the transition period indicates that the central government considered a far-reaching land reform in pastoral areas as too risky from a political point of view. Hence, collective farms were left with the expectation of providing local employment and services. State procurement of *Karakul* pelts also continued, even though it is less attractive by far than keeping local sheep for meat. The rangeland production systems, where only about 10% of the population resides, had been paid less attention also partly because of weak demand from land users and lack of support from politicians, as well as due to low investment returns and unfavorable geographical conditions. In view of emerging socio-economic and environmental constraints in drylands and limited income opportunities, the state encourages an increased number of animals in households without allocating specific grazing land rights to them. In contrast, irrigated agricultural systems received massive public support because they had higher demand from the resource end users and stronger economic incentives and political benefits. For example, in irrigated cotton and wheat areas, in which water resources are also classified as CPR, the system experienced more favorable community-driven water management reforms with established Water User Associations (Ghazouani, Rap, and Molle 2012). These contrasting changes in transition reforms suggest that transition-period agricultural policies were primarily concentrated in the areas where immediate development effects would affect a larger proportion of the population.

The analysis of political economy challenges also demonstrates other determinants that created little demand for both private and community-based rangeland management reforms in Uzbekistan, such as the low profit margin from rangelands, unfavorable geographic conditions and outdated infrastructure in remote pastures in the context of rising input prices and the lack of a historical legacy of traditional land ownership, among other factors. As suggested by Swinnen and Heinegg (2002), there is also a positive correlation between changes in political regime and land reforms (see Figure 3.3). This indicates why any substantial reforms in dryland pastoral systems towards community-based management were mostly overlooked in most of the NIS. Thus, we acknowledge that the pastoral sector is still in a transformation stage in those NIS, and that major structural changes and institutional reforms are yet to come.

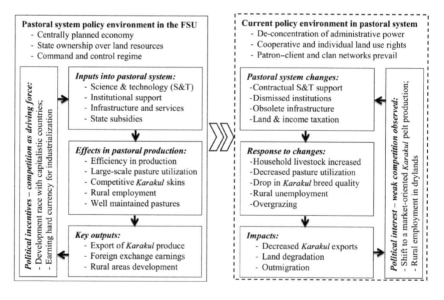

Figure 3.4 Comparing pastoral system changes before and during the transition period

Source: Authors.

As our interview data show, the scientific achievements and institutional models of the FSU, as well as modern approaches, are largely overlooked in the development of rangeland systems in the current transition period. On the one hand, this was due to the fact that the former central planning approaches demonstrated their weakness in the period of the Soviet economic stagnation of the 1980s and were no longer viable after the independence period. The new political leaders of the NIS sought for better solutions than market economy paradigms could suggest. On the other hand, new economies simply could not afford to fund huge research programs, whereas reform priorities, as mentioned earlier, were primarily concentrated in support of the other sectors of the economy in order to avoid a deepening of the post-Soviet period recession and food shortages (UNDP 2010b).

Our historical analysis clearly demonstrates that research experiments and agricultural innovation combined with smart institutional structures and services played an important role in the development of productive livestock systems in the massive rangelands throughout the region, whereas environmental degradation was addressed effectively up to the end of the 1970s. In combination with the political economy factors presented here, the current regimes could still use the unique past experience positively and apply a science-based approach to achieve a fairly successful management of the CPRs.

Conclusions

This chapter contributes to the current state of knowledge on CPR management in the post-socialist economies, where some countries have adopted better institutional reform structures for the commons than others. A strong political will and economic incentives are found to be the major factors in taking reform decisions supported by strong agricultural research and innovation. Our findings also underline the importance of paying attention to the country-specific political economy challenges of rangeland systems in transition countries, whereas reform decisions may be focused towards priority areas other than CPRs. We have discussed why rangelands were neglected in comparison with arable systems, and where cash crops attracted more political and economic incentives and channeled substantial support for community-based water resource management reforms. Our historical analyses also demonstrate important lessons from successful institutional models and effective science-based approaches applied in the past that can be useful for the development of current CPR management systems.

Notes

1 'The Statistics Division of the FAO', http://faostat.fao.org/
2 *Karakul* is a traditional lamb genotype, historically originated from *Karakul* district in Bukhara Region, with high endurance to dryland conditions (Valiev 1980).
3 Estimated from internal budget reports of *Uzbek Karakuli* for 2011.

References

Arasimovich, K. 1929. *Очерки Экономики Каракулеводства Узбекской ССР / Economic Overview of the Karakul Sector in Uzbek SSR*. Народный К. Samarkand: Издание Наркомзема УзССР.

Babaev, A. 1977. Principal Problems of Desert Land Reclamation in the U.S.S.R. In: M. Glantz (ed.) *Desertification: Environmental Degradation in and around Arid Lands*. Boulder, CO: Westview Press.

Babushkin, O., T. Sumochkina, and M. Sitnikova. 2007. *Комплексная Оценка Каракулеводческих Пастбищ Узбекистана / Complex Assessment of Karakul Pastures in Uzbekistan*. Edited by A. Abdullaev. Tashkent: Uzhydromet.

Chayanov, A. 1925. *Организация Крестьянского Хозяйства / Organization of the Peasantry Economy*. Moskva Tse. Moscow: Kooperativnoe Izdatelstvo.

Clark, W. 1993. *Crime and Punishment in Soviet Officialdom: Combating Corruption in the Political Elite 1965–1990*. Contempora. New York: M.E. Sharpe Inc. Print. Congress Catalogue.

Crewett, W. 2012. Improving the Sustainability of Pasture Use in Kyrgyzstan. *Mountain Research and Development* 32 (3) (August): 267–74.

FAO. 1964. Range Study Tour in the Soviet Union. Study Tour Report by Kenneth Pearse. Rome, Italy: FAO.

Farmanov, T. 2005. Economic and Institutional Assessment of Pastoral System. Case-Study of Forish District in Jizzakh Province, Uzbekistan. Report of National Consultant on Economic and Institutional Aspects. Tashkent.

Fedorovich, B. 1950. *Лик Пустыни / A Realm of Desert*. Moscow: Goskultprosvetizdat.

Ferguson, R. 2003. *The Devil and the Disappearing Sea*. Vancouver, Canada: Raincoast Books.

Filtzer, D. and P.R. Gregory. 2006. *The Political Economy of Stalinism: Evidence from the Soviet Secret Archives*. Slavic Review Vol. 65. Cambridge, UK: Cambridge University Press.

Gaevskaya, L. and F. Krasnopolin. 1957. Rangelands of Samarkand Province and its Use in Karakul Sheep Husbandry. In: *Agricultural Issues of Zerafshan Basin*. Tashkent: AN UzSSR.

Gaevskaya, L. and N. Salmanov. 1975. *Rangelands of Deserts and Semi Deserts of Uzbekistan*. Tashkent: FAN.

Ghazouani, W., E. Rap and F. Molle. 2012. Water Users Associations in the NEN Region IFAD Interventions and Overall Dynamics. Rome, Italy: International Fund for Agricultural Development (IFAD).

Gilmanov, T. 1996. Ecology of Rangelands in Central Asia and Modeling of their Primary Productivity. In: *Workshop Proceedings: 'Assessment of Livestock Production in Central Asia'. 27 Feb–1 Mar. 1996. Small Ruminant Collaborative Research Support Program of ICARDA and USAID*. Tashkent: University of California, Davis.

Gregory, P.R. 2008. *Политическая Экономия Сталинизма / Political Economy of Stalinism: Evidence from the Soviet Secret Archives*. 2nd edition. Cambridge, UK: Cambridge University Press.

Gupta, R., K. Kienzler, C. Martius, A. Mirzabaev, T. Oweis, E. de Paul, M. Qadir, et al. 2009. Research Prospectus: A Vision for Sustainable Land Management in Central Asia. Tashkent.

Hayami, Y. and V. Ruttan. 1985. *Agricultural Development: An International Perspective*. Baltimore, MD: Johns Hopkins University Press.

Holland, M. 2010. Mid-term SLM Project Evaluation Report. Materials of the UNDP Project 'Achieving Ecosystem Stability in Degraded Land in Karakalpakstan and Kyzylkum Desert'. Tashkent.

Holmirzaev, I. 2013. Мониторинг Состояния Пастбищного Животноводства В Узбекистане (Monitoring of Pastoral Animal Production Conditions in Uzbekistan). In: *Institutional Aspects of Rational Pasture Use and Conservation / Yaylovlardan Oqilona Foydalanish va Muhofaza Qilishning Institucional Masalalari).* Tashkent: National University of Uzbekistan named after Mirzo Ulugbek, pp. 97–100.

IFPRI/ICARDA. 2008. Economic Analysis of Sustainable Land Management Options in Central Asia. Tashkent.

Kayumov, A. 1977. Пастбища Узбекистана / Pastures of Uzbekistan. *Сельское Хозяйство Узбекистана / Agriculture of Uzbekistan* 2: 50–1.

Kayumov, F., H. Kuvnakov and T. Farmanov. 1993. *Каракулеводческий Подкомплекс И Его Роль В Формировании Конъюнктуры Рынка / Karakul Subsector and its Role in Market Formation.* Tashkent: Uzinformagroprom.

Kerven, C., J. Channon and R. Behnke. 1996. Planning and Policies on Extensive Livestock Development in Central Asia. Working Paper 91. London: Overseas Development Institute.

Kerven, C., B. Steimann, L. Ashley, C. Dear and I. ur Rahim. 2011. Pastoralism and Farming in Central Asia's Mountains: A Research Review. 1. Bishkek: University of Central Asia.

Khudayberdyev, N. 1976. *В Единой Семье Народов-Братьев: Развитие Производительных Сил Узбекистана За 50 Лет / In United Family of Brotherhood Nations: Development of Production Strengths of Uzbekistan in 50 Years.* Tashkent: FAN.

Kreutzmann, H. 2012. Pastoral Practices in High Asia. *Advances in Asian Human-Environmental Research* 323–36. doi:10.1007/978-94-007-3846-1; http://link.springer.com/10.1007/978-94-007-3846-1

Lerman, Z. 2005. The Impact of Land Reform on Rural Household Incomes in Transcaucasia and Central Asia. Discussion Paper No. 9.05. Rehovot: Hebrew University.

Lerman, Z. 2008. Agricultural Development in Uzbekistan: The Effect of Ongoing Reforms. Discussion Paper No. 7.08. Rehovot: Hebrew University.

Lobanov, P. 1953. *Сельскохозяйственная Энциклопедия / Agricultural Encyclopedia.* Издание тр. Moscow: Гос.изд. с/х лит.

Morozova, O. 1946. *Пастбищное Хозяйство В Каракулеводстве Средней Азии / Pastoral Stewardship of Karakul Production in Central Asia/.* Edited by A. Dmitrieva. Moscow: Международная книга.

Nechaeva, N., N. Mordvinov and I. Mosolov. 1943. *Пастбища Каракумов И Их Использование / Pastures of Kara-Kums and their Utilization.* Academy of Ashgabad: Turkmen FAN.

Nishanov, N. 2013. Utilization of Rangelands in Central Asia: Challenges and Solutions. In: *Institutional Aspects of Rational Pasture Use and Conservation / Yaylovlardan Oqilona Foydalanish va Muhofaza Qilishning Institucional Masalalari.* Tashkent: National University of Uzbekistan named after Mirzo Ulugbek, pp. 35–40.

Otakulov, U. 2013. Pasture Protections and Biodiversity Conservation Are Important Factors of Environmental Sustainability / Yaylovlarni Muhofaza Qilish – Bioxilma-Xillikni Saqlash, Ekologik Barqarorlikni Ta'minlashning Muhim Omilidir. In: *Institutional Aspects of Rational Pasture Use and Conservation / Yaylovlardan Oqilona Foydalanish va Muhofaza Qilishning Institucional Masalalari.* Tashkent: National University of Uzbekistan named after Mirzo Ulugbek, pp. 7–8.

Panteleev, I. 1954. Развитие Отгонно-Пастбищного Животноводства В Каспийской Низменности, Кара-Кумах И Кара-Калпакской АССР И Использование Земельных Фондов В Этих Районах / Development of Transhumant Pastoral Approaches of Livestock Production. Moscow.

Paroda, R., S. Beniwal, R. Gupta, Z. Khalikulov and A. Mirzabaev (eds). 2007. From Issues to Actions: Final Report of the Expert Consultation on Regional Research Needs Assessment in Central Asia and the Caucasus, 7–9 March. GFAR, CACAARI, ICARDA-CAC, Tashkent, 14 pp.

Rahimov, R.M. 2012. Evolution of land use in pastoral culture in Central Asia with special reference to Kyrgyzstan and Kazakhstan. In: V. Squires (ed.) *Rangeland Stewardship in Central Asia: Balancing improved livelihoods, biodiversity conservation and land protection.* Dordrecht: Springer, pp. 51–69.

Razzakov, F. 2009. *Коррупция В Политбюро: Дело «красного Узбека» / Corruption in Politburo: Case of 'Red Uzbek'.* Moscow: Eksmo, Algoritm.

Robinson, S., C. Wiedemann, S. Michel, Y. Zhumabayev and N. Singh. 2012. *Pastoral Tenure in Central Asia: Theme and Variation in the Five Former Soviet Republics.* In: V. Squires (ed.) *Rangeland Stewardship in Central Asia: Balancing improved livelihoods, biodiversity conservation and land protection.* Dordrecht: Springer, pp. 239–74.

Rozelle, S. and J. Swinnen. 2009a. The Political Economy of Agricultural Reform in Transition Countries. In: V. Beckmann and M. Padmanabhan (eds) *Institutions and Sustainability.* Dordrecht: Springer Netherlands, pp. 27–41.

Rozelle, S. and J. Swinnen. 2009b. Political Economy of Agricultural Distortions in Transition Countries of Asia and Europe. Agricultural Distortions Working Paper 82. Washington DC: World Bank.

Sergeeva, G. 1951. Improvement of Karakul Sheep Breeding Rangelands of Uzbekistan. *Karakulevodstvo I Zverovodstvo* 4: 77–8.

SNC. 2009. Second National Communication of Uzbekistan for UNCCC. Unpublished report for UNCCD, Bonn. Tashkent.

Sovetkina, M. and E. Korovin. 1941. *Введение В Изучение Пастбищ И Сенокосов Узбекистана / Introduction into Pastures and Hayfields Study in Uzbekistan.* Академия Н. Tashkent: UzFAN.

Sovnarkom. 1945. *О Государственном Плане Развития Животноводства В Колхозах И Совхозах На 1945год / On the State Plan of Livestock Development in Kolkhoz and Sovkhoz for 1945.* Novosibirsk: В Совнаркоме СССР и ЦК ВКП(б).

Squires, V.R. and Lu Qi. 2018. Greater Central Asia: its peoples and their history and geography. In this volume, pp. 251–72.

Steinfeld, H., P. Gerber, T. Wassenaar, V. Castel, M. Rosales and C. De Haan. 2006. Livestock's Long Shadow. FAO Report Vol. 3. Rome, Italy: FAO.

Swinnen, J. and A. Heinegg. 2002. On the Political Economy of Land Reforms in the Former Soviet Union. LICOS Discussion Paper 115/2002. Leuven, Belgium: LICOS Centre for Institutions and Economic Performance.

Swinnen, J. and S. Rozelle. 2006. *From Marx and Mao to the Market.* New York: Oxford University Press.

Teshaboev, M. and O. Sharipov. 2013. Yaylovlardan Oqilona Foydalanish va Muhofaza Qilishning Institucional Masalalari / Institutional Aspects of Rational Pasture Use and Conservation. In: *Institutional Aspects of Rational Pasture Use and Conservation / Yaylovlardan Oqilona Foydalanish va Muhofaza Qilishning Institucional Masalalari.* Tashkent: National University of Uzbekistan named after Mirzo Ulugbek, pp. 4–7.

Umarov, N., N. Shakirov and H. Asilbekova. 2013. Экологическое И Мелиоративное Состояние Используемых Земель, Рекультивация Нарушенных И Загрязненных Земель Пастбищ / Economic and Meliorative State of Land Use, Recultivation of Damaged and Contaminated Pastoral Lands. In: *Institutional Aspects of Rational Pasture Use and Conservation / Yaylovlardan Oqilona Foydalanish va Muhofaza Qilishning Institucional Masalalari.* Tashkent: National University of Uzbekistan named after Mirzo Ulugbek, pp. 16–20.

Undeland, A. 2005. Kyrgyz Livestock Study: Pasture Management Report. Washington DC.

UNDP. 2005. Report on Livestock and Pasture Management. UNDP Project: Establishment of the Nuratau-Kyzylkum Biosphere Reserve as a Model for Biodiversity Conservation in Uzbekistan. Tashkent.

UNDP. 2010a. *Livestock Production in Uzbekistan: Current State, Challenges and Prospects.* Tashkent: Nasaf Publishing House.

UNDP. 2010b. *Food Security in Uzbekistan.* Tashkent: UNDP Uzbekistan.

UNDP. 2011. Economic Assessment of Sustainable Livestock Production and Combatting Desertification Actions in Dryland Regions of Uzbekistan. Joint Project Report by UNDP/GEF/Government of Uzbekistan. Tashkent.

UNDP-GM. 2007. *Traditional Land Management Knowledge in Central Asia.* Almaty: S-Print.

Valiev, U. 1980. *Каракулеводство Афганистана / Karakul Production of Afghanistan.* Moscow: КОЛОС.

Wall, C. 2008. Argorods of Western Uzbekistan. Knowledge Control and Agriculture in Khorezm. ZEF Development Studies 9. Berlin: Lit Verlag.

Wehrheim, P., A. Schoeller-Schletter and C. Martius. 2008. Continuity and Change: Land and Water Use Reforms in Rural Uzbekistan. Socio-Economic and Legal Analyses for the Region Khorezm. Studies on the Agricultural and Food Sector in Central and Eastern Europe 43. Halle: Leibniz-Institut für Agrarentwicklung in Mittel- und Osteuropa (IAMO).

Part II

Sustainable land management

A dream or an economic and ecological imperative

The focus in these three chapters is on sustainability, with an explanation of the multifaceted nature of it. Attention is given to how to assess it and how to monitor change in key indicators.

Squires and Feng open the discussion by trying to answer the questions about what we seek to sustain and why. The principle of inter-generational equity and the adoption of a 'land ethic' that embraces land stewardship are dealt with here.

Feng, Squires and Lu explore the principles and implementation of measures to achieve sustainable land management in Greater Central Asia in a Land Degradation Neutral (LDN) context. LDN is a core policy for the UNCCD and is reinforced by the UN Development goals enunciated in late 2015.

Akramkhanov et al. discuss barriers to achieving sustainable land management in Greater Central Asia, with special reference to the five former Soviet republics.

4 Sustainable land management

A pathway to sustainable development

Victor R. Squires

GUEST PROFESSOR, INSTITUTE OF DESERTIFICATION STUDIES, BEIJING

Haiying Feng

INSTITUTE OF ADMINISTRATION AND MANAGEMENT, XINING, QINGHAI

Introduction: some definitions and challenges

> Creating a shared vision of a sustainable and desirable future is the most critical task facing humanity today. This vision must be of a world that we all want, a world that provides permanent prosperity within the Earth's biophysical constraints in a fair and equitable way to all of humanity, to other species, and to future generations. (Robert Costanza and Ida Kubiszewski, 2014)

The Greater Central Asia (GCA) region (as defined earlier) is an arena for intelligent interstate cooperation where there is growing geopolitical interest. The GCA region as a typical reflection of the consumer-oriented attitude towards the use of natural resources must turn itself into an example of the survival of society on the basis of a new balance with the environment. Nothing is more important to our shared future than making the transition to a sustainable world. One of the most fundamental problems confronting humans at present is how to meet the basic needs and goals of all peoples of this vast region without simultaneously destroying the resource base, i.e. the "environment" from which ultimately these needs must be met. This is why nations have turned to the idea of sustainable development, which was defined by the Brundtland Commission (WCED, 1987) as:

> development to meet the needs of the present without compromising the ability of future generations to meet their own needs ... as a process of change in which the exploitation of resources, the direction of investments, the orientation of technological development are made consistent with future as well as present needs.

The concept of sustainable development arose as a contrast with traditional development founded upon a program of economic growth. However, what is involved here is more than just another new economic program; nor is it simply another concept from the fields of environmental protection or nature conservation. Rather, sustainable development marks an attempt to formulate a program that will integrate different spheres of human activity that had mostly been seen as separate in earlier times. Three generally recognized dimensions of sustainable development have been devised: ecological, social and economic. However, the basis for them is moral reflection regarding humankind's responsibility for nature. Is this all? Perhaps we should also include technical, legal and political dimensions (Pawłowski, 2008). Furthermore, these three dimensions (ecological, social and economic) have been denoted as pillars of sustainability, which reflect that responsible development requires consideration of natural, human and economic capital, or, colloquially speaking, the planet, people and profits. The metaphor of balancing the three pillars does not appropriately account for the complex inter-relationships between human activities and the environment as conceptualized in theories on human–environment systems (Brunckhorst and Trammell, 2016).

Amongst other things, the concept of sustainable development requires the preservation and enhancement of the productive capacity of natural resources; equitable land distribution; a redistribution of wealth to the poor; and the maintenance of the natural ecological processes. FAO (1991) defined sustainable agriculture as:

> the management and conservation of the natural resource base, and the orientation of technological and institutional change in such a manner as to ensure the attainment and continued satisfaction of human needs for present and future generations. Such sustainable development (in agriculture and forestry sectors) conserves land, water, plant and animal genetic resources, is environmentally non-degrading, technically appropriate, economically viable, and socially acceptable.

This definition acknowledges the key role that humans play now and in the future in sustainable agriculture. Until there is consensus about what sustainable development is, however, it will be difficult to implement it. Pawłowski (2008) posed the question "How Many Dimensions Does Sustainable Development Have?" The question is still relevant today.

What do we wish to sustain? Many possibilities exist. Do we want to sustain:

  (i) the rural population and community structure at existing levels;
 (ii) the biological and ecological integrity of the region;
(iii) the financial viability (household incomes) of farmers and pastoralists/herders; or
(iv) the people and their traditional lifestyle?

Once a decision is made as to which of these (singly or in combination) is to be the main aim, then the action taken to achieve this aim can be better specified. But what does sustainable development mean in practice?

Essentially, sustainable development is a set of strategies and tools designed to achieve the following:

- integrate conservation and development;
- ensure satisfaction of basic human needs;

- achieve equity and social justice;
- provide for social self-determination and cultural diversity; and
- maintain ecological integrity.

Each of the above is a major goal in itself and a condition for achieving the others, thus underlining the interdependence of the different dimensions of sustainability and the need for an integrated, interdisciplinary approach to achieve growth that is sustainable. Many groups of researchers are trying to define sustainability indicators and to devise methods to monitor them under field conditions. Sustainability indicators can be of two kinds: physico-biotic or socio-economic (Brunckhorst and Trammell, 2016; van Wyk et al., 2016).

Will the future work?

The world is changing rapidly. Degradation of ecosystem function, productive landscapes, fresh water and social function, along with accelerating climate change, has presented humanity with enormous global sustainability challenges for the 21st century. Water-related climate risks cascade through food, energy, urban and environmental systems. Growing populations, rising incomes and expanding cities will converge upon a world where the demand for water rises exponentially, while supply becomes more erratic and uncertain. If current water management policies persist, and climate models prove correct, water scarcity will proliferate to regions where it currently does not exist, and will greatly worsen in regions where water is already scarce (World Bank, 2016).

There are three basic drivers of change.

Population growth. Although slowing, global population growth is still rising and the highest rates are in those countries which are least able to sustain them. Further analysis of the population growth pattern shows that three categories of nations exist: (i) those in which population growth is below replacement level (e.g., parts of Europe); (ii) those at replacement fertility (e.g., North America); and (iii) those above replacement level (e.g., much of Asia, Africa and South America). This means that populations in those countries will double in less than 25 years (World Bank, 2014). The subject of population growth is complex and controversial and must be approached with respect for various religious and ethical values and cultural backgrounds.

Changes in water availability and variability can induce migration and ignite civil conflict. Food price spikes caused by droughts can inflame latent conflicts and drive migration. Where economic growth is impacted by rainfall, episodes of droughts and floods have generated waves of migration and statistical spikes in violence within countries. In a globalized and connected world, such problems are impossible to quarantine. And where large inequities prevail, people move from zones of poverty to regions of prosperity, which can lead to increased social tensions. This is why water management will be crucial in determining whether the GCA achieves the Sustainable Development Goals (SDGs) and aspirations for reducing poverty and enhancing shared prosperity.

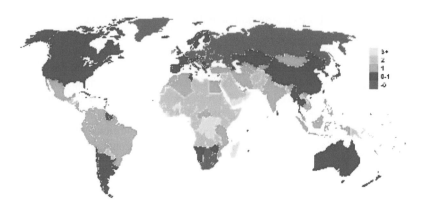

Figure 4.1 Human population growth rate in percent, with negative zero percentage, including the variables of births, deaths, immigration and emigration, as listed on the CIA Factbook (Wikimedia Commons)

Table 4.1 The annual rate of population increase over recent years for each of the GCA countries (expressed as the net gain)

Country	2011	2012	2013	2014
Afghanistan	3.0	3.1	3.2	3.2
Kazakhstan	1.4	1.4	1.4	1.5
Kyrgyzstan	1.2	1.7	2.0	2.0
Mongolia	1.7	1.8	1.8	1.8
Tajikistan	2.2	2.3	2.3	2.3
Turkmenistan	1.3	1.5	1.6	1.6
Uzbekistan	2.7	1.5	1.5	1.6

World Bank data.

Water is the common currency which links nearly every SDG, and it will be a critical determinant of success.

Sustainable development is the key for survival in the 21st century. Natural resources are finite and cannot be used with impunity because we are the custodians of these resources and have a responsibility to pass them to the next generation. This monumental task requires several major commitments, and most important of them is to arrest the population explosion which is occurring in parts of GCA. Natural resources like air to breathe, food to eat, water to drink, and fossil fuel to maintain this lifestyle are being overexploited. An unrestrained consuming culture will accelerate the path to an undesired situation. This situation will have more dire consequences in resource-limited ecosystems such as drylands. Given the severe scarcity of water, ever increasing population and soil salinization, out-of-the-box solutions for the provision of food and clean energy are required to spare meagre freshwater resources for conventional agriculture.

Economic growth. Industrial production had already grown 50-fold during the 20th century and the estimates are that it could well increase by another 350% by the year 2030 (PWC, 2015). If present production trends and population growth continue, the Earth's carrying capacity will be exceeded before long (Holechek, 2013; Smith, 2010; Bawden, 2017).

Technological innovation. The pace of change defies belief. Just consider the changes in the past decades. The Industrial Revolution, important though it was, has been overshadowed by the so-called "Information Revolution". Development of technologies (such as computers, digital communication, microchips) in the second half of the 20th century has led to a dramatic reduction in the cost of obtaining, processing, storing and transmitting information in all forms (text, graphics, audio, video). The main feature of the information revolution is the growing economic, social and technological role of information. Mobile phones bring up-to-date information about market prices, weather and other important data. Computer-based models can simulate real life and are used to derive alternative scenarios.

Social factors affecting the future

Trends are occurring which will have profound effects on society and, in turn, on the struggle to achieve sustainable development. The Rio+20 conference in its publication *The Future We Want* set out goals and objectives that, if adopted, would make for a better world. More recently, the set of Sustainable Development Goals was enunciated (UN, 2015).

The Sustainable Development Goals and targets are integrated and indivisible, global in nature and universally applicable, taking into account different national realities, capacities and levels of development and respecting national policies and priorities. Targets are defined as aspirational and global, with each government setting its own national targets guided by the global level of ambition but taking into account national circumstances.

Urbanization is a major factor, especially in less developed countries and newly industrialized countries. By the year 2020, almost half the world's population (over 3 billion people) will live in urban areas, and while urban population growth in developing countries as a whole has been slowing, three-quarters of a billion people were added to urban areas in the decades just before and after the end of the 20th century. China has an ambitious program to shift to an urban society and has mandated that 80 million will move to urban areas by 2030.[1] This shift is in response to changing methods in agriculture and, to a large extent, to the fact that the demographic profile has shifted. Yet, it is imperative that agriculture be modernized if there is to be food security (Bruins and Bu, 2006; Squires, Hua and Wang, 2015). The change from subsistence agriculture to a full market economy is inevitable, but it begs the question "will there be enough jobs?" (PWC, 2015).

Human investment patterns, family structure and education are also changing globally. Families are getting smaller as the burden of dependents on parents of working age increases (see below for a discussion of the impact on welfare for the elderly, especially in rural areas). Women have assumed greater economic responsibility of families and there is a reduction in the quantity of resources invested in the next generation. Populations are aging and in many developed countries fears have arisen about the viability of retirement welfare funds as the number of workers is reduced as the number of retirees increases rapidly. In developing countries, the lack of any welfare scheme to support aged rural workers has promoted the introduction of pension and other benefits to the elderly in the wake of the demise of the age-old notion of filial obligation that emphasized taking care of elderly parents. In modern society, there are many migrant workers, and some in big countries such as China (or like many from Tajikistan who travel far to work in Russia) cannot find the time or the travel funds to return to their home village more than once every few years. The rural pension/welfare scheme has been successful in China although the level of support in terms of monetary value is low.

Social stability, violence and disorder are worrying trends. There are doubts about the stability of cities because of the rapid population increases, and the slow pace of infrastructure development and provision of services such as health, education and welfare. There are many teenagers who are restless and mobile. Regrettably, the job market is saturated and this leads to restlessness, violence and lawlessness. Because many in the teenage age bracket are unskilled, they are "stranded" in a technological society that has few uses for manual labor at the levels which present themselves.

Equity patterns are also a cause for continuing concern. The gap between the bottom one-fifth and the top one-fifth of society is widening. Equity considerations loom large. Intergenerational equity is also a factor, bearing in mind the definition of sustainable development as including the provision for future generations. We could be like Groucho Marx and say "Why should I care about future generations – what have they ever done for me?", or instead consider what might be the fate of our children's children. However, we should act in way that ensures that they all have opportunities to be at least as well-off as we are (PWC, 2015).

Environmental factors affecting the future

Resource depletion. Everything that we need, want, use, abuse or consume comes from nature. Will it last? Ironically enough, the debate about resource depletion has shifted from concerns about non-renewable resources (e.g., fossil fuels and minerals) to fears that we will run out of renewable resources. There is a growing scarcity of renewable resources such as fresh water, food and fiber, forest products, fish and shellfish. As populations rise we will need more of these resources. The need for efficient food production has never been greater. One in seven humans is undernourished. Urbanization and biofuel production are reducing land availability, and climate change, lack of water (Figure 4.2) and soil degradation are decreasing harvests. Over the past decade, cereal yields per hectare have fallen in one-quarter of countries. Meanwhile, developing nations and the growing world population are demanding more animal protein. The rush to develop the world has

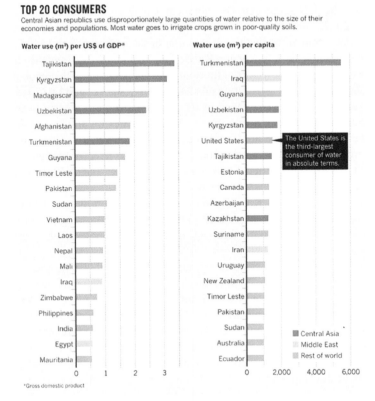

TOP 20 CONSUMERS
Central Asian republics use disproportionately large quantities of water relative to the size of their economies and populations. Most water goes to irrigate crops grown in poor-quality soils.

Figure 4.2 Chart showing water use by various countries in Central Asia (and elsewhere for comparison)

damaged the natural systems that sustain renewable resources. Accelerating land degradation constitutes a risk for sustaining the environment as well as for the livelihood of man. A holistic approach to the planning, development and management of land resources is required to meet the human and environmental needs and to combat degradation. However, it is not only the conservation of nature that is connected with the ecological dimension of sustainable development. Indirectly it relates to the more general matter of the shaping of spatial order, and so also to the proper creation and maintenance of areas inhabited by human beings. This is touching all other dimensions, especially social relations between people.

Examples of this depletion are reflected in:

- land degradation and soil loss;
- spread of deserts;
- build-up of pollutants in the air, water and soil;
- damage to watersheds; and
- loss of biodiversity.

The post-2015 policy environment

An outcome of the UN-sponsored Development Conference held in Rio de Janeiro, Brazil, 20–22 June 2012 was the publication of the document *The Future We Want*. The document embodied the declaration of the conference and sought to encapsulate the conclusions reached in Rio de Janeiro and provide guidance to all UN member countries. One aspect of this that receives attention in this book is the notion of Zero Net Land Degradation (ZNLD). Wang, Squires and Lu (2018) examine the concepts and the likelihood of its operationalization. Several years have passed and there has been action on some of the issues highlighted in the document. The 2015 Paris Climate Conference and several initiatives from the other UN Conventions on Biodiversity and Desertification have been signs of progress. The UN (2015) published *Transforming Our World: The 2030 Agenda for Sustainable Development*. It is a blueprint for the future.

In their book, Costanza and Kubiszewski (2014) encapsulate the issues that underpin sustainability. They have produced a book aimed at addressing concerns about the future of humanity and the future of the planet as a whole but there are lessons for the GCA region.

Sustainable use places the focus on two groups of disenfranchised people: the poor of today and the generations of tomorrow. The rapid degradation of the natural resource base of the rural poor is significantly worsening their poverty. Yet, many of the threats to the environment in the developing world occur as a result of poverty. Indeed, rural poverty and degradation of the environment are mutually reinforcing. When people's survival is at stake, they are forced to overstock fragile grazing lands, cut trees for firewood and overuse ground waters (FAO, 2012).

Few aspects of development have been found to be so complex as the need to reconcile anti-poverty and pro-environment goals. The policy linkages and choices to be made have yet to be articulated. One pivotal point is that no long-term strategy of poverty alleviation can succeed in the face of environmental forces that promote persistent erosion of the natural resources upon which we all depend. A significant portion of the GCA's poor live in rural areas. Rural communities play an important role in the economic development of many GCA countries. There is a clear need to revitalize the agricultural and rural development sectors, notably in developing countries, in an economically, socially and environmentally sustainable manner. It is important to take the necessary actions to better address the needs of rural communities through, *inter alia*, enhancing access of agricultural producers, in particular small producers, women, indigenous peoples and people living in vulnerable situations, to credit and other financial services, markets, secure land tenure, health care and social services. Education and training, knowledge transfer, and appropriate and affordable technologies, including for efficient irrigation, reuse of treated waste water, water harvesting and storage, are vital to any program to implement sustainable land management (SLM) as a set of mainstream rural practices. Rural women are critical agents for enhancing agricultural and rural development and food security and nutrition. Use of traditional sustainable agricultural practices, including traditional seed supply systems, water harvesting and many practices of indigenous peoples and local communities, should receive more attention and support.

The functioning of markets and trading systems can be strengthened by increasing public and private investment in them as part of fostering sustainable agriculture, land management and rural development. Key areas for investment and support include: sustainable agricultural practices; rural infrastructure, storage capacities and related technologies; research and development on sustainable agricultural technologies; developing strong agricultural cooperatives and value chains; and strengthening urban–rural linkages. There is a need to significantly reduce post-harvest and other food losses and waste throughout the food supply chain. It is necessary to promote, enhance and support more sustainable agriculture, including crops, livestock, forestry, fisheries and aquaculture, which improves food security, eradicates hunger and is economically viable, while conserving land, water, plant and animal genetic resources, biodiversity and ecosystems, and enhancing resilience to climate change and natural disasters. It is recognized that there is a need to maintain natural ecological processes that support food production systems. The key role that ecosystems play in maintaining water quantity and quality is recognized and actions within the respective national boundaries to protect and sustainably manage these ecosystems should be supported under the aegis of SLM.

Global and climate change

The economic, social and political conditions have changed dramatically and profoundly since the end of the 20th century and the changes are far reaching. Since the industrial revolution, pressures of change on ecological systems and social systems have greatly accelerated in pace, impact and spatial extent. The industrial revolution provided an expanding capacity of machinery, engineering and efficiencies for large-scale resource consumption and energy conversion. Changing landscape patterns are a reflection of responses and feedbacks of social–ecological interactions that drive change in natural resource capacity, ecosystem health and ecosystem service delivery (Forman, 1995; Antrop, 2005; Weber et al., 2012). Loss of biodiversity, ecological function and services, large-scale land degradation, erosion, and stream and lake pollution have all accelerated in pace and extent over the last millennium. Ecological functions operating across landscapes are compromised and struggling at best. Some are completely degraded or deceased. Humanity now faces many vast and growing challenges from local to global scales – challenges that are likely to decrease future sustainability and quality of life and increase the vulnerability of social and ecological systems responding to future change pressures. Single, narrowly focused "silver bullet" fixes or "command and control" responses of government rarely work (Ostrom and Cox, 2010).

Rapid technological, social and economic changes drive increased complexity by applying growing pressure on water and other natural resources such as arable land. Rapid population growth, increased economic activity and improved standards of living typically lead to expanding pollution and a rise in water demand. In order to meet it, water resources are more intensively developed; but in water-scarce countries, water supply can often be insufficient to cover all needs, leading to competing uses and consequently to the emergence of conflicts for water appropriation. Climate change increases this challenge by adding further uncertainty and variability. Rainfall patterns will exhibit shifts in timing and spatial distribution (increased rainfall variability with more frequent and intense extreme events such as floods and droughts).

Never before has the quest to balance the needs of people, the environment and the economy been so important. Many acknowledge the need to further mainstream sustainable development at all levels, integrating economic, social and environmental aspects and recognizing their interlinkages, so as to achieve sustainable development in all its dimensions. Many also affirm the need to achieve sustainable development by: promoting sustained, inclusive and equitable economic growth, creating greater opportunities for all, reducing inequalities and raising basic standards of living; fostering equitable social development and inclusion; and promoting integrated and sustainable management of natural resources and ecosystems that supports, *inter alia*, economic, social and human development while facilitating ecosystem conservation, regeneration and restoration and resilience in the face of new and emerging challenges.

While the concept of "sustainability" has been widely embraced by governments and business, the world has continued to move in increasingly unsustainable directions, from continued dependence on fossil energy to rising greenhouse gas emissions, and erosion of biodiversity. We know what the right thing to do is, but somehow we cannot make the changes that are needed, and so we continue with business as usual, producing business-as-usual results. Ultimately, the solutions to the problems of the 21st century will come from understanding the tremendous value that the environment provides, and reflecting that value within decision making at every level so that society as a whole thrives. While business-as-usual approaches can be sub-optimal, spending on sustainable outcomes must be balanced with anticipated benefits. There is a need to communicate with stakeholders in a transparent process which provides a robust view of how various options compare over a wide range of possible future conditions, using a language that everyone understands: money. People's opportunities to influence their lives and future, participate in decision making and voice their concerns are fundamental for sustainable development. We must underscore that sustainable development requires concrete and urgent action. It can only be achieved with a broad alliance of people, governments, civil society and the private sector, all working together to secure the future we all want for present and future generations.

Many people, especially the poor, depend directly on ecosystems for their livelihoods, their economic, social and physical well-being, and their cultural heritage. For this reason, it is essential to generate decent jobs and incomes that decrease disparities in standards of living to better meet people's needs and promote sustainable livelihoods and practices and the sustainable use of natural resources and ecosystems. Sustainable development (of which SLM is subset) must be inclusive and people-centered, benefiting and involving all people, including youth and children. Gender equality and women's empowerment are important for sustainable development in Greater Central Asia. Urgent action on unsustainable patterns of production and consumption, where they occur, remains fundamental in addressing environmental sustainability, and promoting conservation and the sustainable use of biodiversity and ecosystems, regeneration of natural resources, and the promotion of sustained, inclusive and equitable growth in the GCA region.

Implementation of sustainable development will depend on active engagement of both the public and private sectors. The active participation of the private sector can contribute to the achievement of sustainable development, including through the important tool of public–private partnerships. We must foster national regulatory and policy frameworks that enable business and industry to advance sustainable development initiatives. Businesses should take into account the importance of corporate social responsibility through engagement in responsible business practices. Land users, including small-scale farmers and fishers, pastoralists and foresters, can make important contributions to sustainable development through production activities that are environmentally sound, enhance food security and the livelihood of the poor, and invigorate production and sustained economic growth. There are different approaches, visions, models and tools available to each country in GCA, in accordance with its national circumstances and priorities, to achieve sustainable development in its three dimensions: social, economic and environmental (the Triple Bottom Line [TBL]). The TBL is an accounting framework that incorporates three dimensions of performance: social, environmental and financial (Pawłowski, 2008). This differs from traditional reporting frameworks as it includes ecological (or environmental) and social measures that can be difficult to assign appropriate means of measurement. Sustainable management is rooted in how sustainability and the Triple Bottom Line directly impact people and the planet. Well before Elkington (1994) introduced the sustainability concept as the "triple bottom line", environmentalists wrestled with measures of, and frameworks for, sustainability (Daley, Cobb and Cobb, 1989). Academic disciplines organized around sustainability have multiplied over the last 30 years. People inside and outside academia who have studied and practiced sustainability would agree with the general definition of the TBL. The TBL:

> captures the essence of sustainability by measuring the impact of an organization's activities on the world … including both its profitability and shareholder values and its social, human and environmental capital. The trick isn't defining TBL. The trick is measuring it. How do you begin to accurately measure social, environmental and financial results in most GCA countries with little infrastructure and limited resources. Fortunately progress has been made. (Talberth, Cobb and Slattery, 2006)

A new approach (Hardisty, 2014) has emerged to evaluate whether a new measure is a viable option. Environmental and economic sustainability assessment (EESA) is a new way to make decisions that meet the challenges of the 21st century. Incorporating elements of lifecycle analysis, risk assessment, cost–benefit analysis and comprehensive sensitivity analysis, EESA provides a fully quantitative, objective and rational way to include all of the social, environmental and economic issues relevant to a decision into one comprehensive analysis.

A number of theoretical models of sustainability were pursued from the late 1980s, which culminated with the Trefoil diagram of social, environmental and economic integration, also called the "people, planet, and prosperity" or "triple bottom line" model (Parkin, 2000; Pope, Annandale and Morrison-Saunders, 2004). More sophisticated models have emerged recently, including the "embedded" or "Russian doll" model, which overlaps instead of intersecting the three dimensions (O'Riordan, Cameron and Jordan, 2001) and the "prism" model, which adds governance as the fourth dimension of sustainability (Spangenberg, 2004).

The term "social-ecological" started to be used by Berkes and Folke (1998) to emphasize the intricate linkages between humans and nature and to stress the artificial and arbitrary delineation between social and ecological systems (Folke et al., 2005). Indeed, many human activities take place in and are dependent on river basins. Social-ecological systems (SES) are by definition complex: they involve a multitude of actors, sectors and scales which are interdependent, and their evolution over the coming decades is uncertain, which makes it hard to have a global understanding of systems and changes. This complexity inherent in SES increases the complexity in their management and governance. Such SES are also referred to as "complex-adaptive systems", because they are characterized by the possession of self-organizing capacities which are responsive to pressures of change. Such adaptive capacities of natural and human systems are critically important responsive mechanisms for dealing with risk, vulnerability and buffering pressures of change (Brunckhorst and Trammell, 2016). The social environment may experience degradation in just the same way as the natural environment. The environment in question comprises a large number of factors, including customs and traditions, culture, spirituality, interpersonal relations and living conditions. Even human relationships with nature have their social dimension, since all the different kinds of activity oriented towards the environment are mediated via the socio-cultural models in place in a given society.

Reversal of desertification and land degradation and drought mitigation and adaptation

The economic and social significance of good land management is well recognized. Soil, particularly its contribution to economic growth, biodiversity, sustainable agriculture and food security, eradicating poverty, women's empowerment, addressing climate change and improving water availability, is vitally important (Cardesa-Salzmann, 2014). Desertification, land degradation and drought are challenges of a global dimension and continue to pose serious barriers to the sustainable development of all countries, in particular GCA countries. Control of desertification, land degradation and mitigation of drought, including the preservation and development of oases (both natural and artificial), restoring degraded lands, improving soil quality and improving water management are high priorities. Success in these aspects will contribute to sustainable development and poverty eradication. There is a need for urgent action to reverse land degradation as part of the strategy to mainstream SLM in national planning and in national budgets.

Mountain ecosystems, so prominent in GCA, play a crucial role in providing water resources to a large portion of the GCA population; fragile mountain ecosystems are particularly vulnerable to the adverse impacts of climate change, deforestation and forest degradation, land use change, land degradation and natural disasters. Mountain glaciers are retreating for the most part (Hagg, 2018) and getting thinner, with increasing impacts on the environment and human well-being. Mountains are often home to communities, including indigenous peoples and local communities, who have developed sustainable uses of mountain resources. They are, however, often marginalized, and continued effort will be required to address poverty, food security and nutrition, social exclusion and environmental degradation in these areas. The sustainable use of biodiversity should contribute to tangible benefits for local people. There is growing support for mainstreaming the consideration of the socio-economic impacts and benefits of the conservation and sustainable use of biodiversity and its components, as well as ecosystems that provide essential services, into relevant programs and policies at all levels, in accordance with national legislation, circumstances and priorities. Investments should be encouraged through appropriate incentives and policies, which support the conservation and sustainable use of biological diversity and the restoration of degraded ecosystems, consistent and in harmony with the Biodiversity Convention and other relevant international obligations.

In sum, proposals for the restoration of damaged ecosystems, e.g. desertification control, reflect the dynamic interactions between the perception of benefits (ecosystem services) and the anticipation of a more desirable flow of benefits, the meanings and values associated with those benefits, the behaviors that result, and the policies and governance processes designed to introduce and facilitate the transformation of benefits. Seen from this perspective, ecological restoration is an adaptive social process embedded in a complex social-ecological system designed ultimately to more broadly align, enable or transform behaviors to ensure the success of restoration to a preferred social-ecological state. The preferred state is one in which meanings, benefits and behaviors are aligned. The relationship between behavioral change and ecological restoration can be understood through the conceptual framework presented in Figure 4.3, which depicts ecological restoration as a social-ecological system. The state of the system is shaped by behaviors that are linked to both resource users and public infrastructure. In this context, the intention of desertification control is to establish a preferred social-ecological state, e.g. a productive and stable rangeland (**1**). To achieve this, resource users must bring about a shift in the meanings that are currently attached to the resource to new shared and prioritized meanings (**2**). The negotiated meanings can then be applied in the determination of the preferred allocation and configuration of benefits (**3**) and in the deliberations of those (private and public) who develop public infrastructure (physical and regulatory) (**4**). The negotiation of meanings and benefits and the discussions around public infrastructure contribute to collective learning, building social capital and a collective identity (**5**) that commits stakeholders, individually and collectively, to the reformed public infrastructure necessary to transform behaviors (**6**) and bring about the desired change in the social-ecological state (see **1**). Because the state of the social-ecological system is under the influence of external forces (**8**), the process is iterative, adapting to accommodate emerging insights and changing circumstances.

We must conclude that the relatively simple principle of sustainable development has led to the formulation of complex action strategies relating to the various dimensions of human activity. These strategies have not always been amenable to implementation in reality, and some of those that have been brought into effect do not operate properly. It is true to say that the problems of the environmental crisis still include many unknowns. To a great extent this reflects the fact that it is difficult to foresee the real long-term consequences of human activity. Science is rather

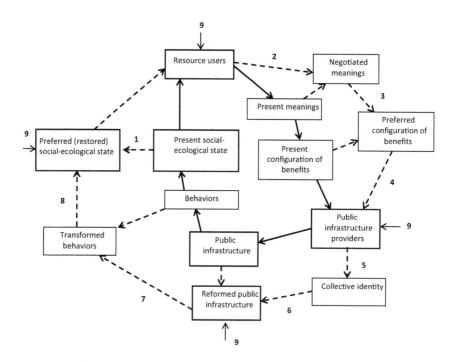

Figure 4.3 A framework depicting desertification control as a social-ecological system wherein behaviors (and changes in behaviors) are directed by the perception of benefits and associated meanings. It also emphasizes the institutional processes designed to prioritize and reflect negotiated meanings for a desired configuration and distribution of benefits

Source: From van Wyk et al. (2014), adapted from Anderies, Jannsen and Ostrom (2004).

difficult to steer and the attachment of priority status to certain disciplines that seem of the greatest importance at a given moment does not necessarily bring benefits.

The rehabilitation of ecosystems must be integrated with transformations of social-ecological systems in their regional landscape or "place" context. Understanding both the characteristics and the circumstances of social-ecological systems change will help identify "leverage" points where SES restorative transformations might more easily allow shifts towards sustainable futures.

Economic driving forces, ecological constraints and policy options for sustainable land use

Land users in GCA countries will be greatly affected. Over the past 30 years, the world has moved towards much greater economic specialization, globalization and complexity at the expense of diversification and local self-sufficiency. Religious beliefs, the need for family labor, lack of freedom for women, lack of education on birth control and lack of contraceptives largely explain the high birth rates in most GCA countries. Conversion of farmland to other uses, loss of the glaciers in the Himalayas, depletion of groundwater and its contamination, soil erosion, climatic instability and population increases are threats. Impending peaks in the extraction rates of oil and phosphorus add to the risk of famine.

The following questions have been raised in this chapter and throughout the book:

- How do we measure and monitor change?
- What are cost-effective techniques and tools to do this monitoring?
- Are there alternative uses for the land?
- What are the biodiversity implications of a change in land use?
- How do we ensure involvement and participation by local people?
- Can forage and water resources be sustained?
- Is pastoralism still a viable option?

Measuring and monitoring change

Indicators of desertification and land degradation are being sought on many fronts, from remote sensing using satellites (Feng et al., 2018; Aralova et al., 2018) to ground-based measurements of plant and soil characteristics (Squires, 2007; Squires, Zhang and Hua, 2010; Squires, 2016). Perhaps the greatest impediment to developing sensitive indicators for measuring ecosystem health is the absence of accurately defined thresholds between the three broad categories of *healthy*, *at risk* and *unhealthy* ecosystems. Determining thresholds can be a complex process. Historical records for most sites in GCA are rarely available, and when available they usually give information only on the biological and physical state of an ecosystem at a point in time. A feature of sensitive indicators that needs to be considered is the range of environments over which an indicator is effective.

A primary requirement for any useful indicator is that its response to changes in ecosystem health be similar over ecotypes in similar sites within a region. This aspect represents a research priority of the highest order.

Alternative land use options in GCA countries

The people of GCA face a number of problems, not the least serious of which are those of the Aral Sea and Caspian Sea (Bekchanov et al., 2018; Saiko, 1998, 2001a, 2001b) and widespread land and water degradation (Squires, 2012; Behnke, 2006; Chen et al., 2014). Clearly, there is a need for the development of sound concepts and technologies necessary to design, build, implement, operate and use comprehensive spatial resource management information as a basis for developing solutions to resource management problems in the GCA region. But there is also a need to adjust the socio-political framework. Land tenure, use rights and other related matters loom large (Robinson et al., 2012; Halimova, 2012). In short, there is a need to create what is referred to by bilateral and multilateral donors as "an enabling environment" that creates conditions favorable to institutional change. The paradigm associated with arresting and reversing land degradation has shifted from science and technology to community action as the single most important component of attempts to combat desertification. The present trend is to involve the farmers, pastoralists and other natural resource users (and managers) who are affected by land and water degradation in the program to arrest and reverse the problems being encountered.

Land users have few options open to them (Squires, Zhang and Hua, 2010). For herders who depend on a highly variable (in both space and time) resource base there are only two basic options available to them. They can either reduce demand for forage/fodder or increase the supply. The next major shift has to be a move from "component thinking" to "systems thinking" – moving toward a whole enterprise approach. This involves, *inter alia*, adjustment to the structure of cropping patterns, shifts in the flock/herd structure, and efforts to conserve fodder such as hay and silage. Protection of rangelands and pasturelands at their crucial stages of development (just after green-up in spring and at seed-set in autumn) is a key intervention. These and other matters are dealt with at length in the books by Squires, Hua, Zhang and Li (2010) and by Squires (2012).

Sustainable land management (SLM) technologies

More than 50 of the most promising and ready to disseminate SLM technologies and approaches were reviewed by national teams in five Central Asian countries. To collect SLM, the Kazakhstan team analyzed reports of research institutes that were completed over the last ten years within the "national program of applied research in agriculture". The Kyrgyzstan team managed to analyze a number of earlier projects conducted by various organizations in the country to extract SLM that showed high potential for dissemination. Organizations – such as GIZ, ICARDA, HELVETAS and CIDA – as well as rural advisory services, NGOs and several farmers were among the main sources of information. Moreover, research institutes under the Ministry of Agriculture and Melioration, such as the Kyrgyz Research Institutes of Irrigation and Farming, and the National Agrarian University, contributed their own SLM. Of many previous projects implemented in Tajikistan, the most relevant was by the WOCAT team that compiled a large database of local SLM, particularly those originating from communities and small-scale producers, representing traditional knowledge. To complement what was collected earlier, the team added SLM technologies that were within the organizations under the Tajik Academy of Agricultural Sciences. The Turkmenistan team collected and described information for 20 SLM options that mainly dealt with irrigated agriculture. Most of the information was drawn from the Turkmenistan Academy of Sciences and the Academy's collaborating organizations. SLM from Uzbekistan were compiled from the proceedings of an annual innovation fair held at the national level, as well as from research institutes under the Scientific Production Center of the Ministry of Agriculture and Water Resources. The national team also utilized other sources of information, such as United Nations Development Program (UNDP) projects and the Small Grants Program of the Global Environment Facility (GEF). All teams also selected from the WOCAT database options that were found suitable for conditions across the five "stans" in Central Asia. The prioritization of selected SLM was conducted at two levels: national and regional. Consultations with representatives of research institutes, farmers and farmers' organizations, NGOs and rural advisory services and decision makers were carried out at national-level workshops organized in each country to present collected SLM technologies.

The set of 11 criteria adapted from partner organization GIZ, used in their earlier projects and agreed by national teams, was used to prioritize and select several SLM per target agro-ecosystem for further discussion at the regional-level prioritization workshop. The SLM approaches and technologies presented by country teams addressed a broad range of issues that are equally important in each ALS in the various agro-ecosystems, and in the relevant countries. In order to reach agreement from each of the participating partners, it was decided to form SLM packages consisting of core technologies for each target around one core technology to enable the adaptation of the packages to local conditions across the five countries. Similarity maps for the five countries were generated. Based on the formulated SLM packages, the country teams agreed on a set of criteria to develop similarity maps. The similarity maps will be used to facilitate SLM dissemination by visually presenting potential areas for upscaling, as well as to identify target areas in which to disseminate particular SLM packages in all four target agro-ecosystems. One of the knowledge gaps identified by the national teams during the inception workshop was the impact of climate change on agriculture. To address these gaps, this activity aimed at assisting in the development of procedures for the calibration of downscaled climate change scenarios in Central Asia *sensu stricto* and an explanation of the potential influence of future climate scenarios on agriculture, especially SLM. This should help their understanding of the downscaling of climate change scenarios to their region and to select SLM options that help farmers to adapt their practices. Historical data on the main meteorological parameters from 22 Central Asian meteorological stations collected during previous ICARDA projects were compiled into a database to aid in the synthesis and consolidation of best practices on SLM. These can be farmer innovations, examples of sustainable management of water, land and forest resources, sustainable farming methods, pasture use, improved livestock and crop production, and promising scientific developments concerning climate change. Champions, those who have achieved in their fields of specialization relevant to SLM, are also of interest.

Biodiversity implications of land use change

Biodiversity is a multifaceted phenomenon involving the variety of organisms present, the differences among them, and the communities, ecosystems and landscape patterns in which they occur. Many factors affect the biodiversity of plants and animals (including birds, insects and other invertebrates). Grazing (defoliation and trampling) is a major factor that is important in areas used for animal husbandry in GCA. Although pivotal in land and water management, responses to pressure are difficult to predict. It is imperative to maintain natural systems to enhance global environmental benefits, including biodiversity conservation, carbon sequestration and ecosystem services such as water flow. These objectives are expected to be achieved through encouraging SLM practices. This implies an ecosystem approach to land management at a landscape scale (Tongway and Ludwig, 2002) across land systems that are used for primary production (herding and farming). Such efforts in GCA aim to promote SLM and combat and reverse land degradation. At the practical (field) level, this calls for participatory, integrated ecosystem approaches to land management in globally significant areas for biodiversity corridors and "hotspots" in the Pamirs, Tian Shan, Altai Shan and Qilian Shan in particular, but also in desert and wetland communities throughout GCA (Zhao and Squires, 2010; Squires and Safarov, 2013).

Land use change (abandonment of irrigated cropland, urban expansion, infrastructure developments such as highways, pipelines, railways and urban expansion) are all threats to biodiversity either directly through habitat destruction or through interruptions to migration corridors for plants and animals.

Involvement of and participation by local people

Land resource management, in its narrowest sense, is the actual practice of using land by local human populations in a sustainable way (Geerken et al., 1998). People are the instruments and beneficiaries, as well as the victims, of all development activities. Their active involvement in the development process is the key to success. Furthermore, unless we keep foremost in our minds the need to continue to improve the welfare of people, environmental programs will certainly fail. The poor in particular tend to be hardest hit by environmental degradation and the least well-equipped to protect themselves. Yet, at the same time, the poor cause much environmental damage out of short-term necessity, ignorance or lack of resources. Lack of easily accessible entry points to embark on more sustainable ways of making a living is a key problem. The challenge is to make participation more than an empty catchword. Practical progress is required at three levels. Firstly, those potentially affected by development projects need to be more involved at the design stage. Secondly, local knowledge needs to be better used in the design and implementation of programs. Thirdly, we need to build our capacity to assess social impacts of policies and investments – a particularly important, but difficult, task, requiring a different skill mix and a different way of doing business (van Wyk, Breen and Freimund, 2014; Watkins et al., 2013). It is essential to facilitate dialogue in the planning process and to get further cooperation between people and institutions by inviting all stakeholders to a participatory decision-making process.

What future for livestock husbandry: whither pastoralism in GCA?

At a time when traditional lifestyles are under threat throughout the world, it seems appropriate to ask what future there is for traditional (non-commercial) range/livestock production systems. The issues have been taken up by Squires and Sidahmed (1998), Strong and Squires (2012) and Kreutzmann (2012). It is clear from these publications that concerns about the sustainability of the resource base have emerged. Many people are left to speculate if the environmental damage done to land and water resources will bring about the demise of traditional pastoralism that operates in a system that is characterized by low or unreliable production, complex semi-natural systems, large-scale management units and greater economic risk.

We can sum up the situation in the following key points:

- *Traditional lifestyles are under threat* throughout the GCA. Environmental, social, political and economic changes will impact on pastoralism. The strongholds of nomadic (Kuchi in Afghanistan/Iran) and semi-nomadic peoples (e.g. Mongolia and Qinghai-Tibet Plateau) are undergoing rapid evolution as market-based systems overtake them.
- *Pastoralism in the arid zone has always been severely constrained.* The key constraints that have plagued pastoralism in the past include both the technical (animal health and nutrition) and socio-political (land tenure, policy issues, religion) and economic (marketing and the emergence of the cash economy) (Wang et al., 2010; Hua and Squires, 2015).
- *Emerging problems include the conservation of the resource base*, including biodiversity issues (Squires and Safarov, 2013), the globalization of the world economy, the breakdown of tradition and the growing impacts of climate change (Hagg, 2017).

On a global basis, livestock production based on pure grazing systems is relatively unimportant, and also grows at the slowest rate. The production of pastoral systems grows at 1%, mixed crop/pasture systems at 3% and industrialized livestock production at more than 7%. It is clear that the balance is changing and intensification is on its way (Michalk et al., 2010), although in some areas horizontal expansion may still be an option. In the light of these observations and from an analysis of the literature, we must conclude that traditional nomadic pastoralism will become less and less important. There will be those who wish to utilize (exploit) the otherwise unusable forage and water resources of the GCA region, but social, economic and political pressures will see the demise of this way of life (Squires, 2012).

Changing climatic conditions, such as frequent droughts, put even more pressure on the system. With so many challenges coming together, it is important to analyze whether pastoralism in itself can be considered a sustainable production system that in principle can cope with these challenges and thus deserves support from policy, or whether the pastoralist production system has a fundamental misfit with today's challenges, in the sense that it is detrimental to the world's scarce resources.

According to Reid, Fernandez-Gimenez and Galvin (2014), rangelands are at a crossroads. Although traditionally defined as lands for grazing wildlife and livestock, some scholars are expanding the way they define this term to include the diverse ways people use rangelands and the varied ecosystem services they provide. Rangelands are now thought of as linked social-ecological systems, each with a unique ecological, historical, political and cultural context, where the social system and ecosystem adapt to each other, and the resilience of this coupled system depends, in part, on this capacity to change. Most global rangelands are still used (and sometimes owned) in common, even in countries with widespread private rangeland, and this communal use can be and often is sustainable, provided that strong local institutions governing their use are in place.

Herders around the world solve the pastoral paradox (the need for both secure but flexible access to land) with institutions that regulate common, quasi-private and private ownership or use to avoid both the tragedy of open access (misnamed the tragedy of the commons), where there are no rules of use, and the tragedy of enclosure on privatized and fragmented land. Growth of populations and consumption, the spread of markets, the commodification of nature, globalization and climate changes are all accelerating the transformation of land use and land tenure in rangelands around the world, causing loss, expansion, fragmentation and reaggregation of rangelands as pastoralists solve the pastoral paradox in different ways. Rangeland degradation is not as widespread as previously thought, but it does occur, particularly in wetter rangelands. Only more long-term and broad-scale co-research, integrating experiential and scientific knowledge, will allow us to clearly distinguish the effects of grazing, climate and land use, as well as when and where degradation is fleeting or permanent. Recent learning and experimentation with both community- and market-based institutions are moving us closer to achieving a socially just, economically viable and environmentally sustainable management of rangelands and the biodiversity they support. Challenges remain, particularly how to ensure that new institutions share benefits equitably, how to manage and monitor spatially extensive systems that often require herd mobility across scales, and how to determine the environmental returns of investing in conservation stewardship practices. Key questions are: As the future brings changes in climate, landownership, new uses (mining, energy) and ranching populations, who are the winners and losers? This applies to pastoral peoples and also to nonhuman species in rangelands. What is driving rangeland degradation – livestock, climate change, or other land use(s)? Where and when do these causes of change operate most strongly and why? What will rangelands look like in the future? Will they be full of turbines to harness wind energy and solar arrays to trap sunlight, be empty of people, or be in some other configuration or combination? How can our current rangeland knowledge be applied to new land uses? What are the best ways to reaggregate fragmented lands and in which situations?

Note

1 12th Five-year Plan.

References and further reading

Anderies, J. M., Janssen, M. A. and Ostrom, E. 2004. A framework to analyze the robustness of social–ecological systems from an institutional perspective. *Ecology and Society* 9: 18 [online]. http://www.ecologyandsociety.org/vol9/iss1/art18/

Antrop, M. 2005. Why landscapes of the past are important for the future. *Landscape and Urban Planning* 70 (1–2): 21–34.

Aralova, D., Kariyeva, J., Menzel, L., Khujanazarov, T., Toderich, K., Halik, U. and Gofurov, D. 2018. Assessment of land degradation processes and identification of long-term trends in vegetation dynamics in the drylands of Greater Central Asia. In this volume, pp. 131–54.

Babaev, A.G. 1999. *Desert Problems and Desertification in Central Asia: The researches of the Desert Institute*. Springer, Heidelberg, 293 pp.

Bawden, R. 2017. Global Change and its Consequences for the World's Arid Lands. In: M.K. Gaur and V.R. Squires (eds) *Climate Variability Impacts on Land Use and Livelihoods in Arid Lands*. Springer Nature, New York.

Bekchanov, M., Djanibekov, N. and Lamers, J.P.A. 2018. Water in Central Asia: a cross-cutting management issue. In this volume, pp. 211–36.

Behnke, R. 2006. *The Socio-Economic Causes and Consequences of Desertification in Central Asia. Proceedings of the NATO Advanced Research Workshop, Bishkek, Kyrgyzstan*. Springer, Dordrecht.

Berkes, F. and Folke, C. 1998. Linking social and ecological systems for resilience and sustainability. In: F. Berkes, C. Folke and J. Colding (eds) *Linking social and ecological systems: management practices and social mechanisms for building resilience*. Cambridge University Press, Cambridge, UK, pp. 1–25.

Bruins, H.J. and Bu, F. 2006. Food security in China and contingency planning: The significance of grain reserves. *Journal of Contingencies and Crisis Management* 14 (3): 114–24.

Brunckhorst, D.J. 2015. The Emergence of Landscape Governance in Society-Environment Relationships. In U. Fra Paleo (ed.) *Risk governance. The articulation of hazard, politics and ecology*. Berlin: Springer, pp. 337–48.

Brunckhorst, D.J. and Trammell, E.J. 2016. Restorative futures for socio-ecological systems in changing times. In: V.R. Squires (ed.) *Ecological Restoration: Global challenges, Social Aspects and Environmental benefits*. Nova Science Publishers, New York, pp. 259–80.

Busby, F.E. 1995. Sustainable use and management of the world's rangelands. In: S.A.S. Omar, M.A. Razzaque and F. Alsdirawi (eds) *Range management in arid zones*. Keegan Paul, London, pp. 1–5.

Cardesa-Salzmann, A. 2014. Combating Desertification in Central Asia: Finding New Ways to Regional Stability through Environmental Sustainability? *Chinese Journal of International Law* 13 (1): 203–31.

Chen, B., Zhang, X., Tao, J., Wu, J., Wang, J., Shi, P., Zhang, Y. and Yu, C. 2014. The impact of climate change and anthropogenic activities on alpine grassland over the Qinghai-Tibet Plateau. *Agric. For. Meterol.* 189–90: 11–18.

Ci, L.J. and Yang, X.H. 2010. *Desertification and its control in China*. Springer/Higher Education Press, Beijing.

Costanza, R. and Kubiszewski, I. (eds). 2014. *Creating a Sustainable and Desirable Future: Insights from 45 Global Thought Leaders*. World Scientific Publishing Co., Singapore.

Daly, H.E., Cobb, J.B. and Cobb, C.W. 1989. *For the Common Good: Redirecting the Economy towards Community, the Environment, and a Sustainable Future*. Beacon Press, Boston MA.

Dukhovny, V.A. and de Schutter, J. 2011. *Water in Central Asia: Past, Present, Future*. Baton Boca, CRC Press.

Elkington, J. 1994. Towards the Sustainable Corporation: Win-Win-Win Business Strategies for Sustainable Development, *California Management Review* 36 (2): 90–100.

FAO. 1991. *Sustainable Agriculture and Development in Asia and the Pacific*. Regional Document No. 2. FAO/Netherlands Conference on Agriculture and Environment, S-Hertogenbosch, The Netherlands, 15–19 April.

FAO. 2012. *Irrigation in Central Asia in figures. AQUASTAT Survey-2012*. FAO Water Reports 39. FAO, Rome. http://www.fao.org/docrep/018/i3289e/i3289e.pdf

Feng, Y., Yan, F. and Cao, X. 2018. Land degradation indicators: development and implementation by remote sensing techniques. In this volume, pp. 155–78.

Folke, C., Hahn, T., Olsson, P. and Norberg, J. 2005. Adaptive governance of social-ecological systems. *Annual Review of Environment and Resources* 30: 441–73.

Forman, R.T.T. 1995. *Land Mosaics: The Ecology of Landscapes and Regions*. Cambridge University Press, Cambridge.

Geerken, R., Ilawi, M., Japa, M., Kaufmann, H., Roeder, H., Sankary, A.M. and Segl, K. 1998. Monitoring desertification to define and implement suitable measures towards sustainable rangeland management. In: V.R. Squires and A.E. Sidahmed (eds) *Drylands: Sustainable use of rangelands into the twenty-first century*. IFAD Series: Technical Reports. IFAD, Rome, pp. 31–44.

Hagg, W. 2018. Water from the mountains of Greater Central Asia: a resource under threat. In this volume, pp. 237–48.

Halimova, N. 2012. Land Tenure Reform in Tajikistan: Implications for Land Stewardship and Social sustainability: A Case Study. In: V. Squires (ed.) *Rangeland Stewardship in Central Asia: Balancing improved livelihoods, biodiversity conservation and land protection*. Springer, Dordrecht, pp. 305–30.

Hansmann, R., Mieg, H.A. and Frischknecht, P. 2012. Principal sustainability components: empirical analysis of synergies between the three pillars of sustainability. *International Journal of Sustainable Development & World Ecology* 19 (5): 451–9.

Hardisty, P.E. 2014 *Environmental and Economic Sustainability*. CRC Press, Boca Raton.

Holechek, J.L. 2013. Global Trends in Population, Energy Use and Climate: Implications for policy development, rangeland management and rangeland users. *The Rangeland Journal* 30 (5): 117–19.

Hua, L.M. and Squires, V.R. 2015. Managing China's pastoral lands: current problems and future prospects. *Land Use Policy* 43: 129–37.

Jazairy, I., Alamgir, M. and Panuccio, T. 1992. *The State of World Rural Poverty: an inquiry into its causes and consequences.* IFAD and New York University Press, New York, 514 pp.

Kreutzmann, H. 2012. *Pastoral practices in High Asia: Agency of 'development' effected by modernisation, resettlement and transformation.* Springer, Dordrecht.

Michalk, D.M., Hua, L.M., Kemp, D., Jones, R., Takahashi, T., Wu, J.P., Nan, Z, Xu, Z. and Han, G. 2010. In: V. Squires, Hua Limin, Zhang Degang and Li Guolin (eds) *Towards Sustainable Use of Rangelands in North West China.* Springer, Dordrecht, pp. 301–24.

O'Riordan, T., Cameron, J. and Jordan, A. 2001. *Reinterpreting the precautionary principle.* Cameron May, London.

Ostrom, E. and Cox, M. 2010. Moving Beyond Panaceas: A Multi-tiered Diagnostic Approach for Social-ecological Analysis. *Environmental Conservation* 37 (4): 451–63.

Parkin, S. 2000. Sustainable development: the concept and the practical challenge. Proceedings of the Institution of Civil Engineers. *Civil Engineering* 138: 3–8.

Pawłowski, A. 2008. How Many Dimensions Does Sustainable Development Have? *Sustainable Development* 16: 81–90.

Pope, J., Annandale, D. and Morrison-Saunders, A. 2004. Conceptualising sustainability assessment. *Environmental Impact Assessment Review* 24: 595–616.

PWC. 2015. *The world in 2050: Will the shift in global economic power continue?* PricewaterhouseCoopers, London. https://www.pwc.com/gx/en/issues/the-economy/assets/world-in-2050-february-2015.pdf

Reid, R.S., Fernandez-Gimenez, M.E. and Galvin, K.A. 2014. Dynamics and Resilience of Rangelands and Pastoral Peoples Around the Globe. *Annual Review of Environment and Resources* 39 (1): 217–42.

Robinson, S., Weidemann, C., Michel, S., Zhumabayyev, Y. and Singh, N. 2012. Pastoral tenure in central Asia: Theme and variation in the five former Soviet Republics. In: V.R. Squires (ed.) *Rangeland Stewardship in Central Asia: Balancing Livelihoods, Biodiversity Conservation and Land Protection.* Springer, Dordrecht, pp. 239–74.

Saiko, T.S. 1998. Geographical and socio-economic dimensions of the Aral Sea crisis and their impact on the potential for community action. *J Arid Environ* 39 (2): 225–38.

Saiko, T. 2001a. The Caspian Sea region: inter-state problems of a shared water body. In: *Environmental Crises: Geographical Case Studies in Post-socialist Eurasia.* Prentice Hall, Harlow, UK, pp. 217–40.

Saiko, T. 2001b. Desiccation in the Aral Sea: The hidden costs of irrigation. In: *Environmental Crises: Geographical Case Studies in Post-socialist Eurasia.* Prentice Hall, Harlow, UK, pp. 242–68.

Smith, L.V. 2010. *The World in 2050.* Penguin Group, New York.

Spangenberg, J.H. 2004. Reconciling sustainability and growth: criteria, indicators, policies. *Sustainable Development* 12 (2): 74–86.

Squires, V.R. 2007. Detecting and monitoring impacts of ecological importance in semiarid rangelands. In: A. El-Beltagy, M.C. Saxena and Tao Wang (eds) *Human and Nature – Working Together for Sustainable Development of Drylands.* Proceedings of the eighth International Conference on Development of Drylands, 25–28 February 2006, Beijing, China. ICARDA, Aleppo, Syria, pp. 718–23.

Squires, V.R. 2012. *Rangeland Stewardship in Central Asia: Balancing Livelihoods, Biodiversity Conservation and Land Protection.* Springer, Dordrecht, 458 pp.

Squires, V.R. 2016. Restoration of Wildlands including Wilderness, Conservation Reserves and Rangelands. In: V.R. Squires (ed.) *Ecological Restoration: Global Challenges, Social Aspects and Environmental Benefits.* Nova Science Publishers, New York, pp. 131–50.

Squires, V.R. and Safarov, N.M. 2013. Diversity of plants and animals in mountain ecosystems in Tajikistan. *Journal of Rangeland Science* 4 (1): 43–61.

Squires, V.R. and Sidahmed, A.E. 1998. *Drylands: sustainable use of rangelands into the twenty-first century.* IFAD Series Technical Reports. IFAD, Rome, 470 pp.

Squires, V.R., Hua, L.M. and Wang, G.Z. 2015. Food security: a multi-faceted and multi-dimensional issue in China. *Journal of Food, Agriculture and Environment* 13 (2): 24–31.

Squires, V., Zhang, D. and Hua, L. 2010. Ecological restoration and control of rangeland degradation: rangeland management interventions. In: V. Squires, Hua Limin, Zhang Degang and Li Guolin (eds) *Towards Sustainable Use of Rangelands in North West China.* Springer, Dordrecht, pp. 81–99.

Squires, V., Hua, L., Zhang D. and Li, G. (eds). 2010. *Towards Sustainable Use of Rangelands in North West China.* Springer, Dordrecht, 353 pp.

Strong, P.J.H. and Squires, V.R. 2012. Rangeland-based Livestock: A vital subsector under threat in Tajikistan. In: V. Squires (ed.) *Rangeland Stewardship in Central Asia: Balancing improved livelihoods, biodiversity conservation and land protection.* Springer, Dordrecht, pp. 213–27.

Talberth, J., Cobb. C. and Slattery, N. 2006. *The Genuine Progress Indicator 2006: A Tool for Sustainable Development.* Oakland CA.

Tongway, D. and Ludwig, J. 2002. Desertification, Reversing. In: L. Rattan (ed.) *Encyclopedia of Soil Science.* Marcel Dekker, New York, pp. 343–5.

Tongway, D.J. and Ludwig, J.A. 2011. *Restoring disturbed landscapes: Putting principles into practice.* Island Press/Society for Ecological Restoration International, Washington DC.

UN. 2015. *Transforming Our World: The 2030 Agenda for Sustainable Development.* UN, New York.

van Wyk, E., Breen, C. and Freimund, W. 2014. Meanings and robustness: propositions for enhancing benefit-sharing in social-ecological systems. *International Journal of the Commons* 8 (2): 576–94.

Wang, F., Squires, V.R. and Lu, Q. 2018. The future we want: putting aspirations for a land degradation neutral world into practice in the GCA region. In this volume, pp. 99–112.

Wang, M.P., Zhao, C.Z., Hua L.M. and Squires, V. 2010. Land Tenure: Problems, Prospects and Reform. In: V. Squires, Hua Limin, Zhang Degang and Li Guolin (eds) *Towards Sustainable Use of Rangelands in North West China.* Springer, Dordrecht, pp. 255–84.

Watkins, C., Massey, D., Brooks, J., Ross, K. and Zellner, M.L. 2013. Understanding the mechanisms for collective decision making in ecological restoration: an agent-based model of actors and organizations. *Ecology and Society* 18 (2): 32.

Weber, M., Krogman, N. and Antoniuk, T. 2012. Cumulative effects assessment: linking social, ecological and governance dimensions. *Ecology and Society* 17 (2) Art. 22 [online].

World Bank. 2014. Population: Annual growth rates. http://data.worldbank.org/indicator/SP.POP.GROW

World Bank. 2016. *High and Dry: Climate Change, Water and the Economy.* Washington DC, World Bank.

World Commission on Environment and Development (WCED). 1987. *Our Common Future.* Oxford University Press, Oxford.

Zhao, C.Z. and Squires, V.R. 2010. Biodiversity of plants and animals in mountain ecosystems. In: V. Squires, Hua Limin, Zhang Degang and Li Guolin (eds) *Towards Sustainable Use of Rangelands in North West China.* Springer, Dordrecht, pp. 101–26.

5 The future we want

Putting aspirations for a land degradation neutral world into practice in the GCA region

Wang Feng, Victor R. Squires and Lu Qi

INSTITUTE FOR DESERTIFICATION STUDIES, CHINESE
ACADEMY OF FORESTRY, BEIJING

Zero net land degradation(ZNLD): origin, definition and challenges

Origin

Land degradation is a major environmental problem affecting the lives and livelihoods of hundreds of millions of people. Global estimates of total degraded land areas vary from less than 10 million km^2 to over 60 million km^2, of which Asia contributes 40% on average (Gibbs and Salmon, 2015). Vast areas of Central Asia (defined below) and Northeast Asia, particularly in China and Mongolia, are under the threat of desertification and land degradation, and the situation is becoming even more challenging with climate change (Huang et al., 2016).

At the United Nations Conference on Sustainable Development (Rio+20), held 20–22 June 2012, governments adopted the outcome document *The Future We Want*, which recognized "the need for urgent action to reverse land degradation". In view of this, nations will strive to achieve a land-degradation neutral world in the context of sustainable development (Stavi and Lal, 2015; Orr et al., 2017). All stakeholders of the United Nations Convention to Combat Desertification (UNCCD) reaffirmed their resolve under the UNCCD to take coordinated action nationally, regionally and internationally.

They set a goal of maintaining a world where the total amount of degraded land remains constant – i.e., there would no net increase – and that would secure the currently available productive land for the use of present and future generations. Achieving a state of land degradation neutrality involves both reducing the rate of land degradation and offsetting newly occurring degradation by restoring the productivity and provision of other ecosystem services in currently degraded lands; in other words, achieving zero net (rather than zero) land degradation or ZNLD (Lal et al., 2012; Stavi and Lal, 2015), the global-scale derivative of which is a land degradation neutral world (LDNW). For achieving this target, it is urgent to monitor the global extent and rate of land degradation and restore degraded land in arid, semi-arid and dry sub-humid areas.

Definitions

Box 5.1   Some terminology and definitions.

Zero net land degradation (ZNLD): The achievement of land degradation neutrality, whereby land degradation is either avoided or offset by land restoration. Promoting the ZNLD target would secure the currently available productive land for the use of present and future generations.

Land degradation: "loss of the biological or economic productivity and complexity of rainfed cropland, irrigated cropland, or range, pasture, forest and woodlands resulting from land uses or from a process or combination of processes, including processes arising from human activities and habitation patterns, such as: (i) soil erosion caused by wind and/or water; (ii) deterioration of the physical, chemical and biological or economic properties of soil; and (iii) long-term loss of natural vegetation" (UNCCD, Article 1).

Desertification: defined by the UN Convention to Combat Desertification (UNCCD) as involving "land degradation in arid, semi-arid and dry sub-humid areas resulting from various factors, including climatic variations and human activities (UN, 1994)."

Sandification: An environmental change whereby an environment becomes sandy. It is often the stage of soil erosion preceding desertification in which wind (predominantly) sorts and entrains fine particles, leaving coarse sand particles that migrate and accumulate on adjacent grasslands or cropland.

Three challenges

What is the geographical unit that makes the "zero net"?

Should the land degradation (LD) of *one bioregion* or of the *whole nation state* be zero? Within the ecosystem, can there be a little bit of degradation of soils while in another bioregion there is the same "amount" of restoration of the ecosystem? Thus, is the total land degradation within a biome or nation state ecosystem zero? For any given "geographical unit" (ecosystem, bioregion or nation state) that is already degraded, there may be big differences in the severity and extent of LD. Should we take that level as "time zero" or is the starting point set at some earlier point in time when ecosystems were more intact? Is it possible to allow the restoration of forests while degrading drylands within the borders of the nation state country? Can one area be traded off against another area? Or is the zero net land degradation an overall calculation, so that the land restoration that takes place in one bioregion can "neutralize" the land degradation in another? If so, who is going to decide what the trade-offs are? And at what cost? If the target is set at ecosystem and/or national level, then will it bring additional obligations only to affected countries? What will be the obligations of other signatory states under the UNCCD? Developed countries have an obligation to help affected-country parties (ACPs).

What is the benchmark or starting point of ZNLD?

What are the operational definitions of *land degradation* and *restoration*? How are *land degradation* and *land rehabilitation* measured? What is the benchmark or starting point of ZNLD? Is the starting point the current LD status or some other pre-degradation status? The first approach (to work from the current status) might be measurable; the last one promotes restoration actively but would be much harder to achieve, especially without financial and technical support from developed countries. In accordance with the United Nations Convention to Combat Desertification the goal is to take coordinated action nationally, regionally and internationally, to monitor, globally, land degradation and restore degraded lands in arid, semi-arid and dry sub-humid areas. Parties to the UNCCD resolved to support and strengthen the implementation of the Convention and the ten-year strategic plan and framework to enhance its implementation (2008–2018), including through mobilizing adequate, predictable and timely financial resources.

Who is going to measure and monitor the rate of land degradation and land restoration?

What is the unit of measure for land degradation? Is it per hectare or per square km? Will new indices apply to all countries? Who is going to measure and monitor land degradation and land restoration? How do local communities, land users, citizens and/or other stakeholders participate in the decision making over ZNLD initiatives/strategies, about the whereabouts of these areas and the monitoring of those same areas. What are the characteristics of restored/degraded land that are allowed under the ZNLD concepts?

Operationalizing ZNLD in Greater Central Asia

Greater Central Asia (see Squires and Lu, 2018) is the core region of the Asian continent and stretches from the Caspian Sea in the west to China in the east and from Afghanistan in the south to Russia in the north. It is also sometimes referred to as "the 'stans" (as six of the nine countries in GCA have names ending with the Persian suffix "-stan", meaning "land of") and is within the scope of the wider Eurasian continent. In modern contexts, all definitions of Central Asia include these five republics of the former Soviet Union: Kazakhstan (population 17 million), Kyrgyzstan (5.7 million), Tajikistan (8.0 million), Turkmenistan (5.2 million) and Uzbekistan (30 million), with a total population of about 66 million as of 2013–2014. Afghanistan (population 31.1 million) is also sometimes included. But GCA encompasses western China and Mongolia.

Despite some uncertainty in defining the exact borders, the GCA region does have some important overall characteristics. For one, GCA has historically been closely tied to its nomadic peoples and the Silk Road. As a result, it has acted as a crossroads for the movement of people, goods and ideas between Europe, Western Asia, South Asia and East Asia (Squires, Shang and Ariapour, 2017).

During pre-Islamic and early Islamic times, Central Asia *sensu stricto* was a predominantly Iranian region that included the sedentary Eastern Iranian-speaking Bactrians, Sogdians and Chorasmians, and the semi-nomadic Scythians and Parthians. The ancient sedentary population played an important role in the history of Central Asia. After expansion by Turkic peoples, Central Asia and the west of China also became the homeland for many Turkic peoples, including the Kazakhs, Uzbeks, Turkmen, Kyrgyz, Uyghurs and other, now extinct, Turkic nations. This sub-region is sometimes referred to as Turkestan.

Since the earliest of times, Central Asia has been a crossroads between different civilizations. The Silk Road connected Muslim lands with the people of Europe, India and China. This position at the crossroads has intensified the conflict between tribalism and traditionalism and modernization.

To make ZNLD operational, a plan of action is required (Chasek et al., 2015). The plan's proposed steps are scoping (determining the spatial scale and the selected domain for which land degradation neutrality is to be achieved), mapping (classifying the lands by their current use and state of their productivity), prescribing (prescribing management practices relevant to each of the land classes), applying the selected land management practices (for either reducing degradation, restoring productivity, or increasing resilience), and monitoring management and its outcomes (which should go together with an assessment of the monitoring results). The challenges posed by each of these steps are illustrated below.

Scoping scale and domain

Although LDNW implies global scale neutrality, this globality is similar but not identical to the globality of greenhouse gas emissions. Whereas emissions from a local site directly affect global warming, local land degradation directly affects the local land user, but only indirectly affects global food security. Therefore, striving for an LDNW is a cumulative result of striving to increase the number of sites that achieve ZNLD. Thus, countries, organizations or sectors that wish to contribute to an LDNW need to determine the spatial scale and the specific geographic or thematic domain within which they aspire to achieve ZNLD. The selected geographic domain can be an individual farm, a rural community, a watershed, an administrative region or a geopolitical region, and a thematic domain may be an area of specific land used for cultivation, rangeland, agroforestry or sylvi-pastoral, each embedded in a specific geographical category. Once the domain is agreed, mapping degradation within its boundaries can proceed.

Mapping degradation

ZNLD's added value to the UNCCD tradition of "combating desertification" through addressing ongoing degradation ("prevention and/or reduction of land degradation") and restoring already degraded land ("rehabilitation of partly degraded land; and reclamation of desertified land") is in its mobilization of both "prevention and reduction" on the one hand, and "rehabilitation and restoration" on the other, to act in unison to stabilize the amount of productive, non-degraded land by neutralizing any additional degradation. This attribute of ZNLD poses a major challenge to the need to quantitatively offset, through the restoration of degraded land, what has been degraded within a given time period in spite of the invested efforts to curb ongoing degradation. This requires classifying and mapping the lands in the area where land degradation neutrality is to be achieved: i.e. identifying lands whose current use is degrading and those that are already degraded (either abandoned or still in use); the former are candidates for applying measures to reduce degradation, and the latter are eligible for applying restoration measures. A third land type may also be identified: land where its current use is not degrading, yet precautionary measures conferring resilience to future degradation risks can be considered.

Monitoring and assessment

Once the selected domain's lands have been classified and mapped and the appropriate management prescribed for each of the land classes, implementation of the management strategy can begin. Yet, a mechanism is required to evaluate the success of the management strategy as expressed in approaching the domain's land degradation neutrality. This mechanism – monitoring and assessment (M&A) – has already been discussed intensively (e.g. Reed et al., 2011) but rarely applied. Despite an abundance of research efforts on combating land degradation, progress has been hampered by lack of effective monitoring and assessment, not only of the state of the land but also of the performance and impact of interventions (Akhtar-Schuster et al., 2011). This is mostly because, while land users do invest in measures to reduce degradation, they habitually do not invest in monitoring and assessment, since these constitute scientific, technological and financial challenges. For example, technologies and methodologies employed for ground monitoring and assessment at the local level differ from the technologies and methodologies used for monitoring at the regional and global levels. Thus, when institutions do engage in the collection and analysis of information, they address the global scale (Safriel, 2007) rather than the needs of local communities (as required by Article 16 of the UNCCD).

ZNLD practices in China

Laws and polices

China has been continuously fighting land degradation since the 1950s through various ecological restoration policies. China and another 46 countries adopted the United Nations Convention to Combat Desertification in Paris on 17 June 1994. To combat desertification and implement the UNCCD, the Chinese government enacted the Desert Prevention and Control Law of the People's Republic of China in 2002, and approved the *National Plan for Combating Desertification* in 2005. Since the end of the 1970s, China has launched a series of key national ecological engineering projects (Wang et al., 2013) such as the "Three-North" Shelterbelt Project (TNSP, 1978–present), Beijing and Tianjin Sandstorm Source Control Project (BTSSCP, 2001–2010), Grain to Green Project (GGP, sometimes called the Returning Farmlands to Forestland (Grassland) Project, 2003–present), and the Natural Forest Conservation Project (NFCP, 1998–present). The TNSP aims at planting shelterbelt forest in arid and semi-arid areas located in the northern part of China; it is scheduled to continue until 2050 and is the most extended and longest-running afforestation plan in the history of the world (Li et al., 2012). The BTSSCP aims to prevent sand dust storms through vegetation restoration in north-eastern China. The GGP and NFCP, respectively, target the conversion of farmland into forestland and grassland and the protection of natural forest through logging bans. Both of them are being implemented throughout most of China but at different levels of inputs of resources (money and personnel).

Monitoring desertification and observation network

Monitoring of the status and dynamics of land desertification and sandification is the fundamental work for combating desertification of China. In line with the Desert Prevention and Control Law of the People's Republic of China and the UNCCD, China began the first national desertification survey in 1994 with repeat surveys planned at five-year intervals, in 1999, 2004, 2009 and 2014. The desertification areas at the time of the five surveys were, respectively, 2.622, 2.674, 2.636, 2.624 and 2.612×10^6 km^2. The latest monitoring results show that, compared with the situation in 2009, the desertification area decreased by 12,120 km^2 (an annual reduction of 2424 km^2), while the net reduction in areas affected by sandification was 9902 km^2 (an annual average reduction of 1980 km^2). The monitoring results indicate that both the desertification area and the sandification area in China have kept on decreasing over the course of three inventories conducted in the past 20 years. Despite the difficulty of preventing and controlling land degradation, obvious achievements have been made.

Furthermore, in order to strengthen monitoring and research in the desertified region, the State Forestry Administration of China has established a national desertification monitoring system and the China Desert Ecosystem Research Network (CDERN), which consists of 43 research stations across the arid, semi-arid and dry sub-humid areas in North China and the karst desertification areas in South-west China. Finally, China has established a databank, which will play an important role in the future.

Perspective

China is committed to go beyond the "land-degradation neutral" objective. The strategic goal of China is that the desertified land (about 5.3×10^5 km^2) caused by human activity will be returned to pre-destroyed status by 2050. Achieving this goal will be a challenge, and China will have to exert more effort. Raising public awareness of, and scientific focus on, the likelihood of severe effects of desertification is the first step in achieving this goal. The government must combat desertification through scientific means without disturbing the water balance in these areas, e.g. by protecting local vegetation in desertification-prone lands and planting suitable vegetation according to local conditions in desertified lands, or just leaving it to recover without human disturbance. The excessive population growth of human and livestock is a challenge to the limited ecological carrying capacity in desertification-prone areas (Squires et al., 2009). Outward migration with a carefully considered relocation policy is an important choice for desertified areas that are judged to be beyond ecological carrying capacity (Tashi and Foggin, 2012). In addition, developing water-saving irrigation techniques, protecting surface soil and groundwater, optimizing future land utilization patterns, and changing structures of agricultural and animal husbandry are effective ways to ensure sustainable development and achieve the Millennium Development Goal (MDG) in arid and semi-arid areas.

Conclusions

It is clear from a perusal of the various national action plans (NAPs) of affected party countries in the GCA region and from the country profiles elsewhere in this volume (Yang et al., 2018) that work is proceeding to combat desertification, reforest land and implement other measures. The most widely agreed definition for Sustainable Land Management (SLM) states that it is a dynamic and evolving concept that aims to maintain and enhance the economic, social and environmental value of all types of lands (including forests), for the benefit of present and future generations (Squires and Feng, 2018).

While SLM is an essential component of any effort to halt land degradation, there is increasing recognition that conservation and sustainable use measures currently in place are no longer sufficient to stem the loss of ecosystem services and achieve ZNLD on a GCA-wide scale. Commitment to ZNLD will require a re-think of NAPs and technical and financial support from the international community. The Chinese "Belt and Road" initiative is a further example of how China is committing to tackling land degradation problems in countries along the route of the ancient Silk Road. Annex 1 provides a statement about the "Belt and Road Initiative".

Annex 5.1 The "Belt and Road" Initiative Joint Action Initiative for Combating Desertification

Desertification and land degradation cause poverty and hunger, undermining the living environment, triggering social conflict and hindering sustainable development. They are a severe challenge to human survival and development worldwide. These challenges of desertification will affect more seriously the livelihoods of farmers and smallholders, in particular women and children living in the degraded land.

The Silk Roads have been important routes for trade and cultural exchanges in human history. Desertification and land degradation are important factors restricting the sustainable development of countries affected by the "Belt and Road" initiative.[1] To promote sustainable development, countries along the Belt and Road need to cooperate with each other and with relevant international organizations. Under the principle of wide consultation, joint contribution and shared benefits, countries can promote the integration of their respective development strategies for joint action and attach importance to ecological environment protection and restoration to effectively tackle desertification and land degradation.

Vision of the Belt and Road and Sustainable Development Goals

The "Vision and Actions on Jointly Building the Silk Road Economic Belt and 21st-Century Silk Road", released by the Chinese government, emphasizes the concept of "ecological civilization" and enhancing cooperation related to the ecological environment, biodiversity and climate change, and jointly building a green Silk Road.

Last year, 17 global Sustainable Development Goals (the SDGs) were adopted by world leaders meeting at the United Nations in New York. One of these goals, SDG 15, calls for efforts to "protect, restore and promote sustainable use of terrestrial ecosystems, sustainably manage forests, combat desertification, and halt and reverse land degradation and halt biodiversity loss".

In particular, Target 15.3 in SDG 15 on land degradation neutrality will be significant for mobilizing support to ensure the link between healthy and productive land and the achievement of many critical development goals, particularly related to climate change. In this regard, at the 12th session of the Conference of the Parties to the United Nations Convention to Combat Desertification (UNCCD COP12), held in October 2015 in Ankara, the parties agreed to achieve land degradation neutrality through voluntary national targets.

Consensus for action

We recognize that a sound ecological environment is the basis for sustainable development in countries along the Belt and Road, and crucial for achieving the vision of the Belt and Road Initiative. The protection and rational use of natural resources, along with the promotion of green economic development and the restoration of land in areas affected by desertification, in combination with eco-logical improvements and desert-culture (local adaptive industry) development, are effective ways to achieve sustainable development in countries along the Belt and Road.

We recognize that, in addressing desertification and land degradation, mankind has accumulated traditional knowledge and practical skills to combat desertifica-tion and land degradation.

We recognize that joint action to combat desertification and to restore degraded land is in line with the vision of the Belt and Road Initiative and meets the devel-opment needs of countries along the Belt and Road. It is also an important way to achieve the LDNW goal of the SDGs. We call upon countries along the Belt and Road to work together proactively to combat desertification, reverse land degra-dation and open a new chapter in green development.

Principles of action

The joint action is to comply with the principle of "peace and cooperation, openness and inclusiveness, mutual learning and mutual benefit", with people engaged at all levels, in particular land users at the community level, in a participatory process.

Framework of action

To effectively curb desertification, reverse land degradation and build a green Silk Road for improving livelihood and alleviating poverty, we propose the following actions:

(1) Take integrated ecosystem management approaches, maintain and strengthen ecosystem stability and biodiversity of desert, steppe, pasture and oasis along the Belt and Road, and improve the ability of ecosystems to adapt to climate change;
(2) Encourage the development of a green economy in dryland areas, advocate integrated land management, coordinate to reflect ecological and economic functions, including through sustainable agriculture, utilize solar energy, wind power and other clean energy, and protect desert landscapes wherever possible;
(3) Establish an integrated eco-protection system of towns (oases) and roads/highways and promote sustainable land and water management in inland river and lake basins;
(4) Promote preparedness and mitigation of drought and the control of shifting sand dunes in dust and sandstorm-originating areas, and improve early warn-ing and emergency response capacity;
(5) Strengthen ecological conservation/protection and restoration in and around the globally significant natural and/or cultural heritage of countries along the Belt and Road.

Cooperation priorities

Joint research, technical exchange, information sharing, technical training and demonstration projects are encouraged in the following key areas:

(1) Monitoring and planning: establish and improve the monitoring and early warning assessment system for desertification, drought, land degradation and sandstorms, including the creation of a land resource atlas and scientific framework for the assessment of land restoration and its impacts on agricultural, pastoral and agroforestry (silviculture) developments and on biodiversity, and prepare a joint action program for combating desertification and restoring the degraded land along the Belt and Road.

(2) Policies and regulations: improve national policies and regulations, encourage and support desertification control projects that help to improve the ecological environment and promote development, carry out policy dialogue, and develop assessment indicators for combating desertification, restoring degraded land and drought mitigation, as well as ecological management standards. In the planning and implementation of these policies, special attention will be given to the active participation of women.

(3) Desert natural heritage list: develop a "List of Important Desert Natural Heritage Sites along the Belt and Road" in joint efforts with international organizations such as the secretariat of the United Nations Convention to Combat Desertification (UNCCD), International Union for Conservation of Nature (IUCN), United Nations Educational, Scientific and Cultural Organization (UNESCO) and World Desert Foundation.

(4) Technology transfer and demonstration: exchange and promote adequate technologies for sand fixation, degraded land vegetation restoration, afforestation for windbreak and sand fixation, mining field reclamation, water conservation and water harvesting afforestation technology in arid areas; promote and utilize practical technologies for soil improvement, saline governance, sustainable agriculture and animal husbandry, as well as water-saving irrigation.

(5) Information sharing: establish internet-based sustainable land management, an ecological restoration directory and information-sharing platform; build a platform to share information on plant germplasm with high economic and/or ecological value, such as sand fixation, drought resistance and salinity tolerance; establish a national technical inventory and technological demand list to promote technology exchange and sharing.

(6) Capacity building: launch projects for experience exchange, training and demonstration of desertification/land degradation control and restoration technology through various ways; carry out specific technical training to enhance the regional, national and local capacity for combating desertification/land degradation.

Cooperation mechanism

Cooperation will be undertaken through existing bilateral and multilateral mechanisms to promote joint action, with voluntary participation.

Explore the feasibility of establishing a dialogue mechanism for joint action, and hold occasional forums as needed, in the form of side events at the Conference of the Parties to the United Nations Convention to Combat Desertification (UNCCD COP), to assess progress and propose further recommendations for joint action. The revolving chairmanship should ensure that each member state gets a chance to lead.

Financing

Participants in the joint action will mobilize resources together by:

(1) Jointly raising funds in the form of voluntary contributions or shared obligations;
(2) Studying the possible establishment of a special open-ended public offering/fund that relies on existing funds to attract additional resources from society;
(3) Applying for projects from existing bilateral and multilateral channels to support joint action, in accordance with the rules and procedures of international financial institutions and international organizations;
(4) Welcoming voluntary contributions or commitments to support follow-up activities from interested stakeholders.

Partnerships

Establish a wide range of partnerships making full use of existing bilateral, multilateral and regional cooperation mechanisms, promoting cohesion and integration of strategies and policies formulated by countries along the Belt and Road and by international organizations, showing common concern and providing support for resource mobilization activities.

Encourage private sector and civil society organizations to participate in joint action.

Relevant countries and partners will further enhance the conservation and restoration of the ecosystem, including soil fertility, in areas affected by desertification/land degradation, heighten the level of green economic development in dryland areas, and provide ecological guarantees, technical support and cooperation momentum for the sustainable development of the countries along the Belt and Road.

Note

1 A modern large-scale revival of trade corridors (road, rail) that more or less follows the main route of the ancient Silk Road that linked Asia with Europe via Central Asia.

References and further reading

Akhtar-Schuster, M., Thomas, R.J., Stringer, L.C., Chasek, P. and Seely, M. 2011. Improving the enabling environment to combat land degradation: Institutional, financial, legal and science-policy challenges and solutions. *Land Degrad. Develop.* 22: 299–312.

Akramkhanov, A., Djanibekov, U., Nishanov, N., Djanibekov, N. and Kassam, S. 2018. Barriers to Sustainable Land Management in Greater Central Asia: with special reference to the five former Soviet Republics. In this volume, pp. 113–30.

Chasek, P., Safriel, U., Shikongo, S. and Fuhrman, V.F. 2015. Operationalizing Zero Net Land Degradation: The next stage in international efforts to combat desertification? *J. Arid. Environ.* 112: 5–13.

FAO. 2003 *World agriculture: towards 2015/2030: An FAO perspective.* FAO, Earthscan, London.

FAO. 2006. *The State of Food Insecurity in the World 2006: Eradicating world hunger – taking stock ten years after the World food Summit.* FAO, Rome.

Gibbs, H.K. and Salmon, J.M. 2015. Mapping the world's degraded lands. *Appl. Geog.* 57: 12–21.

Glenn, E.P., Squires, V.R., Olsen, M. and Frye, R. 1993. Potential for carbon sequestration in the drylands. *Water, Air and Soil Pollution* 70: 341–55.

Grainger. A. (ed.). 2013. *The Threatening Desert: Controlling Desertification.* Earthscan, London.

Hannam, I.D. with Boer, B.W. 2002. *Legal and Institutional Frameworks for Sustainable Soils: A Preliminary Report.* IUCN, Gland, Switzerland and Cambridge, UK.

Hansen, M.C., De Fries, R.S., Townshend, J.R.G., Carroll, M., Dimiceli, C. and Sohlberg, R.A. 2003. Global Percent Tree Cover at a Spatial Resolution of 500 Meters: First Results of the MODIS Vegetation Continuous Fields Algorithm. *Earth Interactions* 7 (10): 1–15.

Heshmati, G.A. and Squires, V.R. (eds). 2013. *Combating Desertification in, Asia, Africa and the Middle East: Proven Practices.* Springer, Dordrecht, 538 pp.

High-level Panel on Global Sustainability. 2012. Secretary-General, Presenting Report of Global Sustainability Panel to General Assembly, Urges Unity in Creating 'Future We Want'. 16 March. http://www.un.org/press/en/2012/ga11212.doc.htm

Huang, J.P., Yu, H.P., Guan, X.D., Wang, G. and Guo, R. 2016. Accelerated dryland expansion under climate change. *Nat. Clim. Change* 6: 166–71.

IFPRI. 2010. *Food Security, Farming, and Climate Change to 2050: Scenarios, Results, Policy Options.* IPRI, Washington.

Kreutzmann, H. (ed.). 2012. *Pastoral practices in High Asia: Agency of 'development' effected by modernisation, resettlement and transformation.* Springer, Dordrecht.

Lal, R. Safriel, U. and Boer, B. 2012. *Zero Net Land Degradation: A New Sustainable Development Goal for Rio+20.* Report Prepared for the Secretariat of the United Nations Convention to Combat Desertification, Bonn.

Li, B., Gasser, T., Ciais, P., Piao, S., Tao, S., Balkanski, Y., Hauglustaine, D., Boiisier, J.-P., Chen, C., Huang, M., Li, L.Z., Li, Y., Liu, H., Liu, J., Peng, S., Shen, Z., Sun, Z., Wang, R., Wang, T., Yin, G., Yin, Y., Zeng, H. and Feng, Z. 2016. The contribution of China's emissions to global climate forcing. *Nature* 531: 357–61. doi:10.1038/nature17165

McDonagh, J. and Bunning, S. 2009. *Field manual for local level land degradation assessment in drylands.* FAO, Rome.

McKinsey Global Institute. 2011. *Resource revolution: Meeting the world's energy, materials, food, and water needs.* McKinsey Global Institute. http://www.mckinsey.com/insights/energy_resources_materials/resource_revolution

Millennium Ecosystem Assessment (MEA) Board. 2005. *Living beyond our means: Natural assets and human well-being.* UN Millennium Goals. UN, New York.

Nachtergaele F. and Petri, M. 2008. *Mapping land use systems at global and regional scales for land degradation.* FAO, Rome.

Nellemann, C., MacDevette, M., Manders, T., Eickhout, B., Svihus, B., Prins, A.G. and Kaltenborn, B.P. (eds). 2009. *The environmental food crisis – The environment's role in averting future food crises. A UNEP rapid response assessment.* United Nations Environment Programme and GRID-Arendal, Cambridge, UK and Arendal, Norway.

OECD and FAO. 2011. *OECD-FAO Agricultural Outlook 2011.* OECD Publishing, Paris. http://dx.doi.org/10.1787/agr_outlook-2011-en

Orr, B.J., Cowie, A.L., Castillo Sanchez, V.M., Chasek, P., Crossman, N.D., Erlewein, A., Louwagie, G., Maron, M., Metternicht, G.I., Minelli, S., Tengberg, A.E., Walter, S. and Welton, S. 2017. *Scientific Conceptual Framework for Land Degradation Neutrality. A Report of the Science-Policy Interface.* United Nations Convention to Combat Desertification (UNCCD), Bonn, Germany.

PBL. 2009. *Trends in global CO$_2$ emissions 2012 report: Background Studies.* PBL, Netherlands Environmental Assessment Agency, The Hague, 39 pp.

Project Catalyst. 2011. Abatement opportunities for non-CO2 climate forcers to complement CO$_2$ reductions and enable national environmental and social objectives. Briefing paper, May 2011. ClimateWorks Foundation, San Francisco.

Reed, M.S. et al. 2011 Cross-scale monitoring and assessment of land degradation and sustainable land management: A methodological framework for knowledge management. *Land Degrad. Develop.* [online at wileyonlinelibrary.com]. doi:10.1002/ldr.1087

Safriel, U.N. 2007. The assessment of global land degradation. In: M.V.K. Sivakumar and N. Ndiaug'ui (eds) *Climate and Land Degradation.* Springer, Berlin.

Squires, V.R. 2013. Replication and scaling up: where to from here? In: V. Squires (ed.) *Rangeland Stewardship in Central Asia: Balancing Livelihoods, Biodiversity conservation and Land protection.* Springer, Dordrecht, pp. 445–459.

Squires, V.R. and Lu, Q. 2018a. Greater Central Asia: its peoples and their history and geography. In this volume, pp. 3–22.

Squires, V.R and Feng, H.Y. 2018b. Sustainable land management: a pathway to sustainable development? In this volume, pp. 75–98.

Squires, V.R., Shang, Z.H. and Ariapour, A. 2017. *Rangelands along the Silk Road: Transformative adaptation under climate and global change.* Nova Science Publishers, New York.

Squires, V.R., Lu XinShi, Lu Qi, Wang Tao and Yang YouLin. 2009. *Rangeland Degradation and Recovery in China's Pastoral Lands.* CABI, Wallingford, 264 pp.

Stavi, I. and Lal, R. 2015. Achieving Zero Net Land Degradation: Challenges and opportunities. *J. Arid Environ.* 112: 44–51.

Tashi, G. and Foggin, M. 2012 Resettlement as Development and Progress? Eight Years On: Review of Emerging Social and Development Impacts of an 'Ecological Resettlement' Project in Tibet Autonomous Region, China. *Nomadic Peoples* 16 (1): 134–51.

Tongway, D.J. and Ludwig, J.A. 2011. *Restoring Disturbed Landscapes: Putting Principles into Practice.* Island Press, Washington DC, 189 pp.

UNCCD. 2012. *Zero Net Land Degradation A Sustainable Development Goal for Rio+20 to secure the contribution of our planet's land and soil to sustainable development, including food security and poverty eradication.* UNCCD Secretariat Policy Brief, Bonn.

Wang, F., Wang, D., Pan, X., Shen, C. and Lu, Q. et al. 2013. Combating desertification in China: Past, present and future. *Land Use Policy* 31: 311–13.

Welton, S., Biasutti, M. and Gerrard, M.B. 2014. *Legal and Scientific Integrity in Advancing a 'Land Degradation Neutral World'.* Sabin Center for Climate Change Law, Columbia Law School, New York.

World Bank. 2011. *Rising Global Interest in Farmland: Can it yield sustainable and equitable benefits?* World Bank, Washington DC.

Yang, Y., Jin, L.S., Squires, V., Kim Kyung-soo and Park Hye-min. 2011 *Combating Desertification and Land Degradation: Proven Practices from Asia and the Pacific.* UNCCD, Seoul, ROK, 168 pp.

Yang, Y., Pak Sum Low, Yang Liu and Jia Xiaoxia. 2018. Mitigation of desertification and land degradation impacts and multilateral cooperation in Greater Central Asia. In this volume, pp. 179–207.

6 Barriers to sustainable land management in Greater Central Asia

With special reference to the five former Soviet republics

Akmal Akramkhanov

ICARDA, TASHKENT, UZBEKISTAN

Utkur Djanibekov

UNIVERSITY OF BONN, GERMANY

Nariman Nishanov

ICARDA, TASHKENT, UZBEKISTAN

Nodir Djanibekov

LEIBNIZ INSTITUTE OF AGRICULTURAL DEVELOPMENT IN TRANSITION ECONOMIES (IAMO), HALLE GERMANY.

Shinan Kassam

ICARDA, CAIRO, EGYPT

Introduction

Humans depend on land resources for provisions, and the sustainability of land capital to provide sustenance becomes invaluable with population growth. Agriculture is one of the major economic sectors of the five 'stans', contributing about 5% to 20% of their GDP (World Bank, 2014a). The diversity of reforms in agriculture produced a variety of challenges for sustainable development in each country of the region. An increasing number of reports suggest a growing understanding among diverse groups of stakeholders that the inappropriate exploitation of land resources results in land degradation, and a decrease or complete loss of productivity. Despite the changes in crop portfolio, farm structures, land reforms, agricultural productivity, resource use and agricultural policies, the agricultural sector has remained important in terms of employment, rural livelihoods, food security and government earnings and therefore for economic development in the region.

There are several important agro-ecosystems in the GCA region characterizing production systems. Two predominant agro-ecosystems are rangelands and irrigated areas. Measured by size, rangelands are the largest production systems. For example, rangelands occupy approximately 260 million ha (Mha) or 65% of the total land mass in the five 'stans' (FAO, 2006). Compared to this vast area, around 8 Mha of irrigated croplands are quite modest in terms of size, which is in sharp contrast to their importance for livelihood security: about 75% of the rural population depend on irrigated agriculture. Due to area's arid climate, most of the regional agricultural production therefore narrowly depends on irrigated agriculture, as is exemplified by Uzbekistan, Turkmenistan and Xinjiang in the far west of China. The irrigated system is crucial not only for food and nutritional security, but also for the environmental and social stability of the entire GCA region.

Over 90% of water withdrawal in the five 'stans' (with the exception of Kazakhstan at 66%) is allocated to the agriculture sector (FAO, 2013), and is mainly delivered through pumping water along surface canals. Due to imperfect irrigation management practices and aging irrigation and drainage infrastructure, water use efficiency has become extremely low, causing severe and significant environmental damage.

Throughout the GCA region there are huge soil degradation challenges, including soil salinity in irrigated lands and soil erosion in rainfed and mountainous areas (Pender et al., 2009). The latter, in the form of wind erosion, resulted in the degradation of 11 Mha of rainfed areas in Kazakhstan alone (Pender et al., 2009). Croplands in irrigated areas are affected by secondary salinization due to elevated groundwater tables related to irrational irrigation water application for crops as well as poor and insufficient number of drainage systems (Gupta et al., 2009, Bekchanov et al., 2018). Regionally, 40%–90% of irrigated lands are affected by secondary salinization (Qadir et al., 2009), with more than 30%–40% in Tajikistan and Kyrgyzstan, around 60% in Uzbekistan, and about 90% in Turkmenistan (Ahrorov et al., 2012). As a result, one of the disturbing issues is the dependence of about one-fourth of the populations in Kazakhstan and Uzbekistan on degraded lands, which is 2.5 times higher than in other 'stans' and far above the world average (HDR, 2010).

The 'problem tree' in Figure 6.1, summarized by Gupta et al. (2009), demonstrates primary causes of land degradation applicable to all agroecosystems in the GCA region. Gupta et al. (2009) grouped causes into: (i) mismanagement and overuse of natural resources; (ii) insufficiency of economic infrastructure and market mechanisms; and (iii) insufficient development of capacity and weak inter-sector coordination. It has been shown that each group of causes results in undesirable outcomes, such as: degradation of land and water resources adversely affects the integrity of the natural ecosystem; problems associated with economic infrastructure and market mechanisms aggravate rural poverty; and weak institutional structures, research capacity and public–private partnerships influence land degradation processes. As a result, they can easily lead to negative consequences, such as: (i) increasing poverty and negative population shifts; (ii) reduced food security and life expectancy and increases in healthcare costs; (iii) social, economic and political instability; and (iv) deterioration in environmental quality (CACILM, 2006).

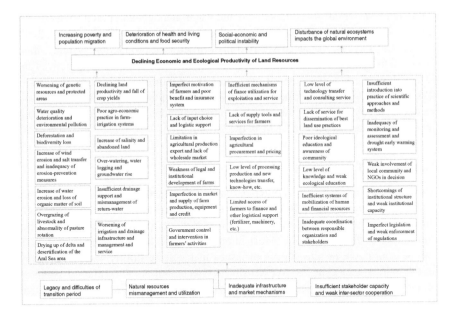

Figure 6.1 Land degradation 'problem tree' and its influence on human wellbeing
Source: Gupta et al. (2009).

Some estimates have already been indicating alarming levels of productivity decline, such as the World Bank (1998) reporting a fall in agricultural yields by 20%–30% across the five 'stans' region since independence. In economic terms, soil salinization annually accounts for US$2 billion of agricultural production loss (World Bank, 1998). Mirzabaev et al. (2015) estimated annual costs of rangeland degradation due to poor rangeland management to have equaled US$4.6 billion between 2001 and 2009. The poorest parts of the region have the most severe impact by land degradation (Mirzabaev et al., 2015). In addition, latest predictions reveal that the temperature rise in GCA will be above global mean values (World Bank, 2014b; Hagg, 2018), which in turn will exacerbate the ongoing desertification processes. Estimates suggest that the irrigation water demand is likely to increase by up to 25% by the middle of the century, while water availability may decline by up to 30%–40% during the same period (World Bank, 2014b; Hagg, 2018; Bekchanov et al., 2018). Concurrently, increasing population, intensive irrigation agriculture practices, and people's heavy reliance on livestock as a cash reserve will put even greater pressure on land cover and soil carbon stock. The degradation of rangelands throughout the region will also reduce the capacity to sequester carbon, significantly diminishing the potential to contribute to climate change mitigation.

The introduction of SLM is crucial to change the baseline course of action (business as usual), which leads to the outcomes mentioned above, and it is necessary to upscale successful SLM experience and lessons (Squires, 2012a). Although a variety of sustainable land and water management practices/technologies has been tested in the region, their adoption remains low (Pender et al., 2009; Gupta et al., 2009). To successfully transform current production systems, there is a need to analyze the barriers that inhibit the introduction of SLM practices and to develop targeted measures to overcome these barriers. Therefore, the aim of this study is to understand the existing barriers to adoption of such practices and technologies by reviewing the existing literature and experiences of national and international researchers on SLM in the GCA region.

Conceptual framework

The success of adopting SLM practices can be addressed through interdisciplinary interlinkages between the four capacities, namely institutional, knowledge, economic and technology (Figure 6.2). This complex interplay of different capacities can hinder or facilitate SLM. Most importantly, this interplay shows that SLM can be contingent on the variety of issues that will be further discussed below. Better market access, access to extension services, farmers' information exchange about SLM, land tenure security and livestock ownership, as well as lower household sizes and lower dependency ratios, are the key factors incentivizing SLM adoption in the transitional economies (Mirzabaev et al., 2015). This framework can be useful for understanding these factors (e.g. as presented by Mirzabaev [2016]). The effects of the interplay between SLM capacities are similar across

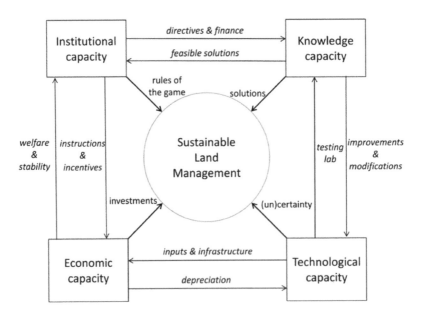

Figure 6.2 Interlinkages between institutional, economic, knowledge and technological capacity to adopt SLM

agro-ecological production systems and countries. The shift to SLM requires a clear understanding of the ability of each of the presented elements to solve the challenges related to adoption and to achieve objectives.

As presented in Figure 6.2, for SLM, the institutional capacity, or the ability of institutions to facilitate the adoption of best practices and technologies, includes the (in)formal rules of the games designed through national plans and norms determining the adoption of technologies/practices. As such, the institutional capacity, being contingent on state objectives via such laws and norms, involves national ministries and, thus, directly links to the capacity of the research and development community to test and implement technology. For the research and development community, institutional capacity provides clear signals for addressing certain solutions, or, in other words, serves the 'local context'. For instance, a lack of economic instruments to reduce irrigation water use in agriculture creates disincentives to adopt water-efficient technologies. Or a poor institutional environment inhibits the formation of farmer cooperatives or user associations, and hinders the adoption of expensive SLM technologies that could be feasible through the formation of service cooperatives. Moreover, the legislative basis for SLM and agricultural norms (e.g. fertilizer application or the number of irrigation events) also shape the knowledge capacity via public finance as well as mapping the direction for the research community. In this way, they have a direct impact on the decisions of farmers to experiment and adopt SLM, or on the knowledge-building capacity of national research centers. The knowledge capacity is also related to the ability of the system to disseminate knowledge, and thus depends on the institutional capacity to provide well-functioning extension services (Squires, 2012b).

Furthermore, the institutional capacity is essential in promoting the economic capacity towards SLM. Firstly, this is related to those technologies that require loans for technology installation, and thus interact with the issues of land tenure, collateral and general economic incentives for shifting from conventional practices. Yet, (technical) norms and instructions also determine how the economic capacity is utilized, for instance by determining crop rotation or afforestation. As part of institutional capacity, policies towards certain crops or sectoral development (food self-sufficiency, crop diversification, climate adaptation, or export promotion) can also incentivize the adoption of certain types of SLM. By advancing the economic capacity to take up SLM practices/technologies, feedback to the institutional capacity comes in the form of higher welfare and stability.

The knowledge capacity can be understood as the ability of knowledge generators to achieve their own objectives related to SLM via field experimentation, learning-by-doing, the demonstration of innovative solutions, and the selection of solutions that fit the 'local context'. Also, solutions that can advance the available technological capacity are required. The solutions developed within the knowledge capacity can impact the way the land is managed by introducing changes directly to the available technological capacity. In this way, the technological capacity is the ability not only of hard infrastructure and already available technologies to utilize SLM practices, but also of agro-ecological and climatic conditions. It can serve as an experimental lab for advancing the knowledge capacity.

The key questions relating to technological capacity are: (i) How well do the available technologies and infrastructure suit the goals of achieving SLM?; (ii) Can it contribute to the economic capacity to bear the costs of shifting from conventional technologies?; and (iii) Can it hinder further investments? For instance, large areas previously under grain production, which were brought into the practice through enormous state expenditures during the Soviet time in the five 'stans', were abandoned due to the mismatch with the economic capacity of newly established farmers. Similarly, the increased variability of water resources and incidence of drought can affect the economic capacity as well as hinder knowledge generation towards SLM. There is an opposite feedback between the two capacities when the lack of financial resources can degrade the technological capacity. This in turn can lower the capacity of available technology to provide certainty for the adequate management of the resource base.

Barriers

Institutional capacity

There is a lack of a systemic approach in devising new interventions. Various national action plans are developed to guide interventions, but without overall vision or the strategy to which such plans contribute it is difficult to imagine that plans will be employed for any intervention. The traditional commodity-based focus in each land use system may need urgent reassessment and diversification, along with subsequent updates and support of policies, legislation and institutions. Changes in policies (and legislation intended to support the realization of those policies) will inevitably have a knock-on effect on the institutional structures and their mandates.

Changes in policy will, for example, require institutions to adjust their mandates and function, and changes in legislation will require different approaches to implement them. Institutions need to evolve from being managers of land to being facilitators by providing a support system for non-state actors (farmers, local communities and households) to manage land – something that is currently not taking place.

Institutional capacity and infrastructure have been eroded due to stagnated public spending in agriculture. Unfortunately, policies that should support the implementation of SLM practices are developed in a piecemeal approach, often with the help of projects (mostly short term), without real 'buy-in' or the consequent financial support from governments. Issues that repeatedly emerged after the collapse of the Soviet Union include: (i) the energy/fuel needs of the local population; (ii) how to manage extensive grazing lands under the new system of private livestock on public lands; and (iii) how to meet the fodder needs of burgeoning livestock numbers. Fodder supply (including crop residues and by-products) from irrigated agriculture and mono-cropping are not reflected in the policies and strategic planning of different sectors due to absent linkages among sectors.

Various initiatives to reform agricultural sectors essentially targeted major production systems, namely irrigated agriculture concentrated in the south and rainfed agriculture in the north of the region. The remaining land use, such as forestry, rangelands and other arid land use systems, did not receive the same amount of attention and largely remain unchanged and function in a similar way to how it was during the Soviet era. For example, although most of the live-stock is in the hands of rural households or '*dehqan*' farms in Uzbekistan and Tajikistan, access to grazing land by these entities is complicated (Halimova, 2012; Robinson et al., 2012).

There are a few exceptions, though: for example, governance concerning pasture use in Kyrgyzstan, where management is being transferred to village com-munities. Responsibility to define grazing areas and periods, stocking rates, and to determine the fees to use pastures are within the remit of village communities that also take care of the maintenance of vegetation cover in allocated territories. On the other hand, institutions and mechanisms to provide seeding materials to maintain grazing areas are no longer functional or significantly weakened, leaving communities and the future state of pastures in limbo.

Many policies have not evolved since Independence to correspond to national development strategies of transition towards a market system. Such policy mis-match results in poor land use practices due to transformed overall conditions from Soviet times. Other, more recent policies need updating because of dynam-ics in agricultural sector reforms or economic changes since the policies were developed, or because of insufficient institutional commitment at the time they were prepared to allow them to be implemented effectively.

Many countries in the region need institutional capacity to enable efficient extension or advisory services. The structural change in agricultural sectors in the 1990s led to the establishment of thousands of unprofessional farmers and small-holders who need even more knowledge and training in agriculture compared with the Soviet period. Funding for public extension systems in the region can hardly cover the targeted policies on filling the knowledge gap and dissemination of new SLM technologies. So the issue of linking knowledge-generating national research systems to the farmer community is not addressed appropriately. The national authorities might take a coordination role and apply a holistic approach in planning the advisory service provision, including opportunities arising from the international donor community, private sector and civil society institutions. That would lead to the prevention of a fragmented delivery of advisory services and can work as an entry point for the development in the long run of national programs of improved extension and advisory services. Taking into account hori-zontal knowledge dissemination, the local community-based organizations should play a leading role in promoting SLM interventions.

Particular emphasis is on land reform in the region, where countries adopted different paths (Robinson et al., 2012). Kyrgyzstan is considered more advanced with respect to land reforms and its farmers have a large share of the land under private use (Lerman, 2009). The dismantling of large Soviet farms into smaller farms led to a mismatch with the agricultural infrastructure that was planned to operate large-scale farms. In Kyrgyzstan, over 400,000 farms were established and most of them have neither the means nor the capacity to manage land sustainably. This demonstrates that existing or introduced laws do not work in practice or have an impact because there is a lack of capacity on the ground to apply them. Weak institutions impede the coordination of small-scale farmers to share responsibilities in managing resources (Kazbekov et al., 2007). Accordingly, land fragmentation in farmers' plots impedes productivity growth in agriculture.

Moreover, in Tajikistan, Turkmenistan and Uzbekistan, the land reforms are not entirely complete. Initial reforms in irrigated agriculture were in line with the pre-independence movement for *destatization* as opposed to *land reform* (Kassam et al., 2016). The nature of lease agreements and the fragility under which lease agreements were (and are) being honored is indicative of the devolution of direct state intervention in the planning of key strategic crops and not necessarily privatized ownership of land assets (Kassam et al., 2016). In these GCA countries, most of the land is not in private use, and farmers lease land from the state. In such conditions, farm structural change can be triggered not by markets but designated by the state to improve agricultural production. Hence, farmers may not be certain whether they will have land in the next years and consequently will be less willing to make investments in SLM that may start to generate benefits in the long term. For example, land tenure insecurity in Turkmenistan and Uzbekistan leads to farmers preferring not to invest in conservation agriculture (Sommer and De Pauw, 2010). At the same time, in Kyrgyzstan farmers own the land but do not conduct such practices (Gupta et al., 2009), which shows that secure land tenure alone does not lead to the adoption of sustainable land use practices.

Also, due to the fact that the land is state property, the state can impose agricultural production quotas and state procurement policies on farmers. Accordingly, Turkmenistan and Uzbekistan follow the state target crop production policy, which was shaped by the Soviet Union, where the state sets priority to produce certain crops. This has led to a situation where agricultural institutions and infrastructure are oriented towards providing suitable conditions and resources for fulfilling the state crop output target (Spoor, 1999). The state preferentially subsidizes some crops, by giving priority in supplying irrigation water and supporting farmers by providing fertilizers and other agricultural inputs. This leads to an agricultural system that is focused on state-targeted crops. For those farmers who lease irrigated lands, contracts with the state for cotton and wheat permit security in the form of access to productive inputs, subsidized credit and crop insurance mechanisms; yet, the incentives for both classes of farmers are not necessarily conducive to investing in sustainable land use management practices (Kassam et al., 2016). Capitalizing on the advantages of existing state support and infrastructure but in favor of alternative crops corresponding to SLM could be the win–win way forward for national agricultural development programs. The significant role of the state in land use necessitates strong state support and progressive policy reforms.

Economic capacity

Almost 25 years after national independence, the five 'stans' have been making efforts to restructure agriculture through various policies, such as those to address agriculture and water-related problems (Bekchanov et al., 2018). These reforms display large differences in the speed and manner of implementation (Spoor and Visser, 2001; Djanibekov and Wolz, 2015). In all these countries, the transition from a planned to a market economy has been driven by a different mixture of economic objectives (Csaki and Nucifora, 2005), mostly based on the natural resource endowment and the need to be independent, rather than playing a certain role in the overall division of labor. Different production unit sizes were considered to be optimal: large collective production units were replaced by smaller individual, often family-based, farms (Lerman et al. 2004), or the land reforms were nonlinear (Djanibekov et al., 2012). While the large farm continued to dominate in the rainfed grain-producing areas of Northern Kazakhstan, the post-Soviet reforms resulted in a progressive increase in the number of small farms without a comprehensive adjustment of the physical infrastructure or proper incentives to make land and water use in agriculture efficient (Lerman and Sedik, 2009).

Agricultural restructuring has either oriented towards a strategy of export revenue generation (such as via cotton production in Uzbekistan and Turkmenistan) and food self-sufficiency (for all five 'stans') with implications for SLM-related issues. Or it has followed no coherent strategy at all, as in Kazakhstan and Tajikistan for most of the 1990s (Pomfret, 2008a, 2008b). One of the examples concerns the presence of state regulation in land allocation decisions by farmers in the largest part of the irrigated areas in Turkmenistan and Uzbekistan.

Farmers are requested to produce a certain amount of cotton (an exported commodity) and/or wheat (the backbone of the food self-sufficiency policy) and to allocate a significant share of their farmland for these two strategic crops (Pomfret, 2008b). Such an institutional setting influences the flexibility of farmers' decision making and limits their economic capacity (revenues from these two crops) to invest in SLM technologies. Giving more flexibility in decision making can lead to land use diversification and the adoption of SLM practices and technologies (Bhaduri and Djanibekov, 2015; Djanibekov and Khamzina, 2016). In addition, from the state perspective, the investments in institutions and infrastructure were effective in institutionalizing centrally planned agricultural production. However, such investments result in high sunk costs and are irreversible. Hence, rapid changes in the production policy (e.g. via full liberalization) can be costly. In the transition stage in the institutional environment, there is uncertainty in outcomes for the national economy. At the same time, maintaining the state target production of crops such as cotton and wheat is still possible through adopting more sustainable land uses such as organic cotton (Franz et al., 2010) and no-till wheat (Nurbekov et al., 2016) and integrating these crops into agroforestry (Djanibekov et al., 2015), which can reduce the pressure on the environment and improve livelihoods. The gradual reduction of cotton cropland, as observed for Uzbekistan towards an increase in fruit and vegetable production, can be accompanied by the integration of SLM-oriented technologies.

Alternatively, the fragmentation of farms, as observed in Kyrgyzstan, or in many irrigated parts of Central Asia after the suspension of the large-scale cotton–wheat production system of Kazakhstan, Tajikistan and Uzbekistan, demonstrated the growth-inhibiting disadvantages of farm fragmentation (Lerman, 2013). The key reason for this is that the restructuring of institutional capacities to support the newly established system comprising numerous small farmers takes considerably longer than the process of farm restructuring (Djanibekov and Wolz, 2015). The issue here is that the classic services provided within the former Soviet agricultural system, such as large-scale machinery, intensive fertilizer application, centrally managed irrigation facilities or extension services, were not adapted quickly enough to serve the newly established small farms. The economic capacity for SLM has been hindered by the lack of institutional capacity to provide the new producers with suitable access to financial services, marketing, the delivery of supplies and equipment, and extension/advisory and information services. Densely populated regions and farms reliant on irrigation facilities and labor-intensive sectors (e.g. fruit and vegetable or livestock farming) are more prone to these problems (Djanibekov and Wolz, 2015).

Financial support to farmers may trigger and enhance the adoption of SLM in the region. For example, subsidies provided by the Kazakh government of about US$6/ha for using no-till practices have accelerated their adoption and were mainly observed on large land areas under agricultural joint-stock companies (Kazakhstan Farmers Union, 2011). The consideration of farmers' economies of scale in combination with low subsidies and the reduction in costs resulting from such practices can result in substantial benefits. However, subsidies paid for some SLM practices must meet strict requirements and compliance can be difficult to achieve. For instance, farm subsidies in Kazakhstan to establish fruit plantations demand a minimum area of 5 ha, the installation of fruit frames and drip irrigation technologies (Lapeña et al., 2014). This requirement precludes the majority of farmers.

Despite the presence of financial support to invest in SLM, institutional capacity as well as the physical infrastructure can make the available SLM technologies unsuitable. The technical capacity, which also consists of machinery and equipment services, originates from Soviet times and is outdated. Newly imported technical equipment and machinery require high investment and operational costs, which are beyond the economic capacity of small farmers. The import of equipment and machinery is a short-term solution since they need to be adapted to local settings.

Knowledge capacity

There are a number of barriers to changing a business-as-usual course of action, including limited awareness about available and applicable SLM technologies and approaches in the region and the lack of experience in applying such practices in the field. The limited awareness of large- and small-scale farmers is a direct indicator of the current state of knowledge dissemination. Analysis of the existing rural advisory services and public extension systems in the region indicates that bottlenecks remain for the efficient channeling of innovations from the research community to farmers and households. These include limited funding of public extension organizations, short-term focus and limited geographical coverage of donor-funded projects, limited ability of farmers to pay for advisory services, insufficient capacity building of consultants and trainers providing advisory services, an obsolete curriculum often failing to introduce the notion of SLM in most public universities, lack of a developed vertical and horizontal (farmer-to-farmer) knowledge transfer system, and an absence of coordination mechanisms in advisory service provision among international donors with national authorities.

The combination of these barriers inhibits implementation of SLM, particularly in the context of the low-priority forestry and pasture use sectors. At the same time, there are positive examples of pilot initiatives in the field on pasture management, joint forestry and community forest management, crop diversification, and conservation agriculture in rainfed areas that demonstrate principles to be applicable in this arid region. As a rule, successful pilot SLM interventions have limited geographic coverage and remain restricted to tested locations at best, or cease once the project or intervention is completed. It is critical to build the practical experience and know-how of key national and local authorities about how selected approaches and practices can be applied most effectively in the field. This will allow for their up-scaling and widespread application (Squires, 2012c).

A fundamental constraint is the limited skills, practical capacity or experience to implement SLM that can be housed within the broader context of integrated land use planning. Traditional thinking for land use planning is highly centralized and narrowly focuses on a single sector. There are usually no options provided for certain land types: for example, arable land is only for crop production, rangelands must be used for livestock production, forest land only for trees. However, the same land can be multi-functional and, as part of the ecosystem, can be used, for example, to produce fuel wood or fruits, used for grazing or as a hunting ground, for watershed protection, disaster risk reduction, or biodiversity conservation. Where an alternative land use is environmentally sustainable, economically sound, financially profitable and socially acceptable, such an option should be institutionally and legally supported.

In order to obtain support, the notion of environmental sustainability through ecosystem services needs to be better recognized. Often, the main challenge faced by farmers to adopt SLM practices is a perception of delayed economic returns. Yet, many SLM ecosystem services are usually neglected. For example, Djanibekov (2015) conducted surveys in northwestern Uzbekistan and revealed that farmers are not familiar with the range of ecosystem services provided by afforestation on degraded cropland, e.g. land rehabilitation and climate change mitigation, and thus their perceived value is low. In the same study, farmers also mentioned that they would start to adopt such land use practices if they observed their benefits and adoption by other farmers.

Technological capacity

There is a rich knowledge base of innovations that contribute to SLM. Most of the programs and projects predominantly engage in the development or adaptation of SLM technologies and approaches to local conditions, but they often fail to make tested innovations available to the general public. It is difficult to navigate among dispersed information and to synthesize the technological options that are required. At the same time, there are global as well as regional initiatives that compile and take stock of best practices in accessible databases. Out of the many global initiatives, the World Overview of Conservation Approaches and Technologies (WOCAT) can be mentioned as an example. It has systematically collected SLM practices from all over the world, and has assessed and described the collected technologies and approaches in a unified template. Such initiatives provide an opportunity for workable solutions, both simple and complex, to be available and accessible through print and online media.

The Central Asian Countries Initiative for Land Management (CACILM) is a program that has been running since 2006 and has also contributed a collection of tested SLM options to the WOCAT database. A recent undertaking within CACILM on knowledge management established a regional *knowledge-sharing platform* led by ICARDA. This recently established knowledge-sharing platform (cacilm.org) is aimed at collecting, sharing, developing and promoting SLM practices that were both provided by national countries and also drawn from world databases that were found suitable for application in the region.

Table 6.1 Types of SLM technologies in different land use categories in Central Asia

SLM types	Rainfed	Irrigated	Pasture	Mountains
Agroforestry amelioration, agroforestry, reforestation/ improving soil vegetation cover	KZ, KG, TJ, UZ	TJ, UZ	TJ, TM, UZ	TJ, TM, UZ
Cultivation of slopes, erosion prevention	KG, TJ	KG, TJ, UZ	TJ, TM	TJ
Improving methods of sowing/ planting crops and soil tillage	KZ, KG, TJ, TM, UZ	KZ, KG, TM, UZ	UZ	KG, TM, UZ
Improving rangeland/fodder production	KG	KG	KZ, KG, UZ	KG
Increase of soil fertility	KZ, KG, TJ, UZ	KZ, KG, TJ, UZ		KG, TJ, UZ
Increasing capacity of land users/ environmental education	TJ	KZ, TJ, UZ	TJ	KZ, TJ
Management of water demand (improving furrow irrigation and resource-saving irrigation technologies)	KG, TJ, TM	KG, TJ, TM, UZ		KG

Source: CACILM (2016).

Note: KG – Kyrgyzstan, KZ – Kazakhstan, TJ – Tajikistan, TM – Turkmenistan, UZ – Uzbekistan.

Table 6.1 shows seven selected types of SLM technologies that were collected from the five 'stans' and grouped by ecosystems and listed in detail in CACILM (2016). However, the issue is not the availability of SLM options but accessibility and lack of skills and information about these technologies.

In spite of the many advantages of SLM approaches, they are not widely recognized for their potential contribution to the agricultural sector and the welfare of the population. Although various studies and practices about SLM approaches are available in the region, there is still a need to disseminate and advise on SLM practices for farmers. The break-up of the Soviet Union resulted in the stagnation of former extension services for agriculture, and poor access to advisory services. Farmers now do not have the necessary information and support for agricultural production. The lack of knowledge can be observed in the absence of skills related to SLM. Farmers are not aware about potential different land uses and how to conduct appropriate management practices.

To address the recognition and potential of SLM, similarity maps could be developed at the regional scale for Greater Central Asia. Similarity maps identify areas, locations and ecosystems where particular SLM approaches have the potential to be out-scaled (replicated and scaled up) based on environmental criteria. Suitability analysis is the fine-tuning of similarity analyses with more specific data, which results in classifying areas/land as highly, moderately or marginally suitable for particular SLM options.

Conclusions

A focus on natural resource management is crucial for sustainable agriculture in Central Asia. Much has been done in this area in terms of research and knowledge dissemination. ICARDA and a number of other international research centers within the GGIAR network have partnered effectively with national research centers in the generation of relevant technologies and contemporary production practices which are particularly suited to small and marginalized farm households. Yet, an enabling environment for supporting the broad uptake of these has been slow in development. In large part, this has been attributable to a general lack of supporting policy mechanisms, as well as a growing dependence on internationally financed project-based support – with short time horizons – and a lack of attention to sustainable (business-driven) approaches to agricultural innovation.

Secure land tenure is a powerful entry point for addressing issues of land degradation and resilience to climate change, and in ensuring sustainable livelihoods for rural populations. In the absence of this, a number of options exist in order to achieve a second-best solution, and which are likely to yield a range of desired outcomes in relation to investments in sustainable land use management practices as well as agrarian reform more generally.

Desirable outcomes from land privatization include those of efficiency within economic theory; and, when concurrently analyzed through gender-based theories and arguments, for inclusivity, equity and well-being. However, when the privatization of land is not an option for various reasons, there are alternative measures for approaching benchmark indicators for promoting SLM. The ability to trade lease agreements can provide a strong measure of efficiency, in terms of ensuring greater measures of productivity and investment through the placing of market-based economic values on lease rights. Yet, in an environment where land privatization has been delayed, tradeable lease rights may have limited appeal. A second-best option is the consideration of formally institutionalizing the collateral lease agreements. Two immediate beneficial outcomes can arise from the ability to collateralize lease agreements. The first is a sense of security in terms of land use rights, both in terms of lower perceived risk of loss to rights as well as in terms of the incentive to invest in soil health. The second, linked to an incentive to invest in land productivity, relates to an ability to access finance to adopt SLM technologies. Furthermore, the reduced control of farmers' production decisions can facilitate rational crop mix choices that are socially efficient, environmentally sound and nationally strategic.

The problem of the smallness of new farms can be addressed through policies tailored to enable such farms to form voluntary service cooperatives, such as for marketing, accessing credits, joint machinery use and purchasing inputs and equipment.

The challenge for the Greater Central Asian countries remains one of an overwhelming (external) focus on land reform, as opposed to agrarian reform, and leads to a lack of acknowledgment that the countries in the GCA region have taken various development paths. Tested options towards implementing SLM would appear to be available given the solutions within economic (land tenure, enabling environment for service cooperatives and access to credits), knowledge (development of extension services and information flow), and financial capacity contingent on a range of acceptable business-driven approaches to agricultural innovation.

References and further reading

Ahrorov, F., Murtazaev, O., Abdullaev, B. 2012. Pollution and salinization: compounding the Aral Sea disaster. In: Edelstein M.R., Cerny A., Gadaev A. (eds.) *Disaster by Design: The Aral Sea and its Lessons for Sustainability. Research in Social Problems and Public Policy*, Vol. 20. Emerald Group Publishing Limited, pp. 29–36.

Bekchanov, M., Djanibekov, N. and Lamers, J.P.A. 2018. Water in Central Asia: a cross-cutting management issue. In this volume, pp. 211–36.

Bhaduri, A., Djanibekov, N. 2014. Potential water price flexibility, tenure uncertainty and cotton restrictions on adoption of efficient irrigation technology in Uzbekistan. In: Lamers, J.P.A., Khamzina, A., Rudenko, I., Vlek, P.L.G. (eds.) *Restructuring Land Allocation, Water Use and Agricultural Value Chains: Technologies, Policies and Practices for the Lower Amudarya Region*. V&R Unipress, Bonn University Press, Göttingen, pp. 217–30.

CACILM. 2006. National Programming Frameworks. Central Asian Countries Initiative for Land Management. Prepared by UNCCD National Working Group, Uzbekistan, February 2006.

CACILM. 2016. Database. http://www.cacilm.org/en/visual/table

Csaki, C., Nucifora, A. 2005. Ten Years of Transition in the Agricultural Sector: Analysis and Lessons from Eastern Europe and the Former Soviet Union. Essays in Honor of Stanley R. Johnson. Berkeley Electronic Press, Berkeley.

Djanibekov, N., Wolz, A. 2015. Entwicklungsprobleme in Zentralasien: Das Beispiel landwirtschaftlicher Dienstleistungsgenossenschaften. *IAMO Jahreszahl* 17: 69–77.

Djanibekov, N., Rudenko, I., Lamers, J.P.A., Bobojonov, I. 2010. Pros and cons of cotton production in Uzbekistan. In Pinstrup-Andersen, P., Cheng, F. (eds.) *Food Policy for Developing Countries: Food Production and Supply Policies.* Cornell University Press, Ithaca, NY.

Djanibekov, N., Sommer, R., Djanobekov, U. 2013. Evaluation of effects of cotton policy changes on land and water use in Uzbekistan: Application of a bio-economic farm model at the level of a water users association. *Agricultural Systems* 118: 1–13.

Djanibekov, N., Van Assche, K., Bobojonov, I., Lamers, J.P.A. 2012. Farm restructuring and land consolidation in Uzbekistan: new farms with old barriers. *Europe-Asia Studies* 64 (6): 1101–26.

Djanibekov, U. 2015. A coevolutionary perspective on the adoption of sustainable land use practices: The case of the Amu Darya River lowlands, Uzbekistan. In: Beunen R., Van Assche, K., Martijn, D. (eds.) *Applying Evolutionary Governance Theory.* Springer, Heidelberg, pp. 233–45.

Djanibekov, U., Khamzina, A. 2016. Stochastic economic assessment of afforestation on marginal land in irrigated farming system. *Environmental and Resource Economics* 63 (1): 95–117.

Djanibekov, U., Dzhakypbekova, K., Chamberlain, J., Weyerhäuser, H., Zomer, R., Villamor, G.B., Xu, J. 2015. Agroforestry for landscape restoration and livelihood development in Central Asia. ICRAF Working Paper 186. World Agroforestry Centre East and Central Asia, Kunming, China, 41 pp.

FAO, 2006. *Compendium of Food and Agriculture Indicators.* FAO, Rome.

FAO, 2013. *Irrigation in Central Asia in figures. AQUASTAT Survey-2012.* FAO Water Reports 39. FAO, Rome. http://www.fao.org/3/a-i3289e.pdf

Franz, J., Bobojonov, I., Egamberdiev, O. 2010. Assessing the economic viability of organic cotton production in Uzbekistan: a first look. *Journal of Sustainable Agriculture* 34 (1): 99–119.

Gupta, R., Kienzler, K., Martius, C., Mirzabaev, A., Oweis, T., de Pauw, E., Qadir, M., Shideed, K., Sommer, R., Thomas, R., Sayre, K., Carli, C., Saparov, A., Bekenov, M., Sanginov, S., Nepesov, M., Ikramov, R. 2009. Research Prospectus: A Vision for Sustainable Land Management Research in Central Asia. ICARDA Central Asia and Caucasus Program. Sustainable Agriculture in Central Asia and the Caucasus Series No.1. CGIAR-PFU, Tashkent, Uzbekistan. 84 pp.

Hagg, W. 2018. Water from the mountains of Greater Central Asia: a resource under threat. In this volume, pp. 237–48.

Halimova, N. 2012. Land Tenure Reform in Tajikistan: Implications for Land Stewardship and Social Sustainability: A Case Study. In: Squires, V. (ed.) *Rangeland Stewardship in Central Asia: Balancing improved livelihoods, biodiversity conservation and land protection.* Springer, Dordrecht, pp. 305–32.

HDR. 2010. *Human Development Report. The real wealth of nations: Pathways to human development.* UNDP, New York. http://hdr.undp.org/sites/default/files/reports/270/hdr_2010_en_complete_reprint.pdf

Kassam, S., Akramkhanov, A., Nurbekova, A., Dhabi, B., Nishanov, N. 2016 (in preparation). Exploring agricultural innovation systems in post-Soviet Central Asia.

Kazakhstan Farmers Union. 2011. Stop the Tractor! I Till No More! http://sfk.kz/ru/homepage/?id=9&kid=10 (in Russian).

Kazbekov, J., Abdullaev, I., Anarbekov, O., Jumaboev, K. 2007. *Improved water management through effective water users associations in Central Asia: Case of Kyrgyzstan (No. H040650)*. International Water Management Institute.

Lapeña, I., Turdieva, M., López Noriega, I., Ayad, W.G. 2014. *Conservation of fruit tree diversity in Central Asia: Policy options and challenge*. Bioversity International.

Lerman, Z. 2009. Land reform, farm structure, and agricultural performance in CIS countries. *China Economic Review* 20 (2): 316–26.

Lerman, Z. 2013. Cooperative Development in Central Asia. Budapest. FAO Regional Office for Europe and Central Asia, Policy Studies on Rural Transition, No. 2013–4. FAO, Rome.

Lerman, Z., Sedik, D. 2009. Agricultural recovery and individual land tenure: lessons from Central Asia. FAO Regional Office for Europe and Central Asia, Policy Studies on Rural Transition No. 2009–3, 12 pp.

Lerman, Z., Csaki, C., Feder, G. 2004. *Agriculture in Transition: Land Policies and Evolving Farm Structures in Post-Soviet Countries*. Lexington Books, Lanham MD.

Mirzabaev, A. 2016. Land Degradation and Sustainable Land Management Innovations in Central Asia. In: Gatzweiler, F.W., von Braun, J. (eds.) *Technological and Institutional Innovations for Marginalized Smallholders in Agricultural Development*. Springer, Dordrecht, pp. 213–24.

Mirzabaev, A., Goedecke, J., Dubovyk, O., Djanibekov, U., Nishanov, N., Aw-Hassan, A. 2015. Economics of land degradation in Central Asia. In: Nkonya, E., Mirzabaev, A., von Braun, J. (eds.) *The economics of land degradation and improvement – a global assessment for sustainable development*. Springer, Dordrecht, pp. 261–90.

Nurbekov, A., Akramkhanov, A., Kassam, A., Sydyk, D., Ziyadaullaev, Z., Lamers, J.P.A. 2016. Conservation Agriculture for combating land degradation in Central Asia: a synthesis. *AIMS Agriculture and Food* 1 (2): 144–56.

Pender, J., Mirzabaev, A., Kato, E. 2009. *Economic analysis of sustainable land management options in Central Asia. Final report for the ADB*. International Food Policy Research Institute and International Center for Agricultural Research in the Dry Areas, Washington DC/Beirut.

Pomfret, R. 2008a. Kazakhstan. In: Anderson, K., Swinnen, J. (eds.) *Distortions to Agricultural Incentives in Europe's Transition Economies*. World Bank, Washington DC, pp. 297–338.

Pomfret, R. 2008b. Tajikistan, Turkmenistan and Uzbekistan. In: Anderson, K., Swinnen, J. (eds.) *Distortions to Agricultural Incentives in Europe's Transition Economies*. World Bank, Washington DC, pp. 219–263.

Qadir, M., Noble, A.D., Qureshi, A.S., Gupta, R.K., Yuldashev, T., Karimov, A. 2009. Salt-induced land and water degradation in the Aral Sea basin: A challenge to sustainable agriculture in Central Asia. *Natural Resources Forum* 33 (2): 134–49.

Robinson, S., Wiedemann, C., Michel, S., Zhumabayev, Y., Singh, N. 2012. Pastoral Tenure in Central Asia: Theme and Variation in the Five Former Soviet Republics In: V. Squires (ed.) *Rangeland Stewardship in Central Asia: Balancing improved livelihoods, biodiversity conservation and land protection*. Springer, Dordrecht, pp. 239–74.

Schwilch, G., Liniger, H.P., Hurni, H. 2014. Sustainable Land Management (SLM) Practices in Drylands: How Do They Address Desertification Threats? *Environ. Mangt.* 54 (5): 983–1004.

Sommer, R., De Pauw, E. 2010. Organic carbon in soils of Central Asia – status quo and potentials for sequestration. *Plant and Soil* 338 (1): 273–88.

Spoor, M.N. 1999. Agrarian transition in former soviet Central Asia: A comparative study of Kazakhstan, Kyrgyzstan and Uzbekistan. *ISS Working Paper Series/General Series* 298: 1–29.

Spoor, M., Visser, O. 2001. The state of agrarian reform in the Former Soviet Union. *Europe-Asia Studies* 53 (6): 885–901.

Squires, V. 2012a. *Rangeland Stewardship in Central Asia: Balancing improved livelihoods, biodiversity conservation and land protection.* Springer, Dordrecht, 455 pp.

Squires, V.R. 2012b. Governance and the Role of Institutions in Sustainable Development in the Central Asian Region. In: *Rangeland Stewardship in Central Asia: Balancing improved livelihoods, biodiversity conservation and land protection.* Springer, Dordrecht, pp. 275–304.

Squires, V.R. 2012c. Replication and Scaling Up: Where to from Here? In: Heshmati, G.A., Squires, V.R. (eds.) *Combating Desertification in Asia, Africa and the Middle East: Proven Practices,* Springer, Dordrecht, pp. 445–59.

World Bank. 1998. *Aral Sea Basin Program (Kazakhstan, Kyrgyz Republic, Tajikistan, Turkmenistan and Uzbekistan). Water and Environmental Management Project.* World Development Sources, WDS 1998-3. World Bank, Washington DC. http://documents. worldbank.org/curated/en/1998/05/442217/aral-sea-basin-program-kazakhstan-kyrgyz-republic-tajikistan-turkmenistan-uzbekistan-water-environmental-management-project

World Bank. 2014a. *World Development Indicators 2014.* World Bank, Washington DC.

World Bank. 2014b. *Turn Down the Heat: Confronting the New Climate Normal.* World Bank, Washington DC. https://openknowledge.worldbank.org/handle/10986/20595

Part III

The nature and extent of land degradation in Greater Central Asia

Monitoring is an important part of the sustainable management of the predominantly dryland parts of Greater Central Asia. The three chapters here present the results of monitoring and evaluation using remote sensing and computer-aided techniques.

Aralova et al. provide a synoptic overview of the current status and trends in land cover and land use (two important indicators of sustainable land use) in Greater Central Asia. Changes in the vegetation as assessed by analyzing a time series of NDVI provide a foundation for their conclusions.

Feng, Yan and Cao discuss the value of remote sensing as a tool in the assessment of sustainable land management in Greater Central Asia. They use results from their fieldwork in northern and north-western China to demonstrate the application of the tools to convert satellite imagery to usable maps that have value to land users, land managers and planners.

Yang et al. report on efforts to mitigate the impacts of desertification and land degradation in the nine countries in the region. An overview of multilateral cooperation in Greater Central Asian region is provided.

7 Assessment of land degradation processes and identification of long-term trends in vegetation dynamics in the drylands of Greater Central Asia

Dildora Aralova

DRESDEN TECHNOLOGY UNIVERSITY, INSTITUTE OF PHOTOGRAMMETRY
AND REMOTE SENSING, DRESDEN, GERMANY AND SAMARKAND STATE
UNIVERSITY, LABORATORY "ENVIRONMENTAL PROBLEMS", UZBEKISTAN

Jahan Kariyeva

OFFICE OF ARID LANDS STUDIES, UNIVERSITY OF ARIZONA, TUCSON,
ARIZONA, USA AND ALBERTA BIODIVERSITY MONITORING INSTITUTE,
ALBERTA, CANADA

Lucas Menzel

INSTITUTE OF GEOGRAPHY, HEIDELBERG UNIVERSITY, HEIDELBERG,
GERMANY

Timur Khujanazarov

WATER RESOURCES RESEARCH CENTER, DISASTER PREVENTION
RESEARCH INSTITUTE, KYOTO UNIVERSITY, UJI, KYOTO, JAPAN

Kristina Toderich

INTERNATIONAL CENTER OF BIOSALINE AGRICULTURE, ICBA-DUBAI,
TASHKENT OFFICE AND SAMARKAND STATE UNIVERSITY, UZBEKISTAN

Umut Halik

FACULTY OF MATHEMATICS AND GEOGRAPHY, CATHOLIC UNIVERSITY
OF EICHSTAETT-INGOLSTADT, EICHSTAETT, GERMANY AND KEY
LABORATORY OF OASIS ECOLOGY, COLLEGE OF RESOURCES AND
ENVIRONMENTAL SCIENCE, XINJIANG UNIVERSITY, CHINA

Dilshod Gofurov

SCIENTIFIC RESEARCH INSTITUTE FOR COTTON BREEDING, SEEDING AND
CULTIVATION AGRO-TECHNOLOGIES, KIBRAY, UZBEKISTAN

Introduction

The Greater Central Asia's (GCA's) rangelands[1] are extensive (Squires & Lu, 2018a; Orlovsky & Orlovsky, 2018) and, in common with other semi-arid and arid regions, their utilization is strongly dependent on the primary production and soils from natural ecosystems (Kharin, 2002).

The United Nations Conference on Environment and Development in 1992 reported that desertification and land degradation in arid, semi-arid and dry sub-humid areas were resulting from various factors, including climatic variation and human activities. The dryland rangelands systems' dynamics are driven by precipitation, which can be low, erratic and highly variable in space and time, while also highly dependent on anthropogenic activities such as grazing and converting lands to agriculture. Management of these rangelands needs to be based on an understanding of their essential productivity and variability and also their dynamics and responsiveness to natural and human influences (Feng, Yan & Cao, 2018).

In the period of 1992–2011 (after post-Soviet), an annual rate of land conversion in the five 'stans' was 11.84%.[2] Land use change is a serious problem because rangelands, wetlands and other natural areas were converted to agriculture, e.g., croplands and pastures. During the Soviet era, vast areas were converted to agriculture; while some have been successful, at the same time many areas were put under serious environmental pressure without rehabilitation. This rapid change in land use has triggered and contributed to land degradation processes. The increased area of land susceptible to degradation in the GCA region has increased under the impact of several environmental factors, e.g. a faster rate of glacial retreat in recent years in the region and a shorter return period of droughts in GCA, the north-western part of China and western Mongolia. Climate anomalies have already been influenced by the faster melting of glaciers, one of the main sources of water resources in all these regions; these include, but are not limited to, the Zerafshan glacier, the Abramov glacier, the glaciers on the Tibetan Plateau (Cruz et al., 2007), in the Tian Shan of Xinjiang, China, and along the Qilian Shan in the Hexi corridor of China (Lu, Wang & Wu, 2009). There has been a decrease in mean precipitation in some observed areas in recent decades. Although evidence of climate-related biodiversity loss remains limited, a large number of plant and animal species are reported to be moving to higher latitudes and altitudes as a consequence of observed climate change in recent years (Yoshio & Ishii, 2001; Squires & Safarov, 2013).

The level of degradation of grasslands/rangelands in GCA has deteriorated significantly (Squires, Shang & Ariapour, 2017) and there is an urgent need to set up mechanisms to reverse degradation processes driven by potential anthropogenic and climate change impacts. GCA has a large fraction of the world's drylands, which are known to be vulnerable to climate change (Zhou et al., 2015, Li et al., 2015). Potential threats to arable land resources because of anthropogenic activity and climate anomalies have been recognized.

Although precipitation and temperature patterns are the most important drivers for dryland ecosystems, human activities, such as overgrazing, overplanting and deforestation, combined with natural processes, are the main causes of desertification and land degradation. Climatic stress contributes to an overall loss of valuable biodiversity on a large scale. Sound management of rangelands focusing on carbon stocks is playing an indispensable role and highlights a need for enhanced understanding of nonlinear interactions between vulnerability induced by human impact versus climate-vegetation trends in the GCA arid regions (Shang et al., 2016). However, the carbon stock of saxaul (*Haloxylon aphyllum*) woodlands (the main woody plant in the deserts of Central Asia) has a low potential to accumulate carbon when compared with other ecosystems of the world (Thevs et al., 2013), but this is the only available way to sequester carbon. Particular attention to rangeland management is required in these areas to prevent further land degradation, deforestation and overgrazing and to support the conservation of biodiversity (Squires & Safarov, 2013). Cumulative effects of climatic stress and anthropogenic pressure contribute to increased rates of biodiversity loss locally and regionally. Rising population and increasingly frequent climate anomalies foster initiatives to develop arid territories as 'grain and/or fodder productivity zones' and thus present problems in counterbalancing agricultural development against conservation management. When determining priorities, it is important to consider the value of rangelands for maintaining essential ecosystem functions. These values may rely on conserving key species (defined in terms of their greater influence on the functioning of ecosystems). Furthermore, the conservation of rangeland ecosystems is essential for ensuring the provision of multiple goods and services, including fuels, fiber, food, feed, meat, water, etc., to conserve biodiversity and, in addition, to preserve the significant scenic beauty of landscapes.

Monitoring rangeland vegetation under these harsh conditions, especially in the cold arid regions of the Tibet Plateau and High Asia generally (Kreutzmann, 2012), is a very complex task on the ground, while remote sensing is the most promising technology for monitoring remote areas to assess the extent and quality of vegetation cover and to build a valuable baseline database for support to develop sustainable regional management scenarios. Remotely sensed data can contribute significantly to the mapping, monitoring and management of land resources with respect to developing economic infrastructure (livestock, meat, wool, crops). The data offer synoptic views of vast landscapes, repetitive imaging and accumulated data archives that allow the monitoring of short-term changes and longer-term trends.

Background to the use of remote sensing technology and the rationale for using it

Remote sensing approaches have been extensively used in many scientific outputs: data products such as local area land cover change, global area coverage, monthly composite climate parameters, and Normalized Difference Vegetation Index (NDVI) composites (Huang et al., 2013). There is considerable international experience and documentation of data calibration and standardization procedures, and many have been well tested throughout the world. Because of its frequency, breadth of coverage an d cost, National Oceanic and Aeronautical Administration (NOAA) Advanced Very High Resolution Radiometer (AVHRR) data are an ideal candidate for broad-scale inventory and monitoring. The AVHRR infrared channel (912.5 nm) at 1 km resolution offers information on local land cover (LC) and the 4 km resolution for global area coverage.

Figure 7.1 is an example of vegetation classification at a regional scale based on 12 months of the greenness index (NVDI). The NDVI index takes advantage of the contrast of the characteristics of two bands from a multispectral raster dataset – the chlorophyll pigment absorptions in the red (R) band and the high reflectivity of plant materials in the near-infrared (NIR) band. An NDVI is often used worldwide for monitoring, predicting agricultural production and drought anomalies, assisting in predicting hazardous fire zones, and mapping desert encroachment. The NDVI is preferred for global vegetation monitoring because it helps compensate for changing illumination conditions, surface slope, aspect, and other extraneous factors (Lillesand et al., 2004).

Figures 7.2a, 7.2b and 7.2c illustrates NDVI values obtained and extracted for GCA, and the black lines outline the area of the five former Soviet 'stans' boundaries (Kazakhstan, Turkmenistan, Uzbekistan, Tajikistan, Kyrgyzstan), Western China, and Mongolia. Many dryland sites in these countries are usually bare areas such as sandy deserts or steppes supporting sparse vegetation.

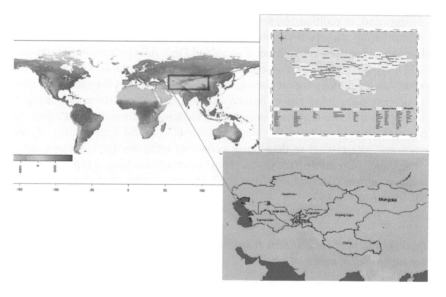

Figure 7.1 Description of Asian vegetation using 12 months of averaged NDVI values (1982–2011) to show the value of archived data that can provide a valuable 'benchmark' against which change can be assessed. The data are at 8 km spatial resolution from GIMMS 3g source. The selected rectangle with the red line is the targeted zone for location of GCA

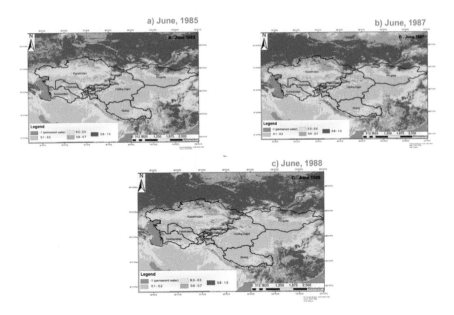

Figures 7.2a–c NDVI values classified as 0.1–0.2 (as low vegetation), 0.3–0.5
(moderate vegetation), 0.6–0.7 (high vegetation) and 0.8–1.0
(evergreen vegetation). Images (7.2a–c) scaled as various years (June
1985, 1987 and 1988) by means of NOAA AVHRR (NDVI3g) data.
Changes in the area with low (yellow) and moderate (brown) vegetation
cover are shown

Each of the vegetation types is identified in the global cover index (Figures
7.2a, 7.2b and 7.2c) and can be characterized in terms of its seasonal greenness
signature. A considerably higher annual production occurs in northern parts, with
growth peaks as boreal forest zones. The bimodal peak in production is displayed
by many of the regional vegetation types, and is worthy of examination in relation
to rainfall records.

The phenologically defined vegetation types can be characterized by total
productivity (using greenness as a proxy for net primary productivity or NPP),
sometimes defined by integration of the area under the curves, or by approximat-
ing the sum of monthly NVDI values, and phenology here defined as the annual
amplitude of greenness. However, change detection techniques do not always fully
account for changes in biophysical parameters and processes (Lambin, 2000),
therefore there was a need for more comprehensive analyses of the change processes for selected ecosystems. In addition, there are data archives which might
be used in long-term studies of Land Use Land Cover (LULC) change dynamics
and trends. The NDVI has been correlated (Feng, Yan & Cao, 2018) with leaf area
index (LAI), biomass of vegetation and NPP.

Study areas

In this chapter semi-arid and arid lands of GCA we focus on the following five former Soviet republics – Turkmenistan, Uzbekistan, Kazakhstan, Tajikistan and Kyrgyzstan – plus Mongolia and part of Western China – Xinjiang and Tibet (Xizang). Figure 7.3 shows all the selected countries with administrative borders and geographic coordinates. The selected western part of China borders with Tajikistan, Kyrgyzstan and Kazakhstan and is near Pakistan and Afghanistan.

The goal of this study is to understand variability, long-term trends and changes in vegetation on selected target areas at a host of spatial scales over the past 30 years. NDVI values have been scaled to values ranging from −1000 to 1000, water pixels are assigned the value of −10000, and masked pixels are −5000 (Figure 7.1).

Year-by-year population has increased in Uzbekistan, Tajikistan and most of the central part of China, whereas the western regions of China (Xizang, Xinjiang) are less densely populated (Table 7.1) due to the harsh, rugged and/or arid terrain conditions. With the majority of the population of these countries living in rural areas, as shown in the population density column, desertification and land degradation are taking a heavy toll on their ability to survive.

Climate change and global warming have already affected some parts of the region, leading to an increase in the number of 'ecological refugees'. The accelerating

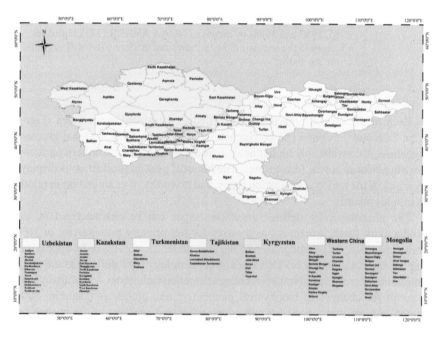

Figure 7.3 Geographic locations of the selected zones (no water sources shown). The relative size of the countries in GCA are shown (see color code)

Source: DIVA-GIS, administrative boundaries.

Table 7.1 Geographical extent, demography of population, growth rate and density (2012)

Country	Area (km²)	Population (2012)	Population density* (per km²) (2012)		Population growth rate (%)
Kazakhstan	2.724900	1.7948816	6.3	(0.47)	1.14
Kyrgyzstan	199900	5.604212	27.8	(0.64)	1.04
Tajikistan	143100	8.052512	55.9	(0.74)	1.88
Turkmenistan	488100	5.171943	10.5	(0.51)	1.41
Uzbekistan	447400	30.185000	67.5	(0.64)	0.93
Mongolia	1.565000	2.796000	1.76	(0.31)	1.37
Xizang	1.228400	n/a	2.42	(0.90)	n/a
Xinjiang	1.664897	20.523090	2.75	(0.47)	1.28
Western China#	2.867269	24.312166	7.145	n/a	n/a

Notes: * Rural population proportion is shown in parenthesis.

Western China is a composite of several provinces and autonomous regions. Estimates are provided.

loss of biological biodiversity over the last decades presents a threat to humans too. For example, the shrinking of the Aral Sea and land degradation in the northern part Karakalpakstan poses risks to human health (Glantz, 1999; Saiko, 1998, 2001; Orlovsky & Orlovsky, 2001).

Overview of climatic and rainfall conditions in core study areas

The dryland zones in GCA (as defined earlier) have mostly a temperate continental arid climate with very hot (dry) summers and cold winters; total precipitation ranges from 0 to 300 mm. A vast majority of the land in the area is low lying at less than <300 m, but there are also some peaks of more than >5000 m. It is important to evaluate how climate has varied and changed in the past for focused regions. The monthly mean historical rainfall and temperature data can be mapped to show the baseline climate and seasonality by month, for specific years, and for rainfall and temperature. The charts (Figure 7.4) show mean historical monthly temperature and rainfall for focus regions during the time period 1901–2015. The dataset was produced by the Climatic Research Unit (CRU) of the University of East Anglia (UEA). Precipitation and temperature patterns are most important for understanding dryland ecosystems, which are strongly dependent on these two factors. Cumulative effects of climatic stress and anthropogenic pressure contribute to accelerate rates of biodiversity loss at the local and regional level.

Gridding precipitation in core study areas

As Figure 7.4 shows, the mean annual precipitation on the lowlands is less than 10–130 mm/year in Xinjiang and Mongolia, and part of Xizang province. Countries around the Aral Sea Basin also have less than 100–130 mm, while north

140   *Dildora Aralova et al.*

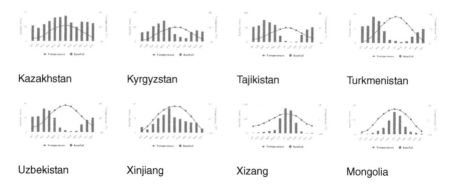

Kazakhstan   Kyrgyzstan   Tajikistan   Turkmenistan

Uzbekistan   Xinjiang   Xizang   Mongolia

Figure 7.4 Average historical monthly rainfall and temperature for the GCA countries (1900–2015)

of the Karakum Desert and part of Turkmenistan are in the range of 130–270 mm/year. Precipitation is higher in the eastern part of Uzbekistan (200–320 mm/year), while in Kazakhstan annual precipitation is 90–200 mm/year. In the mountainous areas in parts of Tajikistan and Kyrgyzstan, it is above 550 mm/year and reaches 500–700 mm/year. Most lowlands and mid altitudes of Uzbekistan and southern Tajikistan are under a Mediterranean climate with winter precipitation

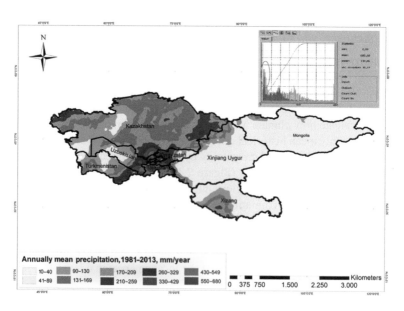

Figure 7.5 Overview of mean annual precipitation trends (mm/year) for selected target zones for the period 1981–2012. Data for Xinjiang and Mongolia are incomplete but some areas receive as little as 50 mm

Source: CRU-TS, vers.3.23.

and summer drought (Le Houerou, 2005). Higher precipitation is demonstrated in Tajikistan. When mapping the precipitation, some gaps in the data were noted during the period 1981–1991; these may be due to the use of uncorrected or raw corrected data for selected target areas.

The impact of environmental drivers such as precipitation can be used to explain the spatio-temporal variability of vegetation phenology in this selected area. Three core landscapes types were identified: desert, steppe, and mountainous zone.

Material and methods

Datasets for historical climatic data

This new version (vers.6) of the database GPCC (Global Precipitation Climatology Center) is for 1901 to 2012, globally at 0.5 degree spatial resolution on land areas. We have selected and extracted for the period of 1981–2011, for preliminary analysis or to predict further solid earth (water bodies excluded) vegetation patterns compared with temperature and NDVI datasets. To help facilitate the use of this database for the monitoring of each country, we applied a GIS and R software package to convert the raw data into the raster format and calculated trend analysis with R programming. Much of the observational data is updated monthly for each country.[3] For precipitation mapping, our approach depends on a resampling method that uses a weighted average of the four nearest cells to determine new cell values under bilinear interpolation.

Satellite-based vegetation indices data (GIMMS NDVI3g)

As the goal of this study is to understand variability, we studied long-term trends and changes in vegetation patterns on selected target areas at a host of spatial scales over the past 30 years. The earliest version of the GIMMS NDVI dataset spans 1981–2006 and the latest period is July 1981 to December 2011 and is termed NDVI3g (third generation GIMMS NDVI from AVHRR sensors), with spatial resolution of 8 km. The data are carefully assembled from different AVHRR sensors and accounts for various issues, such as calibration loss, orbital drift, volcanic eruptions, etc. (Pinzon & Tucker, 2014). To convert the NDVI range of −1 to 1 we used the following formula:

$$NDVI = float\ (raw/10000) \tag{1}$$

Theoretical NDVI range is between −1 and 1, with values around 0 for bare soil (low or no vegetation) and values of 0.7–1.0 or higher for dense vegetation (forest zones). Water = (−0.1), no data = (−0.05) (Figure 7.2b). In the majority of resampling raster data, calculated with our approach, the values and output were calculated for each cell. We then used a 2x2 neighborhood rule of the input raster to check error output. The applied approach does not create any new cell values, so it is useful for resampling categorical land use data. Datasets were resampled; this acts as a type of low-pass filter for discrete data, generalizing it and filtering out anomalous data values.

Results and discussion

Seasonal change dynamics of Asiatic rangelands at landscape scale

Deserts cover more than 30% of Western China's territory (as defined here), and the areas threatened by desertification amount to over 25% of that region's total landmass, while the former Soviet republics (the 5 'stans') plus Afghanistan consist of 295 million hectares (Babaev, 1999). The winter cold deserts of Central Asia cover an area of 2.5 million km^2 (Thevs et al., 2013). Mongolia has about 69% of its land area classified as desert (Dorj et al., 2013; Batkhishig, 2013).

In this part, we describe the trend analysis of precipitation and NDVI dependency for targeted zones. Cross-tabulated analysis has revealed significant evidence of change in the dynamics between the land cover classes of grasslands/rangelands or sparsely vegetated and soil surfaces covered with low vegetation associated with salinization, which is widespread in the dry steppe zones (Toderich et al., 2013), or accumulation resulting from a very low precipitation and absence of moisture during the warming period. For selected zones, the total mean of precipitation is equal to 110 mm/year. A statistical overview for accumulation precipitation describes it as 0–100 mm/year for GCA countries. As precipitation is equal to 100–110 mm/year, the covariance map shows that core values of NDVI are estimated between 0.04–0.34 (low precipitation/low vegetation). A covariance map was developed through a simple kriging technique for guided sampling processing of data averaged for GCA countries. The outcomes of the covariance method used with certain datasets were mapped.

The covariance map (not shown) indicated dependency $NDVI_{value/year}$ + $Precipitation_{mm/year}$ for estimating extensive coverage of vegetation patterns under various precipitation anomalies. Areas with high vegetation density had values between 0.53 and 073 but covered only a small area within the targeted zones. A vast area was occupied by low vegetation densities in the range of 0.05–0.14 values.

The trend towards larger bare areas between plants and sparser vegetation

In the calculation of bare areas and sparse vegetation we had to distinguish between natural and artificial (i.e., man-made) surfaces (infrastructure, roads, mines, settlements). We analyzed population density (Table 7.1) as a factor but noted that for further forecasting such data are not suitable indicators. On the other hand, there was a positive correlation between population dynamics and vegetation change. The objectives were to examine the spatio-temporal patterns of land use/cover dynamics and quantify the rate and direction of these dynamics.

We have concentrated on bare areas and their spread, and on areas with sparse vegetation. Two major land use/cover patterns – namely bare areas and sparse vegetation (>15%) – were identified and we calculated differences in vegetation change between the years 2000 and 2009 using MODIS images (Figures 7.6a and 7.6b). Applying a linear trend analysis for the period 1982–2011 has revealed several hotspot zones with ongoing desertification (Xinjiang, Turkmenistan, Mongolia) and slight ongoing regeneration in the eastern part of Uzbekistan: i.e., results show that desert areas have expanded and now cover approximately 11% of Mongolia. Mining areas are an important factor in the creation of bare land. In total, 38.5 million ha or 24.6% of the whole country in Mongolia is covered by mining leases. Large land areas are destroyed by mining operations and/or by mining surveys and prospecting (Batkhishig, 2013).

Mapping degradation values and factors of potential decreasing of greenness

For estimating degradation values of dryland ecosystems, we selected two criteria for analysis of ongoing processes, based on NDVI category values extracted categories: *bare areas* (Figure 7.6a) and *sparse vegetation* (Figure 7.6b) (Ricotta et al., 1999). Initial results indicated changes of ecosystems (regeneration) in the green belt of Kazakhstan; also, Kyrgyzstan had slightly higher NDVI values compared with Tajikistan and Turkmenistan (Figure 7.6b). Given the potential

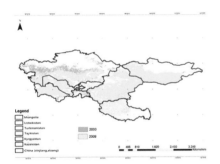

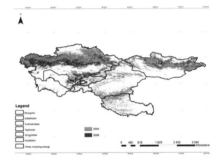

Figure 7.6a Description of bare areas from 2000 (pink) and 2009 (yellow) in the drylands of Greater Central Asia. Modified under MODIS time-series data (2000–2009) with extraction one legend (bare areas) under classification IGBP

Figure 7.6b Description of sparse vegetation (>15% cover) on the base NDVI index from 2000 (red) and 2009 (green) in the drylands of Greater Central Asia. Modified under MODIS time-series data (2000–2009) with extraction one legend (sparse vegetation coverage, where covered solid earth pattern ≤15%) under classification IGBP

relevance of these issues and access to NDVI status, overall agricultural areas (arable land) have sharply decreased (≈4.14%) after the independence of these five 'stans' (1993–2011), but not in China and Mongolia. In most cases in these countries, land abandonment due to high salinity levels is common. Annual changes of land use are equal to ≈0.22% from the total.

The land cover types of drylands (Figure 7.7) were identified under bare lands and we calculated changes for the period 2000–2009; no changes were observed in Mongolia, Western China and Uzbekistan; slight changes were observed in the northern part of Kazakhstan and parts of Tajikistan (Figures 7.6a and 7.7). There was an expansion of sparsely vegetated areas between 2001 and 2009 (Figure 7.6b). Bare areas covered approximately 6.4% of Kazakhstan; 29.3% of Kyrgyzstan; 30.6% of Tajikistan; 23.9% of Turkmenistan; 26.8% of Uzbekistan; 53.9% of Mongolia; and 12.8% of Western China. The calculation is not related to the size of the country. Whilst Kazakhstan is five times bigger than Tajikistan and Kyrgyzstan, areas of serious degradation in Kyrgyzstan and Tajikistan are five times higher compared with Kazakhstan. Both Kyrgyzstan and Tajikistan are located in mountainous zones, and Kazakhstan is located in the open lowlands. In addition, naturally rocky mountain areas are classified as bare areas. As mentioned before, vegetation anomalies observed in the southern part of Mongolia were much higher than in other targeted zones, while the northern area was in the high-density category (Figure 7.7). Vegetation cover of sandy rangelands has a mosaicking spatial structure, determined mostly by the distribution of fixed and mobile dunes.

Land Use and Land Cover (LULC) change

During the analysis period, a considerable decline in vegetation cover was observed in 2001 (Figure 7.8) in Western China (Xinjiang), Turkmenistan and Tajikistan. Change of land cover classes were gradual in the study area in GCA, being more noticeable especially in Turkmenistan and Tajikistan, and in Western China (Xinjiang). Significant negative trends in vegetation greenness continued for large parts of GCA (Figure 7.9). Since the beginning of the 20th century, the situation in Mongolia has been well documented in the *Desertification Atlas of Mongolia* (Institute of Geoecology, 2014), developed in conjunction with the Swiss Agency for Development and Cooperation and the Environment Information Center of Mongolia. The *Atlas* could be a useful model for each of the countries in GCA to follow.

Based on the geographic location of each country, we assumed that our selected sampling zones were adequate to detect changes in the areas with sparse vegetation, with the exception of Kyrgyzstan and Xinjiang region of Western China. Xizang region has large areas classified under the *open grassland*s category, while Xinjiang has significant areas under *irrigated croplands* (Figure 7.7). Agriculture is practiced along the rivers and characterized by monocultures of various cash crops. In rain-fed zones of northern Kazakhstan, dominant crop types include wheat, oats and barley, while along the Amudarya and Syrdarya rivers (southern

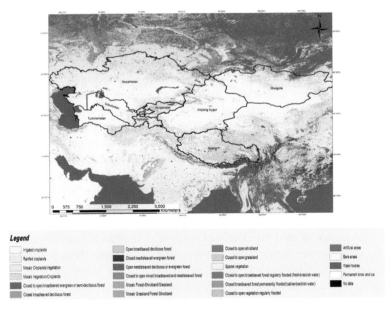

Figure 7.7 Land Use and Land Cover map (LULC) for the former Soviet republics, Western China and Mongolia (2001–2009). Modified under MODIS data with classification scheme of IGBP

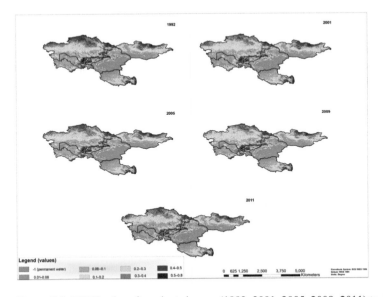

Figure 7.8 NDVI values for selected years (1992, 2001, 2005, 2009, 2011) to describe vegetation change dynamics in GCA. Red, brown and yellow colors indicate low vegetation cover

Source: Time series of bi-monthly NDVI3g data, AVHRR-GIMMS.

Kazakhstan, Uzbekistan and Turkmenistan) vast areas are used for irrigated crops (e.g., cotton and rice) (Figure 7.7 – eastern side). Xinjiang also produces cotton and rice along the Tarim River and in the Junggar basin.

The demographic data show high density population in Uzbekistan and Tajikistan (Table 7.1) and this is strongly correlated with increasing areas of artificial surfaces, especially in Uzbekistan and Western China (Figure 7.9), while artificial surfaces serve as indicators of a country's economic development as well as population growth. Kazakhstan and Uzbekistan have major cities with urban development and construction; meanwhile, there is an indicator factor of permanent vegetation reduction (cutting shrubs, felling trees for construction). According to our data, the total urban area in Western China is 0.16%; for Kazakhstan it is 0.18%, and it is noticeably higher for Uzbekistan (0.24%). Neeti et al. (2012) mentioned that patterns of agricultural intensification, along with urban expansion, have brought changes in seasonal patterns of greenness. Ecological frameworks for assessing land degradation have used changes in the area of artificial surfaces as an indicator of extension of land degradation. Commonly, vegetation is destroyed and coverage and density are reduced for built-up artificial surfaces. These profiles or indicators were used to predict how urbanization and invasive plants could spread by occupying currently natural and untransformed areas with the same characteristics as those currently affected and displaying degradation status.

Over the last two decades, plant responses to climate change have appeared to be somewhat straightforward: increasing temperatures and longer growing seasons may have had an effect but our study was not designed to measure plant productivity. A study published by the US Global Change Research Program suggests that recent higher temperatures lead to the evaporation of moisture from soils, thereby increasing the frequency, intensity and duration of droughts in the region. There have been reports from land users about an alarming decrease in the frequency and depth of snowfall over the past few years as well as the consistent rise in temperatures being witnessed over the recent years.

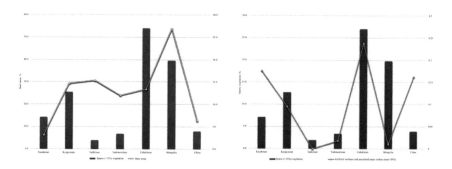

Figure 7.9 Potential association of sparse vegetation with bare areas (left) and artificial surfaces in the GCA countries (right). Modified under legends MODIS–IGBP data. Bars indicate the extent of sparse vegetation (by country) and the line the % of bare land

Geostatistical analysis of vegetation conditions among GCA countries (1982–2011)

The analyses were conducted using geostatistical extensions of the ArcGIS (v.10.x ESRI Inc., Redlands, CA, USA) with simulation layer stacking bi-monthly NDVI3g data and certain datasets of climate (temperature, precipitation). Then we utilized an ordinary kriging/cokriging method for each of the selected regions and combined them. Ordinary kriging is responsible for a random function model of spatial correlation to calculate a weighted linear combination of selected datasets to predict the response for an unmeasured location. This approach is useful to develop predictive vegetation map status for vast areas (in our case, the Central Asia *sensu stricto* and Western China and Mongolia), and these trends clearly described the ongoing status of vegetation conditions at present (1982–2011).

A cumulative analysis of NDVI values shows that vegetation accumulation started going up from 0.07 values and maximized at 0.78 values. A high density of accumulation is from under 0.09 to 0.36 values, meanwhile it means that 0.09–0.36 are the main values for accumulation indicated in these regions. For all the regions we can say that a low vegetation cover prevails in these areas and skewing density accumulated around values of 0.07; while the second tile is not reached even up to values of 0.30. Accumulation of greenness values is very low due to the fact that selected areas are located in arid and semi-arid environments. Also, results illustrate that the vegetation index for areas in Xinjiang, part of Xizang, Kazakhstan, and the northern part of Mongolia's non-boreal zones ranges from 0.07 to 0.14.

Based on the analysis of NDVI time-series and climatic data, we assume that vegetation development is more sensitive to precipitation anomalies, for nearly 70–80% of the arid and semi-arid zones of the land surface in the study area, biomass production is driven by precipitation (amount and distribution). Generally, desert vegetation is dependent on seasonal precipitation (Aralova et al., 2015), which is demonstrated by a rapid green-up and plant growth during the rainy period, but during hot summer periods many semi-shrubs and shrubs lose their leaves to reduce evaporation rates and may again sprout new leaves in the beginning of the cooler autumn season. For these reasons, near-infrared (NIR) and red (R) reflectance will be lowest in the summer period in the natural environments of the arid and semi-arid zones of GCA, especially in Xinjiang. Note that in Mongolia and Xizang (Figure 7.4), there is summer rain in June, July and August and warmer temperatures favorable to plant growth: i.e., the average daily summer temperature in Xizang is 10°C and 20°C in Mongolia.

However, NDVI can be used as an indicator of biomass and greenness (Boone et al., 2000; Chen & Brutsaert, 1998; Kariyeva & van Leeuwen, 2012). For all the study regions we can say that a low vegetation cover prevails in these areas and skewing density accumulated around values of 0.07; while the second tile is not reached even up to values of 0.30. According to Kariyeva et al. (2012), phenological dynamics of irrigated and natural drylands of the five 'stans' showed variability in regional scale vegetation response (phenology) due to altered land

and water use patterns following the 1991 collapse of the USSR and changes of institutional and governmental systems in these countries (Squires & Lu, 2018a).

Assumed changes of vegetation regarding average values of NDVI for the last three decades in GCA countries (1982–2011)

Finer scale analysis reveals trends that are more relevant to human decision making. It is possible that regional economics have played a major role in guiding investment. In areas where irrigation water is available, more money is invested for planting agricultural crops (including greenhouse crops) to meet demand from the expanding urban populations and also to ensure food security. Slight changes of vegetation communities under overgrazing and the long-term use of lands already under pressure (without rotation) caused degradation processes (erosion, salinity, soil structure decline, and loss of vegetative cover) on the rangelands (Qi & Evered, 2008). A time-series analysis that enables a review of trends and provides identification of hotspot areas gives more opportunity to develop strategic plans for soil and vegetation cover to ensure their healthy status. All data derived from remotely sensed imagery will need to be confirmed on the basis of ground truth data (Feng, Yan & Cao, 2018). This is a major challenge for the generally poorly equipped and inadequately financed land management agencies in most of GCA. It also provides an incentive for China in particular to share its experience, baseline database and personnel, and for greater cooperation of information to occur under CACILM II (see Squires & Lu, 2018b; Yang et al., 2018) for a discussion on international and cross-border cooperation.

Potential ecological threats: remnant vegetation – values and threats in GCA

Rapid demographic change and climatic cataclysms (extension of hot dry periods over recent decades) will see raising awareness about the implications of ecosystem and biodiversity losses in GCA countries and the need for trans-boundary cooperation and coordination, especially in the sharing of data derived from the application of remote sensing and other tools. Figure 7.4 demonstrates distinct temporal patterns of bioclimatic synchronies for different land use patterns; Figure 7.8 shows that the northern part of GCA has mostly denser and more intact vegetation cover zones and the southern part demonstrates the inverse of this condition. Many factors must be considered when defining scenarios for changing threats to biodiversity in this southern area: e.g., the fact that many people now live in conditions of poverty and in resource-poor environments. This increases the risk of mass migration as 'ecological refugees'. These developments put increasing pressure on the environment. Higher awareness by authorities (e.g., those responsible for natural resource management, in particular) and access to the tools described in this chapter could help in gaining an understanding of the consequences for both natural and human systems, particularly as they might affect the allocation and availability of resources to these systems.

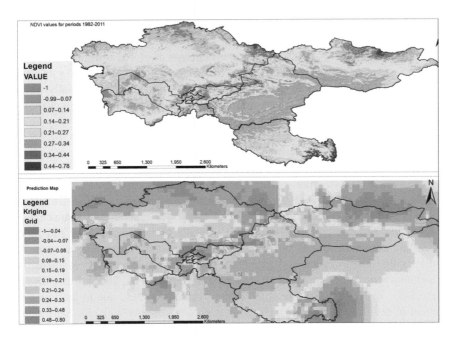

Figure 7.10 Assessment of the current status of vegetation value (top) and predicted changes of vegetation values (below) in upcoming years. Kriging methodology of vegetation with mapping GIMMS NDVI3g and temperature dataset (1982–2011)

In the past, extensive agricultural development resulted in native vegetation being cleared across vast areas of GCA, especially converting desert and steppe zones to an agricultural sector (Bekchanov et al., 2018), which caused serious issues (Orlovsky & Orlovsky, 2018) leading to the Aral Sea Crisis (ASC). Extensive land use change of the fertile lands of Uzbekistan and Turkmenistan and part of Kazakhstan and Tajikistan has occurred. About 30–40% of the native vegetation in arid and semi-arid lands has been removed and there is continuing pressure on remnant vegetation. Native vegetation is no longer commonly seen, and agricultural intensification, including the expansion of irrigated horticulture into areas that traditionally practiced dryland grazing and cropping enterprises, has led to continuous pressure on remnant vegetation. Many studies are already demonstrating other serious potential threats besides the ASC. The Aral Sea Basin continues to be desiccated, on the boundaries of both Kazakhstan and Uzbekistan.

We assumed that core potential threats to remnant vegetation will occur in Xinjiang, China, as population steadily rises and conversion to irrigated cropland accelerates (Figure 7.10). We are led to this conclusion by a centered sampling point (calculated as the difference between estimated trend values (NDVI) and higher negative trends) in the Xinjiang area. There is also a noticeable shift from high vegetation values to very low vegetation values in the central part of Xinjiang. Based on this analysis, we are able to identify the current 'hotspot' zones as the zones in central Xinjiang, after Mongolia (southern side). Some studies assumed that more rain and less snow is also one indicator of global warming issues. In those regions where there are likely to be fewer snow days but more rain days per year, conditions will be more optimal for revitalizing lands, especially stable sands during the period of water shortage (seasonal irrigation and high evaporation), for both the five 'stans' and Afghanistan.

Discussion

Despite such serious environmental problems as the shrinking of the Aral Sea, which lies between Uzbekistan and Kazakhstan, a worsening environment/situation is expected and occurring in Xinjiang, which will experience climate change anomalies and low rainfall (Figure 7.4) annually (during April, May and June; monthly accumulated ≥20 mm). Because the ecosystems are so fragile, some of the applied interpretation and prediction datasets and tools could be unfeasible and impractical, e.g. the Sahel ecosystem (Rasmussen et al., 2001). Currently, in the Sahel ecosystem they are able to detect more greenness due to the involvement of several donors with such projects as 'Great Green Wall' for Sahara and Sahel, which has planted over 3 million trees (3,746,777 as of 8 March 2016), mostly in the Burkina Faso area (Ecosia Gmbh). After several decades of vegetation loss due to climate and human-driven pressure, ecosystems in a vast area in the western part of the Sahel (or West Sahel), which was previously identified as a high desertification zone, are now changing and demonstrating rapid recovery, with greenness after an increasing number of rainy days. As stated by Eklundh and Sjöström (2005), after long-term drought events and when annual precipitation has increased, the general recovery of the vegetation communities is faster in semi-arid environments.

Restoring degraded land is a revolutionary step for those areas in GCA. Due to economic situations, in addition to intensive agriculture activities, there is also a heavy reliance on extracting natural resources: e.g., oil and gas in Kazakhstan, Turkmenistan and Uzbekistan, and mining gold and other minerals, e.g. in Kyrgyzstan (Orlovsky & Orlovsky, 2018; Squires & Lu, 2018a). These industries and the associated infrastructure (mines, pipelines, roads, railways, water diversions, mine waste disposal, etc.) are an additional potential danger and threat to fragile ecosystems.

The GCA countries are at different stages of economic development and technological advancement, and possess varying levels and types of technical expertise, experience and human capital. The natural resource sectors of each country have common as well as unique problems. Policy dialogue, pilot demonstration, capacity building and information sharing of experiences and transfer of technology have begun and are likely to accelerate in pace over the coming decades. Although the extent of global change impacts on natural and human systems depends largely on adaptation capacities, not every societal approach to adaptation is mitigating or focused on sustainability. Projections averaged across a suite of climate and vegetation models show a progressively increasing drought risk across much of the GCA region.

Practical solutions are needed:

- establishment of planting to revitalize ecosystems in the eastern part of the selected area (Xinjiang, southern part of Mongolia, border zones of Xizang (north) with Xinjiang (south part), Turkmenistan (middle part) and Uzbekistan (Aral Sea Basin));
- maintenance and management of native vegetation planting for the long term to meet sustainable objectives for local people, the landscape and wildlife; and
- improved land and transboundary water management practices (river basin management).

The adoption of the global Goals for Sustainable Development (15 goals) to become land degradation neutral by 2030 (Wang, Squires & Lu, 2018) poses many challenges for countries in GCA. Only by getting outside help, such as that to be offered through the recently reinvigorated CACILM/GEF project and from China through the State Forest Administration and APFNET (Yang et al., 2018), will the required capacity building (human, financial and technical) be achieved in a short time.

Acknowledgments

This research was supported by Deutscher Akademischer Austauschdienst (DAAD) for young academics to pursue a degree abroad. We are grateful to Dr. Victor Squires, Adelaide, Australia for his help with the preparation of the English version of this chapter, and also to Professor Elmar Csaplovics (TU-Dresden) for his potential interest in this area and willingness to share accumulated knowledge and experiences among foreign students at the Institute of Photogrammetry and Remote Sensing, TU-Dresden.

Notes

1 Rangelands, in international terminology, refers to natural areas of grassland, shrubland and steppe.
2 Source: FAOSTAT, http://faostat.fao.org
3 Source: KNMI Data Center, Netherlands, https://data.knmi.nl.portal

References and further reading

Aralova, D.A., Toderich, K., & Osunmadewa, B. 2015. Spatial Distribution Patterns of Vegetation Cover in Deserts of Central Kyzylkum with application of Vegetation Indices (VIs). *Journal Arid Land Studies* 25 (3): 265–8.
Babaev, A.G. 1999. *Desert problems and desertification in Central Asia: The researches of the Desert Research Institute.* Springer, Berlin, 293 pp.

Batkhishig, O. 2013. Human impact and land degradation in Mongolia. In: J. Chen, S. Wan, G. Henebry, J. Qi, G. Gutman, S. Ge, & M. Kappas (eds) *Dryland East Asia: Land dynamics amid social and climatic change*. Higher Education Press, Beijing, pp. 265–80.

Bekchanov, M., Djanibekov, N., & Lamers, J.P.A. 2018. Water in Central Asia: a cross-cutting management issue. In this volume, pp. 211–36.

Boone, R.B., Galvin, K.A., Smith, N.M., & Lynn, S.J. 2000. Generalizing El Niño effects upon Maasai livestock using hierarchical clusters of vegetation patterns. *Photogrammetric Engineering and Remote Sensing* 66 (6): 737–44.

Chen, D., & Brutsaert, W. 1998. Satellite-sensed distribution and spatial patterns of vegetation parameters over a tallgrass prairie. *Journal of the Atmospheric Sciences* 55 (7): 1225–38.

Chen, J., Wan, S., Henebry, G., Qi, J., Gutman, G., Ge, S., & Kappas, M. 2013. *Dryland East Asia: Land dynamics amid social and climatic change*. Higher Education Press, Beijing, 420 pp.

Cruz, R.V., Harasawa, H., Lal, M., Wu, S., Anokhin, Y., Punsalmaa, B., Honda, Y., Jafari, M., Li, C., & Huu Ninh, N. 2007. Asia. In: M.L. Parry, O.F. Canziani, J.P. Palutikof, P.J. van der Linden, & C.E. Hanson (eds) *Climate Change 2007: Impacts, Adaptation and Vulnerability. Contribution of Working Group II to the Fourth Assessment Report of the Intergovernmental Panel on Climate Change*. Cambridge University Press, Cambridge, UK, pp. 469–506.

Desertification Atlas of Mongolia. 2014. Institute of Geoecology and Swiss Agency for Development and Cooperation, Ulaanbaatar.

Dorj, O., Enkhbold, M., Lkhamynjin, S., Nijiddorj, Kh., Nosmoo, A., Puntsagamil, M., & Sainjargal, U. 2013. Mongolia: Country features, the main causes of desertification and remediation efforts In: G.A. Heshmati, & V.R. Squires (eds) *Combating Desertification in Asia, Africa and the Middle East: Proven Practices*. Springer, Dordrecht, pp. 217–29.

Eklundh, L., & Sjöström, M. 2005. Analysing vegetation changes in the Sahel using sensor data from Landsat and NOAA. ISPRS Proceedings. http://www.isprs.org/proceedings/2005/isrse/html/papers/

FAO. 1993. Sustainable development of drylands and combating desertification. FAO position paper. FAO, Rome.

Feng, Y., Yan, F., & Cao, X. 2018. Land degradation indicators: development and implementation by remote sensing techniques. In this volume, pp. 155–78.

Fensholt, R., & Proud, S.R. 2012. Evaluation of earth observation based long term vegetation trends: Intercomparing NDVI time series trend analysis consistency of Sahel from AVHRR GIMMS, Terra MODIS and SPOT VGT data. *Remote Sens. Environ.* 119: 131–47.

Glantz, M.H. 1999. *Creeping environmental problems in the Aral Sea basin*. NATO Advanced Research Workshop on Critical Scientific Issues of the Aral Sea Basin. Cambridge University Press, Cambridge, UK, 320 pp.

Heshmati, G.A., & Squires, V.R. 2013. *Combating desertification in Asia, Africa and the Middle East: Proven Practices*. Springer, Dordrecht.

Huang, J., Wang, X., Li, X., Tian, H., & Pan, Z. 2013. Remotely Sensed Rice Yield Prediction Using Multi-Temporal NDVI Data Derived from NOAA's-AVHRR. *PLOS ONE* 8 (8): e70816. https://doi.org/10.1371/journal.pone.0070816

Institute of Geoecology. 2014. *Desertification Atlas of Mongolia*. Mongolian Academy of Sciences and the Environmental Information Center, Ministry of Environment and Green Development, Ulaanbaatar, 134 pp.

Kariyeva, J., & van Leeuwen, W. 2012. Phenological dynamics of irrigated and natural drylands in Central Asia before and after the USSR collapse. *Agr. Ecosyst. Environ.* 162: 77–89.

Kariyeva, J., van Leeuwen, W.J.D., & Woodhouse, C.A. 2012. Impacts of climate gradients on the vegetation phenology of major land use types in Central Asia (1981–2008). *Frontiers of Earth Science* 6 (2): 206–25. http://doi.org/10.1007/s11707-012-0315-1

Kharin, N.G. 2002. *Vegetation Degradation in Central Asia under the Impact of Human Activities*. Springer, Berlin, 182 pp.

Kreutzmann, H. 2012. Pastoral practices in High Asia: Agency of 'development' effected by modernisation, resettlement and transformation. Springer, Dordrecht.

Lambin, E. 2000. Land-cover categories versus biophysical attributes to monitor land-cover change by remote sensing. *Observ. Land Space: Sci. Custom. Technol.* 137–42.

Le Houerou, H.N. 2005. *Atlas of climatic diagrams for the isoclimatic Mediterranean region*. Copymania, Montpellier, 220 pp.

Li, Q., Yang, S., Xu, W., Wang, X.L., Jones, P., Parker, D., Zhou, L., Feng, Y., & Gao, Y. 2015. China experiencing the recent warming hiatus. *Geophys. Res. Lett.* 42. doi:10.1002/2014GL062773

Lillesand, T.M., Kiefer, R.W., & Chipman, J.W. 2004. *Remote Sensing and Image Interpretation*. 5th edition. John Wiley & Sons Inc., New York.

Lu, Q., Wang, X., & Wu, B. 2009. An analysis of the effects of climate variability in northern China over the past five decades on people, livestock and plants in the focus areas. In: V. Squires, X. Lu, Q. Lu, T. Wang, & Y. Youlin (eds) *Degradation and recovery in China's pastoral lands*. CABI, Wallingford, UK, pp. 33–47.

Mohammat, A., Wang, X., Xu, X., Peng, L., Yang, Y., Zhang, X., Myneni, R.B., & Piao, S. 2013. Drought and spring cooling induced recent decrease in vegetation growth in Inner Asia. *Agr. Forest. Meteorol.* 178–179: 21–30.

Neeti, N., Rogan, J., Christman, Z., Eastman, J.R., Millones, M., Schneider, L., Nickl, E., Schmook, B., Turner II, B.L., & Ghimire, B. 2012. Mapping seasonal trends in vegetation using AVHRR-NDVI time series in the Yucatán Peninsula, Mexico. *Remote Sensing Letters* 3 (5): 433–42.

Orlovsky, N., & Orlovsky, L. 2001. White sandstorms in Central Asia. In: Yang Youlin, V.R. Squires, & Lu Qi (eds) *Global Alarm: Desert and sandstorms from the world's drylands*. UN, Beijing, pp. 169–201.

Orlovsky, L., & Orlovsky, N. 2018. Biogeography and natural resources of Greater Central Asia: an overview. In this volume, pp. 23–47.

Pinzon, J., & Tucker, C. 2014. A Non-stationary 1981–2012 AVHRR NDVI$_{3g}$ Time Series. *Remote Sens. 6* (8): 6929–60. doi:10.3390/rs6086929

Qi, J., & Evered, K. 2008. *Environmental Problems of Central Asia and their Economic, Social and Security Impacts*. Springer, Dordrecht, 400 pp.

Rasmussen, K., Fog, D., & Madsen, J.E. 2001. Desertification in reverse? Observations from northern Burkina Faso. *Global Environmental Change* 11: 271–82.

Ricotta, C., Avena, G. et al. 1999. Mapping and monitoring net primary productivity with AVHRR NDVI time-series: statistical equivalence of cumulative vegetation indices. *ISPRS Journal of Photogrammetry and Remote Sensing* 54 (5): 325–31.

Ruecker, G.R., Dorigo, W., Lamers, J.P., Ibragimov, N., Kienzler, K., Strunz, G., ... Symeonakis, E. 2014. Mapping and assessing water use in a Central Asian irrigation system by utilizing MODIS remote sensing products. *Remote Sensing* 6 (9): 012005.

Saiko, T.S. 1998. Geographical and socio-economic dimensions of the Aral Sea crisis and their impact on the potential for community action. *J. Arid Environ.* 39 (2): 225–38.

Saiko, T. 2001. *Environmental Crises: Geographic Case Studies in Post-socialist Eurasia*. Pearson Education, Harlow, UK. 320 pp.

Shang, Z.H., Dong, Q.M., Degen, A.A., & Long, R.J. 2016. Ecological restoration on Qinghai-Tibetan plateau: Problems, strategies and prospects. In: V. Squires (ed.)

Ecological Restoration: Global Challenges, Social Aspects and Environmental Benefits. Nova Science Publishers, New York, pp. 151–76.

Squires, V.R., & Lu Qi. 2018a. Greater Central Asia: its peoples and their history and geography. In this volume, pp. 3–22.

Squires, V.R., & Lu Qi. 2018b. Unifying perspectives on land, water, people, national development and an agenda for future social-ecological research. In this volume, pp. 283–305.

Squires, V.R., & Safarov, N. 2013. Diversity of plants and animals in mountain ecosystems in Tajikistan. *J. Rangeland Science* 4 (1): 43–61.

Squires, V.R., Shang, Z.H., & Ariapour, A. 2017. *Rangelands along the Silk Road: Transformative Adaptation under Climate and Global Change.* Nova Science Publishers, New York.

Thevs, N., Wucherer, W., & Buras, A. 2013. Spatial distribution and carbon stock of the Saxaul vegetation of the winter-cold deserts of Middle Asia. *J. Arid Environ.* 90: 29–35.

Toderich, K.N., Shuyskaya, E.V., Rajabov, T.F., Ismail, S., Shaumarov, M., Yoshiko, K., & Li, E.V. 2013. Uzbekistan: Rehabilitation of desert rangelands affected by salinity, to improve food security, combat desertification and maintain the natural resource base. In: G.A. Heshmati, & V.R. Squires (eds) *Combating desertification in Asia, Africa and the Middle East: Proven Practices.* Springer, Dordrecht, pp. 249–78.

Wang, F., Squires, V.R., & Lu Qi. 2018. The future we want: putting aspirations for a land degradation neutral world into practice in the GCA region. In this volume, pp. 99–112.

Yang, Y., Low, P.S., Yang, L., & Jia, X. 2018. Mitigation of desertification and land degradation impacts and multilateral cooperation in Greater Central Asia. In this volume, pp. 179–207.

Yoshio, M., & Ishii, M. 2001. Relationship between cold hardiness and northward invasion in the great mormon butterfly, *Papilio memnon* L. (Lepidoptera: Papilionidae) in Japan. *Appl. Entomol. Zool.* 36: 329–35.

Zhou, Y., Zhang, L., Fensholt, R., Wang, K., Vitkovskaya, I., & Tian, F. 2015. Climate Contributions to Vegetation Variations in Central Asian Drylands: Pre- and Post-USSR Collapse. *Remote Sens.* 7: 2449–70.

Datasets and websites

ArcGIS (v.10.x, ESRI Inc., Redlands, CA, USA)

CIMP5 data: http://cmip-pcmdi.llnl.gov/cmip5/data_portal.html

Climate Change Knowledge Portal: http://sdwebx.worldbank.org/climateportal/

CRU time series: https://crudata.uea.ac.uk/cru/data/hrg/

Ecosia tree planting initiative, Gmbh: https://www.ecosia.org/

FAOSTAT: http://faostat.fao.org

GIMMS NDVI3g data: https://nex.nasa.gov/nex/projects/1349/

Global Inventory Modeling and Mapping Studies (GIMMS) Satellite Drift Corrected and NOAA-16 incorporated Normalized Difference Vegetation Index (NDVI), Monthly 1981–2006: http://glcf.umd.edu/library/guide/GIMMSdocumentation_NDVIg_GLCF.pdf

GPCC data: http://www.esrl.noaa.gov/psd/data/gridded/data.gpcc.html

R programming packages: https://www.rstudio.com/products/rstudio/download/

8 Land degradation indicators

Development and implementation by remote sensing techniques

Feng Yiming, Yan Feng and Cao Xiaoming

INSTITUTE OF DESERTIFICATION STUDIES, CHINESE
ACADEMY OF FORESTRY, BEIJING

Introduction

The monitoring and evaluation of land degradation is the core component of desertification studies. As a technology of air-to-ground observation, remote sensing
technology (RST) has become an important research approach of the geosphere,
atmosphere, biosphere and cryosphere, because of its broad detection range, fast
data collection, large quantities of information acquired in a short period and at
a low cost. There are fewer restrictions due to better access to a wide range of
ground surface conditions, etc. With the development of remote sensing techniques, information obtained in this way has become an important tool for the
evaluation of land degradation. At the same time, RST provides technical support
and a guarantee of precision for quantitative land degradation monitoring and
evaluation of land degradation in Greater Central Asia (GCA).

Remote sensing monitoring indicators of land degradation

Indicators relating to land use and land cover (LULC) types may be classified as
shown in Table 8.1.

Table 8.1 Classification system of land use and land cover (LULC)

First-level types of LULC	Second-level types of LULC
Cultivated land	Dry land
	Paddy field
	Irrigable land
	Vegetable plot
Woodland	Forest land
	Shrub land
	Open woodland
	Other woodland
	Nursery lot

(continued)

Table 8.1 (continued)

First-level types of LULC	Second-level types of LULC
Meadowland	Natural meadow
	Improved grassland
	Artificial pasture
Land for residential, industrial and traffic use	Land for residential use
	Land for industrial use
	Land for traffic use
	Others
Water area	
Unused land	Sandy land (dunes and sand plains)
	The Gobi Desert
	Waterless, remote grassland
	Saline-alkali soil
	Swampland
	Bare land
	Others

The main vegetation monitoring indicators include vegetation type, vegetation indices, vegetation coverage and vegetation biomass, which are shown in Table 8.2.

The main soil monitoring indicators of land degradation (Table 8.3) include soil type, soil moisture, surface particle size of the soil and extent of soil erosion amount.

Table 8.2 The main vegetation monitoring indicators

Types		Unit
Trees	The main tree species	
	Origin	
	Crown density	
	Biomass	g
Shrubs	The main species	
	Origin	
	Coverage	%
	Biomass	g
Grass	The main species	
	Origin	
	Coverage	%
	Biomass	g
Composite coverage		%
Composite biomass		g
Vegetation index		

Table 8.3 The main soil remote sensing monitoring indicators

Indicators	Unit
Soil type	Description, profile
Soil moisture	%
Surface particle size of the soil	mm
Soil erosion	t/ha

The types of land degradation and the indicators to assess the degree of land degradation are varied, but in this study we followed the technology regulation of the
fourth round of national desertification and sandification monitoring (State Forest
Administration, 2009). These were based on articles written by Li et al. (2005),
and take into consideration the environmental characteristics of the GCA region.
There are four types of land degradation: wind erosion, water erosion, freeze/thaw
and salinization, each of which can be ranked by degradation degree on a scale of
mild, moderate, severe and very severe. Specific indicators are shown in Table 8.4.

Table 8.4 Remote sensing monitoring indicators of land degradation types and degree of
severity

Types of degradation	Degrees of degradation
Wind erosion	For woodland, meadowland and unused land:

(1) Vegetation coverage: for the arid and semi-arid areas, <10%
(scored 60), 10–24% (scored 45), 25–39% (scored 30), 40–54%
(scored 15), >55% (scored 5);

(2) Surface configuration: no sand dunes can be seen from the image
(scored 10), the image shows sand dunes but without shadow and
lines basically (scored 20), the image clearly shows sand dunes
and the shaded area of sand dunes is <50% (scored 30), Gobi,
wind-eroded land, bare land and clearly delineated sand dunes of
which the shaded area is >50% (scored 40).

Land is graded according to the summed score of (1) and (2): not
desertified ≤20, mild 21–35, moderate 36–60, severe 61–85, and
very severe ≥86.
For cultivated land:

- Mild degradation: there are planted tree belts and other protective
measures. Land is cropped normally and the crop is in good condition.
- Moderate degradation: there are planted tree belts. The crop yield is
not as good as in 'Mild'.
- Severe degradation: there are no protective measures. The growth of
the crop depends on natural rainfall. The crop is in bad condition.
- Very severe degradation: there is no guarantee of biomass harvest.

(continued)

Table 8.4 (continued)

Types of degradation	Degrees of degradation
Water erosion	For woodland, meadowland and unused land:

(1) Vegetation coverage: >70% (scored 1), 69–50% (scored 15), 49–30% (scored 30), 29–10% (scored 45), <10% (scored 60);
(2) Slope: <3% (scored 2), 3–5% (scored 5), 6–8% (scored 10), 9–14% (scored 15), ≥15% (scored 20);
(3) The area ratio of the erosion gully: ≤5% (scored 2), 6–10% (scored 5), 11–15% (scored 10), 16–20% (scored 15), >20% (scored 20).

Land is graded according to the summed score of (1), (2) and (3): non-desertification ≤24, mild 25–40, moderate 41–60, severe 61–84, and very severe ≥85.

For cultivated land:

- Mild degradation: slope <5 degree. The area ratio of the erosion gully 6–15%. Crop is in good condition.
- Moderate degradation: slope 5–8 degree. The area ratio of the erosion gully 16–40%. The crop yield is not as good as in 'Mild'.
- Severe degradation: slope 9–14 degree. The area ratio of the erosion gully 41–60%. The crop is in bad condition.
- Very severe degradation: slope ≥15 degree. The area ratio of the erosion gully >60%.There is no guarantee of harvest.

Salinization	For woodland, meadowland and unused land:

- Mild degradation: there is a small amount of saline and alkaline at the surface (≤20%) and vegetation coverage ≥36%.
- Moderate degradation: salinization soil covers 21–41% of the area and vegetation coverage 21–35%.
- Severe degradation: salinization soil covers 41–60% of the area and vegetation coverage 11–20%.
- Very severe degradation: salinization soil covers 61% of the area. There is hardly any vegetation.

For cultivated land:

- Mild degradation: there is a small amount of saline and alkaline at the surface (≤20%). Crop is in good condition.
- Moderate degradation: salinization soil covers 21–41% of the area. The crop yield is not as good as in 'Mild'.
- Severe degradation: salinization soil covers 41–60% of the area. The crop is in bad condition.
- Very severe degradation: salinization soil covers 61% of the area. There is no guarantee of harvest.

Freeze thawing	

- Mild degradation: the surface is covered with small pieces of dry thawing, swamp, bare land and/or quicksand. Vegetation coverage 40–50%. The area of bare land and broken land is <10%.
- Moderate degradation: the surface is covered with pieces of dry thawing, swamp, bare land and/or quick sand. Vegetation coverage 30–40%. The area of bare land and broken land is 10–30%.

- Severe degradation the surface is covered with pieces of bare land with mud-rock flow slopes, quick sand, freeze/thaw slumping ridges and/or ridges/bare slopes. Vegetation coverage 20–30%. The area of bare land and broken land is 30–50%.
- Very severe degradation: the surface is covered with large areas of bare land, mud-rock flow slopes, quicksand, freeze/thaw slumping ridges and/or ridges/bare slopes. Vegetation coverage ≤20%.The area of bare land and broken land >50% (Li et al., 2005).

Techniques of applying remote sensing to monitor land degradation

Techniques relating to LUCC and its change dynamics, including the classification of land objects

Promoted by the International Geosphere-Biosphere Program (IGBP) and the International Human Dimensions Program on Global Environmental Change (IHDP), land use/cover change (LUCC) research quickly became a focus area in the study of global environmental change in the 1990s. The processes in LUCC involve cross-cutting issues that recognize both natural and anthropogenic forces. LUCC is an important topic in the research on global environment change and is at the core of sustainable land use research. Compared with the traditional ground survey methods for determining LUCC, remote sensing has been widely used and rapidly developed because of its advantages (periodicity, reality, and macroscopic scope).

During the course of information extraction of LUCC using remote sensing techniques, the most important and basic work is remote sensing classification according to the LUCC classification scheme. Classification must be carried out according to the same criteria: namely, the characteristics of satellite images with the same physiognomy distribution should have the same classification results. Desertification researchers in the GCA region classify remote sensing images according to different classification methods: supervised and non-supervised classification, decision tree classification and visual interpretation and translation methods.

Supervised classification of imagery derived from RST involves technicians in a learning process based on discriminant function. During the course of training a site is selected in a desertified area, and features such as surface water, mobile dunes, semi-fixed sand and fixed desert may be noted. Certain requirements must be given more attention: land types used in training samples should be consistent with the regional categories of the whole study area. The training samples should be selected in the center of larger regions, which could better highlight the representativeness of the sample. In addition, the numbers of training samples should provide enough information and overcome the influence of various accidental factors. Selection of the training samples should meet the criteria to allow the establishment of a classification determination function for that land type. And relationships among the required sample numbers, classification methods, and the kinds and dimensions of featured space must be considered. For example, the number of training samples for the maximum likelihood method should be at least n+1 (n is the dimension of the feature space), which could assure the non-singularity of the covariance matrix. Supervised classification commonly used for the specific classification method mainly includes the minimum distance classification method and multistage cutting classification, characteristic curve window method and maximum likelihood ratio classification method.

The assumption of the unsupervised classification of remote sensing is that the same object in the same condition has the same spectral characteristic. Compared with supervised classification, the unsupervised classification method does not need the acquisition of prior knowledge of image features. It relies on the spectral and texture information of different land types for feature extraction. According to their similarity, a group of pixels could be classified into a number of categories. The clustering analysis method of unsupervised classification is the hierarchical cluster method and dynamic clustering method, which results in reducing the distance between the same kinds of individual pixels as far as possible. The result of unsupervised classification is to distinguish different categories, but it does not determine the properties of the category. So the category attribute must be determined by visual interpretation or field survey.

Decision tree classification is a method where the characteristics of each pixel value set become the benchmark for layered successive comparison. The method has advantages such as a clear structure, an easy to understand reconciliation, no black box structure, and better interactivity, which has very obvious advantages in its practical application. The decision tree classification method can simulate classification based on a range of data from the root node through to hierarchical rule classification. A decision tree is composed of a root node, a series of internal branches and a final leaf node; each node has only one parent node, but can have two or more child nodes, so as to establish a framework for the tree structure.

Manual visual interpretation delineates different types of pattern according to the characteristics of the sample image (shape, size, texture, etc.) and spatial characterisics (graphics, location and layout, etc.) through a combination of different information, such as geographic distribution and other factors. The method can make full use of personnel interpretation based on local knowledge, has good flexibility, and allows extraction of spatial information. Manual visual interpretation is commonly used in desert ecosystems, but the interpretation of desert features is more time-consuming and results vary from person to person. In order to improve the accuracy and efficiency of remote sensing interpretation in desert ecosystems, one or more additional classification methods can be used according to the actual needs.

LUCC of desert ecosystems in GCA can be divided into three types: one is degradation, where land quality has been reduced although the land cover type does not change: for example, due to overgrazing of grassland, or logging where the forest cover density decreased. The second is the conversion of land cover from one cover type to another, such as land occupied by city or industrial construction, or woodland felled and converted for pasture or arable land. The third is improvement of land cover, such as the establishment of soil improvement, farmland terracing, or the establishment of sown grassland. In desert ecological systems, according to space-based remote sensing and through the use of GIS, the LUCC analysis module can facilitate dynamic monitoring of ecological systems. Currently, LUCC study features types of temporal and spatial dynamic monitoring and analysis. Commonly used methods are as follows.

The transfer matrix

The Markov transfer matrix model plays an important role in the flow between different land types and desertification degree. By means of the transfer matrix, not only can we get a quantitative description transformation between the different degrees of desertification, but we can also reveal the transfer rate between the different degrees of desertification, leading to a better understanding of the temporal and spatial evolution processes of land desertification. The Markov model is a special kind of random motion process, applicable to data in a series collected at specific time intervals. A metastable system from time t to *t+1* reflects a transformation through a series of processes. This transformation time T state depends only on the time t relevant requirements. In LUCC research in the desert ecosystem, we can use any two years of land use changes in the land cover map through superimposition to develop a land use change map. Statistics from the superimposition of the land use change maps of each type of land area change can tell us how much of a given land use type has been converted to other types, and how much area remains for the class that did not change.

Land use change intensity index

The land use change intensity index refers to a region i of the unit area of land use type j, change from period a to period b. The change intensity index formula for calculating the land is as follows:

$$LTI_i = \frac{K_{j,b} - K_{j,a}}{LA_i} \times \frac{1}{T} \times 100\% \qquad 8.1$$

Where LTI_i is land use type j in the space of a unit i, while $k_{j,b}$ and $k_{j,a}$ respectively are land use change intensity index at the beginning of the study a and at the end of the study b of land use type j in the space unit i area; LA_i land area for space unit i; t at the end of the study and initial phase, time (years).

Areal extent of land use change and trend

Land use changes for any given land use category in the region occur as a number of patches and their distribution and areal extent (total size) can be quantitatively expressed.

$$D = N_i/n \times 100\%$$ 8.2

Where D is a land use change along a multi degree gradient, N_i is patch number of the land use change type, n for all the land use types and the number of plaques. Importance degree can be quantitatively expressed. Types of land use change of regional importance are used to determine direction and pace of land use change. This is an important basis for determining IV.

$$IV = D+B$$ 8.3

Where IV is the importance for a land change type; D is multi degree for the kind of the land use changes type; B is the kind of the land using changes in the type area as a percentage of the total area of all patterns. Such information is of value to land managers and policy makers.

Speed of land use change

A single land use dynamic can be quantitatively described as a certain time range within a certain land use change speed, which has a positive role for comparing the land use changes in regional differences and predicting the future land use change trend. The land use dynamic degree model can be used to analyze the dynamic change of land use types, which can reflect the regional LUCC intensity.

$$R = \frac{U_b - U_a}{U_a} \times \frac{1}{T} \times 100\%$$ 8.4

Where U_a and U_b are, respectively, the number at the beginning of the study and at the end of the study for a certain type of land use; t is the length of time; T is the time when the single land use change rate corresponds to the study area.

Regional differences of LUCC

Regional differences of LUCC can be quantified if we know the respective land use change rates. That is, in a certain period of time on a certain area of the land, and knowing the dynamic degree of a given site and the regional land use dynamic degree, we can compare sites 'at risk' relative to the regional land use change rate. Such information, which can be derived from the following equation, is useful to land managers and policy makers.

$$LUR = \frac{R_p}{R_t} = \frac{|U_b - U_a| \times C_a}{U_a \times |C_b - C_a|}$$ 8.5

Where LUR is the relative change rate of land utilization; R_p and R_t are respectively the certain area and the land use dynamic degree; U_a and U_b are respectively the beginning and the end of the monitoring period of a certain land use type. C_a and C_b are the monitoring areas at the beginning and the end of the whole study area. When

LUR>1, the local land use change amplitude is greater than the extent of change as a whole; and LUR<1 shows a lower amplitude than the overall land use change.

Techniques for identifying and extracting the vegetation information

As an important factor in the geographical environment, vegetation is very useful for indicating desertification and development. Vegetation of different types (trees, shrubs, grass, etc.) has a strong and unique spectral reflectance curve.

In desert regions with many types of landforms and topographic features, according to the vegetation spectral curve features, reflectivity of near infrared and red of vegetation can be used to establish the vegetation index model (either linear or nonlinear). The vegetation index can better reflect the vegetation types, growth and biomass and other information, and is an important tool to assess vegetation in desertified areas by remote sensing monitoring.

In order to quantitatively assess the vegetation growth and coverage characteristics, dozens of vegetation indices have been developed. Among them, the most widely used ones are NDVI (normalized difference vegetation index), RVI (ratio vegetation index), DVI (difference vegetation index), SAVI (soil adjusted vegetation index), MSAVI (modified soil adjusted vegetation index) and EVI (enhanced vegetation index).

$$NDVI = (NIR - RED)/(NIR + RED) \tag{8.6}$$

$$RVI = NIR / RED \tag{8.7}$$

$$DVI = NIR - RED \tag{8.8}$$

$$SAVI = (1 + L)\left[(NIR - RED)/(NIR + RED + L)\right] \tag{8.9}$$

$$MSAVI = \left(2NIR + 1 - \sqrt{(2NIR + 1)^2 - 8(NIR - RED)}\right)/2 \tag{8.10}$$

$$EVI = G \times \frac{NIR - RED}{NIR + C_1 \times RED - C_2 \times BLUE + L} \tag{8.11}$$

Where the *NIR, RED* and *BLUE* are the reflectance in the near infrared band, the red band and the blue band; L is the background adjustment parameter; C_1 and C_2 are the fitting coefficient; G is the gain factor. For the calculation of MODIS EVI, $C_1 = 6$, $C_2 = 7.5$, $G = 2.5$. According to the study area range and ground vegetation seasonal characteristics, the use of the vegetation index can quantitatively reflect the variation characteristics of ground vegetation (see Aralova et al., 2018).

Techniques for extracting the vegetation coverage

Vegetation coverage is the percentage of total or individual plant communities on the part of the vertical projection area and quadrat area ratio, which can reflect the degree of dense vegetation and presence or absence of plant photosynthesis. In desertification areas, vegetation coverage in practical application can be obtained mainly by field measurement, or a combination of field and remote sensing. For the combination method, the correlation between the vegetation coverage from field measurement and remote sensing data is analyzed and empirical models are established to quantify vegetation coverage over a large area. The combination method is also called the 'experience model' method. Several measurement methods exist: the 'single band' model, 'band combination model' and 'vegetation index empirical model'. In accordance with the establishment of the empirical model using statistical methods, the field and remote sensing comprehensive measurement method can be divided into the regression model, decision tree and artificial neural network method.

The estimation of vegetation coverage is mainly divided into the vegetation index method and the mixed pixel decomposition method. The mixed pixel decomposition method assumes that a pixel in a remote sensing image may actually be constituted by multiple components; each group of remote sensing sensors observes different information, so remote sensing information (band or vegetation index) can be decomposed to build a pixel decomposition model to estimate the vegetation coverage. The developed mixed pixel decomposition model is mainly a linear model, probability (probabilistic) model, geometrical optics (geometric-optical) model, random geometric stochastic geometric model analysis and fuzzy model. Among the models, the most commonly used is the linear model, which can be defined as:

$$R_{i\lambda} = \sum_{k=1}^{n} f_{ki} C_{k\lambda} + \varepsilon_{i\lambda} \qquad\qquad 8.12$$

Where $R_{i\lambda}$ is spectral reflectance of the ith pixel in λ band; f_{ki} is the k basic components of *ith* pixel accounted for the component value; $C_{k\lambda}$ is spectral reflectance of the K basic group in the λ band; $\varepsilon_{i\lambda}$ is residual error value; n is divided into a set of basic numbers.

The most simple linear pixel decomposition model is the Dimidiate Pixel Model. This pixel model assumes that each pixel consists of only two parts: the vegetation cover and non-vegetation cover. The resulting spectral information falls into two groups (the linear factor component). The respective area occupied by each component in the pixel ratio becomes the weight of each factor. The vegetation coverage percentage of pixels is the coverage of vegetation. Using NDVI as the remote sensing data source for the Dimidiate Pixel Model is a widely used approach which is often referred to as the vegetation index method:

$$F_c = \frac{NDVI - NDVI_{soil}}{NDVI_{veg} - NDVI_{soil}} \qquad\qquad 8.13$$

Attributes previously defined.

Based on the Dimidiate Pixel Model, Gutman and Ignatov (1998) put forward a method using the pixel density model to obtain vegetation coverage. According to the distribution of the vegetation of different pixels, pixels are divided into uniform and pixel density pixels and non-density pixels and mixed density sub-pixels mode.

Techniques for extracting the net primary productivity, etc.

Net primary productivity (NPP) refers to amount of photosynthetically fixed organic carbon minus respiratory consumption and is used for vegetation growth and reproduction by green plants. The NPP per unit area in unit time represents the cumulative amount of organic matter and is an important component of the terrestrial carbon cycle. NPP cannot only be used to reflect the production capacity of vegetation type under the natural environment condition, but can also be used to determine ecosystem carbon sequestration. In the early days of NPP research, according to the statistical relationship between NPP and climate, some people established an NPP climate estimation model. Further work, based on plant growth and development of the basic physiological and ecological process, encouraged some people to combine climate and soil physical data to establish NPP estimates for an ecological process model. With the development of remote sensing and computer technology, remote sensing model and NPP have penetrated into many areas. The relationship between the vegetation index and NPP is routinely calculated. Based on the resource balance theory of the light use efficiency model, NPP estimation on regional and global scales becomes possible.

NPP estimation through the utilization of solar energy is based on the balance of resources point of view, which assumes that the ecological process tends to adjust plant characteristics in response to environmental conditions. In conditions of extreme or rapidly changing environmental factors, if vegetation (or even individual plants) cannot fully adapt to the new environment, NPP will be constrained by the shortage of resources. Therefore, any resource limitation on the growth of plants (such as water, nitrogen, light, etc.) can be used for the estimation of NPP. The relationship formula for NPP and the restriction of resources is as follows:

$$NPP = F_c \times R_u \qquad\qquad 8.14$$

Where F_c is the conversion factor and R_u is the upper limit for absorption of resources. Photosynthetically active radiation (PAR) is the driving force of plant photosynthesis. PAR is a decisive factor in plant NPP and plant absorbed photosynthetically active radiation (APAR) is particularly important. The famous Monteith equation is established on this basis:

$$NPP = APAP \times \varepsilon \qquad\qquad 8.15$$

Where ε is plant energy utilization, which is influenced by water, temperature, nutrients, etc.

With the development of remote sensing technology, the plant's APAR can be estimated using remote sensing information. Therefore, an NPP model based on APAR has shown great potential. Regional or global NPP models include CASA, GLO-PEM, SDBM, and so on. Use of a 'solar energy utilization rate' model to estimate the vegetation's NPP has three major advantages: 1) the model is relatively simple, and can directly use remote sensing to obtain full coverage data, and the experiment can be easily promoted on regional and global scales; 2) the canopy leaves' absorbed photosynthetically active radiation ratio can be obtained through remote sensing, which does not need a field experimental determination; and 3) the exact NPP seasonal and inter-annual dynamics can easily be obtained.

Techniques for identifying and extracting soil information

Remote sensing has been widely used in regional soil classification studies, and has become one of the universal soil digital cartography methods.

Soil classification

In the second general survey of soil in China in 1979, remote sensing interpretation based on aerial photographs or satellite images was widely used. Soil classification by remote sensing is based on the relationship between soil and landscape. The theoretical basis (Jenny formula) is the basic formula of soil forming factors (Jenny, 1941). Soil is the product of the joint action of the parent material, biology, climate, terrain, time, etc. There is a functional relationship between soil properties and soil-forming factors. Remote sensing could effectively extract soil environmental factors, e.g. soil spectral characters, vegetation information, landform, etc. Based on AVHRR (Odeh & McBratney, 1998), Odeh, McBratney and Chittleborough (1995) studied soil classification and soil properties. Different soil types have a different soil moisture, organic material, disposition, organization structure and reflectance spectrum. There are differences in the reflection strength and shape in different bands or reflections for the same type of soil. For example, the high reflectance of alluvium, alkali soil, aeolian sandy soil and gravel soil resulted in a white tone in the remote sensing images. Meadow soil and boggy soil appeared dark, due to the low reflectance in corresponding bands.

Soil classification based on remote sensing is also an effective method to monitor desertification. In desertified areas, the development of soil is weak, and the difference between landscape types is also small. It is difficult to classify soil only by spectral information. Auxiliary information, such as topographic information, is usually used in this research. For example, based on remote sensing data, topographic information, and Jeffries-Matusita distance analysis, Kang et al. (2008) classified the soil in the Ebinur Lake in Xinjiang. The desert soil and saline-alkali soil was classified successfully.

In classification studies, the selection of distinguishing features is very important. Generally, several features are usually used in the classification.

VEGETATION INDEX

The vegetation index could detect the differences of biological natures in different soil types, such as biomass, vegetation cover, etc. NDVI is considered to be one of the most useful vegetation indices.

$$NDVI = \frac{\mathrm{Re}\,f_{nir} - \mathrm{Re}\,f_{ir}}{\mathrm{Re}\,f_{nir} - \mathrm{Re}\,f_{ir}} \tag{8.16}$$

To overcome the saturation in high cover area and the influences of soil in low cover area, EVI is also a useful index.

$$EVI = 2.5\frac{\mathrm{Re}\,f_{nir} - \mathrm{Re}\,f_{ir}}{C1\,\mathrm{Re}\,f_{ir} - C2\,\mathrm{Re}\,f_{blue} + L} \tag{8.17}$$

Where $L = 1$ is the soil modified parameter, and $C1$ and $C2$ are 6.0 and 7.5.

HUMIDITY INDEX (NDMI)

NDMI could be used to classify vegetation water content and soil moisture. It could be calculated by:

$$NDMI = \frac{\mathrm{Re}\,f_{green} - \mathrm{Re}\,f_{SWIR}}{\mathrm{Re}\,f_{green} + \mathrm{Re}\,f_{SWIR}} \tag{8.18}$$

NORMALIZED DIFFERENCE SNOW INDEX (NDSI)

NDSI could be used to extract the snow information.

$$NDSI = \frac{\mathrm{Re}\,f_{visible} - \mathrm{Re}\,f_{SWIR}}{\mathrm{Re}\,f_{visible} + \mathrm{Re}\,f_{SWIR}} \tag{8.19}$$

In addition, surface reflectance, the surface temperature at day and night, texture features and terrain parameters are also considered as effective parameters to classify soil type. Based on the classified features mentioned above, the differences of every soil type in distribution, color, texture, moisture content and biological properties could be distinguished more easily.

Soil moisture

Soil moisture is a key variable of the climate system. It constrains plant transpiration and photosynthesis in several regions of the world, with consequent impacts on water, energy and biogeochemical cycles. Moreover, it is a storage component for precipitation and radiation anomalies, inducing persistence of the climate system. Finally, it is involved in a number of feedback loops at the local, regional and global scales, and plays a major role in climate change projections. Since the 1970s, the inversion of soil moisture using remote sensing has been studied and has developed various research directions. Methods include those set out below.

MICROWAVE REMOTE SENSING METHOD

Satellite microwave remote sensing has been considered as the major method to estimate regional and global soil moisture. There are two methods: passive microwave and active microwave. Passive microwave estimates soil moisture by measuring the land surface brightness temperature in microwave radiation. Active microwave estimates soil moisture by comparing the differences between the energy emitted by radar and the measured energy reflected by the land surface.

Passive microwave   Passive microwave measures the brightness temperature using a microwave radiometer, based on a physical model and an empirical and statistical model to estimate soil moisture. Most algorithms are based on the radiation transfer model, and there are different characteristics in different models. There are three parts in these models: the dielectric model contacting soil moisture and dielectric constant, the roughness predictable model concerning the surface scattering characteristics, and the vegetation canopy model to estimate vegetation extinction. There are also atmospheric models in some radiation transfer models. Passive microwave is considered to be the most successful method to estimate soil moisture.

The energy emitted by microwave is described by microwave brightness temperature (T_B):

$$T_B = t(H,\theta)[RT_{sky} + (1-R)T_{surf}] + T_{atm}(H,\theta) \qquad 8.20$$

Where $t(H,\theta)$ is atmospheric transmissivity; T_{sky} is atmospheric downward radiance; $T_{atm}(H,\theta)$ is the upward radiation measured by the sensors; T_{surf} is the surface temperature heat; R is the reflectance.

Passive microwave (1.4 GHz) measures surface soil moisture (w_s) (2~5 cm). So to 'smooth' soil, the microwave reflectance could be described by a Fresnel equation. The horizontal polarization (h) and vertical polarization (v) under the un-vertical angle of incidence (θ) could be calculated by Fresnel reflectance:

$$R_h = \left| \frac{\cos\theta - \sqrt{d - \sin^2\theta}}{\cos\theta + \sqrt{d - \sin^2\theta}} \right|^2 \qquad 8.21$$

$$R_v = \left| \frac{d\cos\theta - \sqrt{d - \sin^2\theta}}{d\cos\theta + \sqrt{d - \sin^2\theta}} \right|^2 \qquad 8.22$$

Where d is the dielectric constant term, which could be determined by soil moisture, and is also affected by the soil texture and structure.

There are lots of inverse algorithms, and soil moisture could be classified by an empirical algorithm and radiation transfer model. The statistical inversion, artificial neural network and genetic algorithms relate to an empirical algorithm; the AMSE-R algorithm and the surface parameter inversion model (LPRM) relate to the radiation transfer model (Liang et al., 2013).

Active microwave Different soil water contents have different radar echo signals, and, on the basis of this, the relationship between the backscatter coefficients (σ) and soil water content could be established. The common active microwave imaging system is synthetic aperture radar (SAR). This system can acquire an image by sending a pulse signal and then simulating a long aperture signal, so that high surface resolution data could be acquired. According to a statistical approach, based on an empirical correlation analysis between soil moisture and the backscatter coefficient, soil moisture could be estimated by the active microwave remote sensing method. For arid and saturated soil, the nonlinear relationship would be helpful. In this method, soil roughness is an essential parameter (Liang et al., 2013).

We calculate σ of the synthetic aperture radar-related soil moisture (w_s) by the differences of bare soil and the dielectric constant of water. The empirical model can be used to describe this (Ulaby et al., 1996). The backscatter of radar on vegetation surface is characterized by:

$$\sigma = t^2 \sigma_{soil} + \sigma_{veg} + \sigma_{multi} = \exp(-2\tau_c)\sigma_{soil} + \sigma_{veg} + \sigma_{multi} \qquad 8.23$$

Where σ_{soil} is the backscattering of bare soil; t^2 is the bidirectional optical parameter; σ_{veg} is the backscattering of vegetation; σ_{multi} is the multiple scatters between the vegetation and land surface.

At present, there are two active microwave remote sensors used to estimate soil moisture: synthetic aperture radar and microwave scatterometer. However, with the early synthetic aperture radar (ERS-1, RADARSAT-1), it was difficult to estimate soil moisture using only single band, single view and single observation data without *a priori* knowledge. For new synthetic aperture radars (ENVISAT-ASAR, RADARSAT-2, ALOS, TerraSAR-X), multi-band, multi-view and multi-observation have been included to decrease the characteristic parameter attributes in the topsoil except soil moisture. There are many algorithms based on radar to estimate soil moisture, including empirical algorithms and algorithms based on physical principles (Shoshany et al., 2000; Moran et al., 2004; Wang & Qu, 2009). Empirical algorithms are used to invert soil moisture by establishing the regression correlations between the measured soil moisture and backscatter coefficients. Generally, based on the regression analysis, various indices are calculated by backscatter coefficients: e.g., Shoshany et al. (2000) proposed the normalized backscatter moisture index (NBMI):

$$NBMI = \frac{\sigma_{t1} - \sigma_{t2}}{\sigma_{t1} + \sigma_{t2}} \qquad 8.24$$

Where σ_{t1} and σ_{t2} are the backscatter coefficients at different times.

OPTICAL AND THERMAL INFRARED REMOTE SENSING

Triangle method The triangle method (Figure 8.1) explains the two-dimensional spatial distribution between the surface radiative temperature (T_R) and the

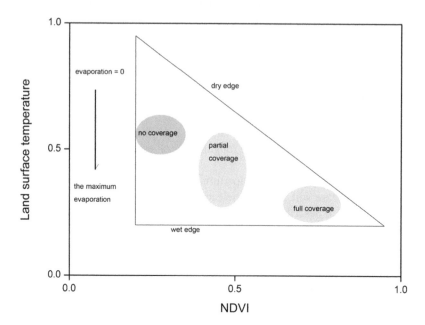

Figure 8.1 The triangle relationship between NDVI and LST

vegetation coverage (F) of the pixels (Carlson, 2007). On the whole, the surface radiative temperature decreases with an increase in vegetation coverage, so the triangle is formed. The narrow vertices in the triangle correspond to the surface radiative temperature of dense vegetation. The simulated results of the soil–vegetation–atmosphere transfer model (SAVT) also verified the triangle, and could be used to estimate soil moisture. For example, Carlson (2007) defined the available water in the topsoil (M_0) as the proportion of the soil moisture and the field water-holding capacity:

$$M_0 = \sum_{i=1}^{i=3} \sum_{j=1}^{i=3} a_{ij} T^{*i} F_r^j \qquad 8.25$$

$$T^* = \frac{T_R - T_{min}}{T_{max} - T_{min}} \qquad 8.26$$

$$F_r = (\frac{NDVI - NDVI_{min}}{NDVI_{max} - NDVI_{min}})^2 \qquad 8.27$$

Where i and j are the simulated surface radiative temperature; T_{max} and T_{min} are the maximum and minimum surface radiative temperature; F_r is the vegetation abundance calculated by NDVI; a_{ij} is the parameter coefficient.

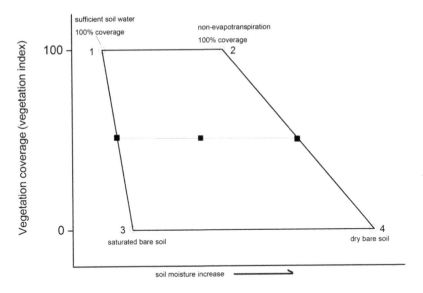

The difference between the temperature of land surface and air

Figure 8.2 The relationship between the temperature difference of the vegetation and land surface air

Source: Moran et al. (1994).

Trapezoidal method At a specific time, the scatters of the difference between land surface temperature and air temperature and vegetation cover could form a trapezoid (Figure 8.2).

The water deficit index (WDI) could be calculated from the trapezoid method:

$$WDI = \frac{\Delta T - \Delta T_{L13}}{\Delta T_{L24} - \Delta T_{L13}}$$ 8.28

Temperature-vegetation dry index The temperature-vegetation dry index (TVDI) was constructed based on the empirical relationship between land surface temperature and NDVI (Sandholt et al., 2002).

$$TVDI = \frac{T_s - T_{min}}{a + bNDVI - T_{min}}$$ 8.29

Where T_{min} is the minimum surface temperature of the wet edges; T_s is the land surface temperature; a and b are the parameters of the simulated dry edges; $T_{s_{max}} = a + bNDVI$, $T_{s_{max}}$ is the maximum land surface temperature for a certain NDVI. To get a and b, the study area must be large enough, and soil moisture needs to vary from wet to dry, and cover should range from bare land to dense vegetation.

Thermal inertia (TI) method   The basic principle of the thermal inertia method is based on the proportion of soil moisture and the thermal inertia. For a homogeneous body, the thermal inertia could be calculated as:

$$TI = \sqrt{\rho c K} \qquad\qquad 8.30$$

Where K is thermal conductivity [J/(m.s.K)]; ρ is density (kg/m³); TI is thermal inertia [Ws$^{1/2}$/(m².K)]. The thermal inertia is the measurement of the ability of the temperature variation caused by impeding external factors. For a certain incident flux, there is a negative relationship between soil temperature changes and thermal inertia. Due to the large specific heat of water in soil, thermal inertia is also affected by soil moisture. So, when inverting the thermal inertia by remote sensing, soil moisture could be estimated according to the relationship between soil moisture and thermal inertia.

Apparent thermal inertia (ATI) is one of the simple approximations. In an area with uniform solar radiation, ATI could be calculated as:

$$ATI = \frac{a(1-\alpha)}{T_{day} - T_{night}} \qquad\qquad 8.31$$

Where a is the experimental coefficient related with the types of soil; α is albedo; T_{day} and T_{night} are the land surface temperatures at day and night. Soil moisture could be estimated by using a constructed linear model, logarithmic model or index model.

The grain size of topsoil

The texture of topsoil is closely related to land degradation. In arid and semi-arid area, wind erosion strongly affects topsoil grain size. Therefore, grain size distribution in the topsoil can potentially be used as an indicator of land degradation. It is thus possible to monitor land desertification by topsoil grain size change in arid and semi-arid areas using the remote sensing technique (Xiao et al., 2005). From the 1970s, the properties of topsoil were closely related to land size and its spectral reflectance. According to Zhu et al. (1989), different extents of desertification have different topsoil texture – the more severe the desertification, the coarser the topsoil grain composition. Salisbury and D'Aria (1992) pointed out that the grain size of topsoil played a significant role in erosion potential and other mechanical properties, and the ratio of ASTER thermal infrared bands 10/14 could be used to estimate particle size. Fu et al. (2002) found that overgrazing can accelerate soil erosion by wind and result in topsoil coarsening.

Detailed knowledge of soil surface texture would dramatically improve the ability to model wind erosion and dust emission in desert soils where wind erosion is strongly controlled by surface grain size (Mahowald, 1999; Marticorena & Bergametti, 1995; Okin et al., 2001). Remote sensing tools that produce quantitative information on soil surface texture would be useful supplements to traditional soil maps for planning purposes. Such tools would also prove valuable in the emerging field of predictive soil mapping (Scull et al., 2003).

RADIATIVE TRANSFER MODELS

Gregory et al. (2004) reported that there was negative correlation between effective grain size of sand in the plume and reflectance; the most significant correlations occurred in the short-wave infrared.

EMPIRICAL MODELS

Based on the reflectance of topsoil grain size in different bands, various indices are established. The empirical models are then established by the relationship between topsoil grain size and the index to invert topsoil grain size in a large area. In 2006, Xiao (2006) proposed a topsoil grain size index (GSI) to monitor desertification. The GSI showed a positive correlation with fine sand content. The results showed that the fine sand content of topsoil increased in most places, indicating a coarsening process of the topsoil in the study area. The fast soil coarsening of degradation is largely caused by human activities. The GSI was designed especially for using Landsat TM/ETM+ data as follows:

$$GSI = (R - B) / (R + B + G) \qquad \qquad 8.32$$

Where R, B and G are the reflectance of the red, blue and green bands of the Landsat TM and ETM+ sensors.

CLASSIFICATION METHOD BASED ON REMOTE SENSING INFORMATION

The classification method used is based on remote sensing information, such as texture information, roughness, elevation, and the spectral index related to topsoil grain size. For example, Yao et al. (2013) estimated topsoil grain size in the Gobi Desert by principal component analysis. In this study, 43 remote sensing and earth-science factors (e.g. the reflectance of every band, DEM, NDVI, GMEI, K-T transformation) were selected to describe topsoil grain size. Based on the principal component analysis, the estimated model was established to classify the topsoil grain size in the Gobi Desert.

Land degradation in Greater Central Asia

Throughout GCA, desertification is a problem (Aralova et al., 2018). In the Aral Sea Basin, from the 1960s, the water yield from Amu Darya and Syr Darya to the Aral Sea showed a continued decrease (Orlovsky & Orlovsky, 2018). In dry climate conditions, the intensive evaporation resulted in reduced water yield, and the water level fell. In 1989, the Aral Sea was divided into two lakes because of the falling water level. Saline-alkaline land without vegetation appeared in the dry lake, and the salt content of the lake water also jumped. There was a large area of vegetation degradation in the delta region into the lake. Studies also point out that a persistent multi-year drought in Central Asia had affected close to 60 million people as of November 2001. To a certain extent, persistent drought caused the land degradation.

Based on the Ts-NDVI triangle space with the MODIS NDVI dataset, Cao et al. (2016) monitored soil moisture in the Mongolian Plateau in 2000–2012. Based on TVDI, the spatio-temporal aridification was studied from the general Ts-NDVI space method. The study pointed out that TVDI showed the water deficit for the region from low (bare soil) to high (full vegetation cover) NDVI values. Drought was widely spread throughout the Mongolian Plateau, and there was aridification in the study period (Figures 8.3 and 8.4).

Based on the Albedo-NDVI space from MODIS, Zhang et al. (2013) studied the spatio-temporal desertification in 2000–2012 in Turkmenistan. That paper

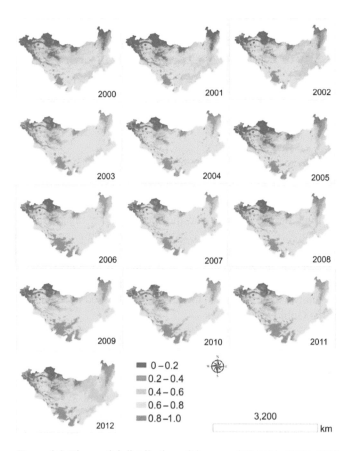

Figure 8.3 The spatial distribution of the annual TVDI in 2000–2012

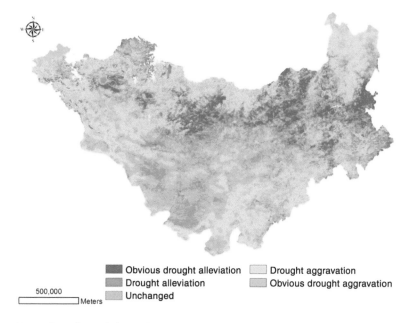

Figure 8.4 The variations (slope) of TVDI in 2000–2012

established the desertification classification index model, and extracted the dynamic information of soil desertification. The paper noted that, during the study period, the total desertification area had decreased, but the severity of desertification increased. The incremental increases in the area of extremely severe desertification and severe desertification lands were 2173.27 km^2 and 43,428.47 km^2 respectively. Moreover, Yan et al. (2014) monitored the desertification in the typical oasis area of Turkmenistan from 2001 to 2012. The paper combined six remote sensing indexes (EVI, MSAVI, TVDI, LST, Albedo, FVC) to monitor the spatio-temporal desertification in the study area. The results showed that the total area of desertification decreased, while the desertification degree was aggravated in this area.

Based on time-series SPOT vegetation remote sensing data, Kuang et al. (2014) used Theil-Sen and Mann-Kendall to calculate land degradation intensity and trend in Central Asia from 1999 to 2012. The paper pointed out that the city zone around two deserts (Kalakum Desert and Kyzylkum Desert) and the saline-alkali land in the west of Kazakhstan are regions suffering from the most serious land degradation. Based on NOAA and MODIS remote sensing data, Zhou et al. (2006) took FVC, MSAVI, Albedo, LST and TVDI as the indices to monitor land degradation in Central Asia in 1990–2005. The paper noted that, in the five 'stans' in Central Asia, the area of vegetation degradation showed a general increase and the ecological environment was degrading.

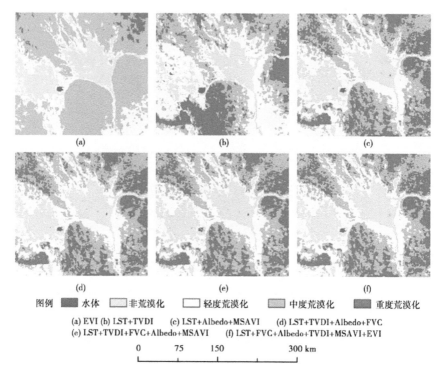

Figure 8.5 The classification based on different combinations of the six indexes

References and further reading

Agrawala, S., Barlow, M., Cullen, H. & Lyon. B. 2001. The drought and humanitarian crisis in central and southwest Asia: a climate perspective. *Drought Network News* 12: 6–12.

Aralova, D., Kariyeva, J., Menzel, L., Khujanazarov, T., Toderich, K., Halik, U. & Gofurov, D. 2018. Assessment of land degradation processes and identification of long-term trends in vegetation dynamics in the drylands of Greater Central Asia. In this volume, pp. 131–54.

Boyd, D.S., Foody, G.M. & Ripple, W.J. 2002. Evaluation of approaches for forest cover estimation in the Pacific Northwest, USA, using remote sensing. *Applied Geography* 22: 375–92.

Cao, X., Feng, Y. & Wang, J. 2016. An improvement of the Ts-NDVI space drought monitoring method and its applications in the Mongolian plateau with MODIS, 2000–2012. *Arabian Journal of Geosciences*. doi:10.1007/s12517-016-2451-5

Carlson, T. 2007. An overview of the 'triangle method' for estimating surface evapotranspiration and soil moisture from satellite imagery. *Sensors* 7: 1612–29.

Fu, H., Wang, Y.R., Wu, C.X. & Ta, L.T. 2002. Effects of grazing on soil physical and chemical properties of Alxa desert grassland. *J. Desert Res.* 22: 339–43.

Gutman, G. & Ignatov, A. 1998. The derivation of the green vegetation fraction from NOAA/AVHRR data for use in numerical weather prediction models. *International Journal of Remote Sensing* 19 (8): 1533–43.

Hansen, M.C., DeFries, R.S. & Townshend, J.R.G. 2002. Towards an operational MODIS continuous field of percent tree cover algorithm: Examples using AVHRR and MODIS data. *Remote Sensing of Environment* 83: 303–19.

Kang, Q., Zhang, Z. & Zhao, X. 2008. A Study of Soil Classification Based on Remote Sensing in Arid Area. *Journal of Remote Sensing* 12 (1): 159–67.

Knorr, W. & Heimann, M. 1995. Impact of drought stress and other factors on seasonal land biosphere CO_2 exchange studied through an atmospheric t racer transport model. *Tellus B* 47 (4): 471–789.

Kuang, W., Ma, Y.G., Ling, H. & Liu, C. 2014. Analysis of land degradation intensity and trend in Central Asia from 1999 to 2012. *Remote Sensing for Land and Resources* 26 (4): 163–9.

Li, S., Yang, P., Wang, Y. & Zhang, C. 2005. Preliminary Analysis on Development and Driving Factors of Sandy Desertification on Ali Plateau. *Journal of Desert Research* 25 (6): 838–44.

Liang, S., Li, X. & Wang, J. 2013. *Quantitative Remote Sensing: concept and algorithm.* Science Press, Beijing, pp. 551–73.

Mahowald, N. 1999. Dust sources and deposition during the last glacial maximum and current climate: A comparison of model results with paleo data from icecores and marine sediments. *Journal of Geophysical Research* 104 (D13): 15895–916.

Marticorena, B. & Bergametti, G. 1995. Modeling the atmospheric dust cycle: 1. Design of a soil-derived dust emission scheme. *Journal of Geophysical Research* 100: 16415–30.

Moran, M., Peters-Lidard, C. & Watts, J. 2004. Estimating soil moisture at the watershed scale with satellite-based radar and land surface models. *Canadian Journal of Remote Sensing* 30: 805–26.

Moran, S.M., Clarke, T.R. & Inoue, Y. 1994. Estimating crop water deficit using the relationship between surface-air temperature and spectral vegetation index. *Remote Sensing of Environment* 49: 246–63.

Odeh, I.O.A., McBratney, A.B. & Chittleborough, D. 1995. Further Results on Prediction of Soil Properties from Terrain Attributes: Heterotropic Cokriging and Regression Kriging. *Geoderma* 67: 215–26.

Odeh, I.O.A. & McBratney, A.B. 1998. *Using NOAA Advanced Very High Resolution Radiometric Imageries for Regional Soil Inventory.* Proceedings of the 16th World Congress of Soil Science, Montpellier, France.

Okin, G.S. & Painter, T.H. 2004. Effect of grain size on remotely sensed spectral reflectance of sandy desert surfaces. *Remote Sensing of Environment* 89: 272–80.

Okin, G.S., Murray, B. & Schlesinger, W.H. 2001. Degradation of sandy arid shrubland environments: Observations, process modelling, and management implications. *Journal of Arid Environments* 47 (2): 123–44.

Orlovsky, L. & Orlovsky, N. 2018. Biogeography and natural resources of Greater Central Asia: an overview. In this volume, pp. 23–47.

Potter, C.S., Randerson, J.T. & Field, C.B. 1993. Terrestrial ecosystem production: A process model based on global satellite and surface data. *Global Biogeochem. Cyc.* 7: 811–41.

Prince, S.D. & Goward, S.N. 1995. Global primary production: A remote sensing approach. *J. Biogeogr.* 22: 815–35.

Purevdorj, T.S., Tateishi, R. & Ishiyame, T. 1998. Relationships between percent vegetation cover and vegetation indices. *International Journal of Remote Sensing* 9 (18): 3519–35.

Quamby, N.A., Townshend, J.R.G. & Settle, J.J. 1992. Linear Mixture Modeling Applied to AHVRR Data for Crop Area Estimation. *International Journal of Remote Sensing* 13 (3): 415–25.

Salisbury, J. & D'Aria, M. 1992. Infrared (8–14 um) remote sensing of soil particle size. *Remote Sensing of Environment* 42: 157–65.

Sandholt, I., Rasmussen, K. & Andersen, J. 2002. A simple interpretation of the surface temperature/vegetation index space for assessment of surface moisture status. *Remote Sensing of Environment* 79: 213–24.

Scull, P., Franklin, J., Chadwick, O.A. & McArthur, D. 2003. Predictive soil mapping: A review. *Progress in Physical Geography* 27 (2): 171–97.

Seneviratne, S.I., Corti, T., Davin, E.L., Hirschi, M., Jaeger, E.B., Lehner, I., Orlowsky, B. & Teuling, A.J. 2010. Investigating soil moisture–climate interactions in a changing climate: A review. *Earth-Science Reviews* 99: 125–61.

Shoshany, M., Svoray, T. & Curran, P. 2000. The relationship between ERS-2 SAR back-scatter and soil moisture generalization from a humid to semi-arid transect. *International Journal of Remote Sensing* 21 (11): 2337–43.

State Forestry Administration. 2009. The technology regulation of the 4th round of national desertification and sandification monitoring, State Forestry Administration, Beijing.

Ulaby, F.T., Dubios, P.C. & van Zyl, J. 1996. Radar mapping of surface soil moisture. *Journal of Hydrology* 184: 57–84.

Wang, L. & Qu, J.J. 2009. Satellite remote sensing applications for surface soil moisture monitoring: A review. *Frontiers of Earth Science in China* 3: 237–47.

Xiao, J., Shen, Y., Ryutaro, T. & Bayaer, W. 2005. Detection of land desertification and topsoil grain size using remote sensing. Geoscience and Remote Sensing Symposium, IGARSS'05. Proceedings.

Xiao, J., Shen, Y., Tateishi, R. & Bayaer, W. 2006. Development of topsoil grain size index for monitoring desertification in arid land using remote sensing. *International Journal of Remote Sensing* 27 (12): 2411–22.

Yan, X., Ding, J., Zhang, Z. et al. 2014. Dynamic remote sensing monitoring of desertifica-tion of a typical oasis in Turkmenistan of Central Asia. *Journal of Natural Disasters* 2 (23): 103–10.

Yang Xiaoping. 1998. Desertification and land use in the arid areas of central Asia. *Quaternary Sciences* 5 (2): 119–27.

Yao, S., Zhang, T., Zhao, C. & Liu, X. 2013. Saturated hydraulic conductivity of soils in the Horqin Sand Land of Inner Mongolia, Northern China. *Environ. Monit. Assess.* 185 (7): 6013–21.

Zhang, Y.J., Tiyip, T., Xia, J. et al. 2013. Desertification monitoring with remote sensing in the Central Asia: a case of Turkmenistan. *Arid Land Geography* 36 (4): 724–30.

Zhou, K.F., Zhang, Q., Chen, X. & Sun, L. 2006. The features and trend of the changes in ecological environment in the arid area in Central Asia. *Science in China Ser. D.* 36 (2): 133–9.

Zhu, Z., Liu, S. & Di, X. 1989. *Desertification and its rehabilitation in China.* China Science Press, Beijing.

9 Mitigation of desertification and land degradation impacts and multilateral cooperation in Greater Central Asia

Yang Youlin

ASIA-PACIFIC REGIONAL COORDINATION UNIT OF UNCCD

Pak Sum Low

ADJUNCT PROFESSOR, INSTITUTE OF SUSTAINABLE DEVELOPMENT AND
ARCHITECTURE, BOND UNIVERSITY

Yang Liu

INSTITUTE FOR DESERTIFICATION STUDIES, CHINESE ACADEMY OF
FORESTRY

Jia Xiaoxia

NATIONAL BUREAU TO COMBAT DESERTIFICATION, STATE FORESTRY
ADMINISTRATION

Basic facts of each affected country in GCA

Table 9.1 is a compilation of pertinent facts about each of the nine countries in Greater Central Asia (GCA). In what follows we give a few highlights for each country to augment Table 9.1. We also provide a key reference that can provide readers with more detailed information.

Afghanistan

Afghanistan is a landlocked country covering a total area of 652,225 km². The climate of the country ranges from continental to subtropical. The Marji Desert of the southwest and the desert of the northern parts, which continues to the banks of Amu-Darya, are characterized by a subtropical climate, while the remaining areas and mountainous parts of the country are characterized by a continental climate.

Of the total population of Afghanistan (Table 9.1) about 80% is dependent on the agricultural and natural resources management sectors. Wheat is the main crop in the country and cultivated by almost all farmers. Successful harvests depends on irrigated and rain-fed land. During the last four decades, Afghanistan has undergone drastic policy changes, from a semi-market-led economy of the

Table 9.1 Basic facts of countries in Greater Central Asia

Country	Location	Land area (km²)	Population	Unique circumstances
Afghanistan	29⁰–35⁰N	652,225	32,527,000	War, and civil strife
China*	27⁰–42⁰N 75⁰–92⁰E	3,017,000	30,000,0000	Part of China's national program
Islamic Republic of Iran	24⁰–40⁰N 44⁰–64⁰E	1,648,000	79,109,000	Rapid population growth and livestock inventories
Republic of Kazakhstan	40⁰–56⁰N 46⁰–88⁰E	2,700,000	17,625,000	Local site pollution caused by industrial and military activities
Republic of Kyrgyzstan	39⁰–44⁰N	199,900	5,940,000	The multiple causes of land degradation and desertification are complex
Mongolia	46⁰N–05⁰E	1,555,116	2,959,000	Mining land area has increased rapidly over recent years
Tajikistan	36⁰–41⁰N	443,100	8,482,000	Climate change, deforestation, inappropriate irrigation and mismanagement of water resources
Turkmenistan	35⁰–48⁰N	488,100	5,374,000	Aral Sea Basin: 80% of land is desert
Uzbekistan	37⁰–46⁰N 56⁰–74⁰E	447,400	29,893,000	Aral Sea Basin, salinity

Note: * Estimates based on Xinjiang, Tibet and Qinghai (principally).

1960 and 1970s to a highly centralized state controlled system in the 1980s, and to a serious lack of proper economic policy and inefficiencies of public institutions of the present day (Emadi, 2012).

The combination of human activities and harsh climatic conditions causes the rapid spread of desertification in many provinces of Afghanistan. The prevailing problems that Afghanistan faces today are those of soil erosion, overgrazing, deforestation and rangeland degradation. All of these contribute to a decline in the numbers of wildlife, biodiversity loss, further land degradation and the destruction of fragile ecosystems. The existing irrigation system is operating at a low efficiency rate of about 25%. At the same time, the agriculture resources have been under pressure with the population growth and return home of refugees.

The prolonged civil war combined with successive years of serious drought has reduced cultivating areas markedly and impacted livestock adversely. The agriculture sector (1998–2006) in Afghanistan was greatly damaged (a reduction of 40–50%). More than 80% of land is potentially subjected to soil erosion.

Soil infertility is common and salinization is accelerated, water table falls dramatically, destruction of vegetation is extensive, and soil erosion by water and wind is prevailing in many parts of the country (Safar and Squires, 2017). It is estimated that 29% of the total area is affected by water erosion, 5% by wind erosion and 3% by salinity. A large amount of land is washed away in floods year by year. To supply sufficient water for agricultural purposes and domestic consumption, tube wells are drilled illegally. This activity is accelerated during the years of drought, which causes severe decline of the water table and a reduction in the efficiency of traditional irrigation systems.

China's western regions

China (Table 9.1) is one of the countries suffering from severe desertification and land degradation over a vast area (Jia et al., 2011). The area of desertification-prone land is about 0.92 million km^2 in the western regions that are the focus of this chapter. The area of shifting sand disaster-affected lands, as a result of wind erosion, is about 0.83 million km^2 in western China. The sand disaster-affected land stretches along an east-to-west axis and eastward from the Tarim Basin in Xinjiang. The area of desertification-prone land caused by freezing and thawing phenomena in alpine mountains is 363,600 km^2, which is distributed mostly on the Qinghai-Tibet Plateau. The area of salinized land is about 173,700 km^2, distributed mainly in the oasis around the Tarim Basin, the alluvial plain at the northern foot of the Tianshan Mountain and in the Qinghai-Tibet Plateau.

Islamic Republic of Iran

Most of the country is covered by arid deserts and semi-arid lands in which the average annual precipitation is less than 250 mm, which is scattered irregularly both in terms of time and area. The potential for evapotranspiration is many times more than precipitation; it amounts to more than 100 times in certain regions of the country. The net result is that vegetation cover is sparse (Ariapour, Badripour and Jouri, 2017). The biological productivity of agricultural lands and natural vegetation has been severely compromised. Desertification, land degradation and drought adversely affect about half of the country's population. Besides a large human population, there is also a large number of livestock that is dependent on the land (Table 9.1). The pressure on the land has resulted in an increase in utilization of marginal areas for both cropping and grazing. This situation, coupled with the natural erodibility of the land and erosive nature of the rainfall in drylands, which comes as flash floods, has caused severe soil erosion, salinization, alkalization and land degradation (Ariapour, Badripour and Jouri, 2017).

Republic of Kazakhstan

Kazakhstan is a landlocked country where the climate is continental. Oil and gas are the leading economic sectors in Kazakhstan. It has considerable agricultural potential with its vast steppe lands accommodating both livestock and grain production. At present, the processes of desertification and land degradation are marked in nearly all administrative areas of Kazakhstan. The prevailing problems of land degradation and sustainable land management (SLM) identified in Kazakhstan include: a) loss of soil fertility due to inappropriate land use practices in rain-fed arable lands; b) inefficient water use, salinization and water logging of irrigated arable lands, caused by deteriorating irrigation and drainage systems and mismanagement of watershed and water resources; c) degradation of pasturelands caused by local overgrazing and overuse of large pasture areas due to the giving up of mobile grazing practices, local livestock concentration and catastrophic decline of wild ungulates' populations; d) forest degradation and deforestation caused by illegal logging and wildfires; e) drying out of large areas of the Aral Sea (Bekchanov et al., 2018) and associated negative consequences; and f) local site pollution caused by industrial and military activities (Bekniyaz, 2011). In many cases these problems have impacts on ecosystem types of global importance and affect neighboring countries (Akiyanova et al., 2017).

Republic of Kyrgyzstan

Kyrgyzstan is a landlocked country bordering Kazakhstan, China, Tajikistan and Uzbekistan (Table 9.1). Land degradation and desertification are serious economic and social problems related to the environment that face Kyrgyzstan. The multiple causes of land degradation and desertification are complex, and vary across Kyrgyzstan's regions (Ridder, Isakov and Ulan, 2017). But to a greater extent deterioration and exhaustion of land resources are the result of admittedly incorrect and destructive agricultural practices, overgrazing, deforestation and the cutting down of bushes, forest degradation, loss of biodiversity and natural disasters.

Kyrgyzstan's economy was, and still remains, primarily agricultural. Its economy was severely affected by the collapse of the Soviet Union and the resulting loss of its vast market. During the period of joining and ratification of the United Nations Convention to Combat Desertification (UNCCD) in 1997, all organizational work was done as the initiative of the Kyrgyzstan Scientific-Research Institute of Irrigation at the Ministry of Agriculture, Water Resources and Processing Industry. The country is rich in expertise and know-how to mitigate DLDD, conserve soil, protect forests and steppes, and manage upstream water. However, Kyrgyzstan faces the pressure of weak economic and financial constraints (Ridder, Isakov and Ulan, 2017).

Mongolia

Mongolia is a landlocked country, bordered by Russia to the north and China to the south, east and west. Mongolia is a highland, and a cold and dry country. It has an extreme continental climate with long, cold winters and short summers, during which most precipitation falls. The extreme south is the Gobi Desert, some regions of which receive no precipitation at all in most years.

Economic activities in Mongolia are traditionally based on animal husbandry and agriculture. Mongolia is ranked as a lower middle income economy by the World Bank. Some 22.4% of the population lives on less than US$1.25 a day. Agriculture in Mongolia contributes about 20.6% of Mongolia's annual Gross Domestic Product (GDP) and employs 42% of the labor force. However, the high altitude, extreme fluctuation in temperature, long winters and low precipitation provide limited potential for agricultural production. The degree of severity of land degradation in Mongolia is very high, with a continuous year-on-year increase for decades. In particular, drift by wind is increased, and its degree deepened. Due to soil loss and erosion damage, agronomy and the farming industry are worsening, with declines in area, productivity, cereal quality and the employment rate of local farmers. The area of mining land has increased rapidly over recent years (Tsed, 2011).

Tajikistan

Tajikistan is landlocked and is the smallest nation in Central Asia by land area (Kurbanova and Squires, 2017). The problem of land degradation and desertification gets increasingly urgent every year. The basic driving factors of desertification are climate change, deforestation, inappropriate irrigation and mismanagement of water resources, overexploitation of agricultural lands and overgrazing (Squires, Kurbanova and Madaminov, 2013). All these factors decrease the livelihoods of the affected population. As a result, the low level of incomes has increasingly negative consequences on land degradation and desertification. Although Tajikistan possesses water resources and high moisture in some areas, part of the territory of Tajikistan is home to deserts. Desertification and land degradation are observed in many natural zones that are connected to Tajikistan's position among the large deserts of the Eurasian continent – the Gobi, the Kara Kum Desert, the Kyzyl Kum Desert and the Taklimakan Desert.

Turkmenistan

Turkmenistan is located in the western part of Central Asia. About 80% is covered by the Kara Kum (Garagum) Desert, the largest desert of Asia, while the remainder is covered by plains, mountains, rivers, lakes, water reservoirs and oases. Turkmenistan has a subtropical desert climate that is severely continental. Summer is long (from May through September), hot and dry, while winter is generally mild and dry (Annagylyjova and Squires, 2017).

Turkmenistan is one of the large desert countries, with nomadic livestock raising and intensive agriculture in irrigated oases. Its economy is predominantly agriculture, which accounts for almost half of its GDP. The territory of Turkmenistan is situated in the desert zone, with a rather fragile ecosystem, and where every unreasonable step in land management could cause irreversible catastrophic consequences. Integrated estimation of the current state of desertification in the territory of Turkmenistan was carried out in 1998 by the Institute of

Deserts Research, Turkmenistan Academy of Sciences, based on space images, observations at field stations, subject maps and statistic materials.

It is estimated that desertification-prone lands and degraded lands cover 446,287 km², a large portion of the land of the country, and they can be classified as the following types:

- Plant cover degradation: 367,522 km², occupies 76.3% of total desertification-prone land;
- Sand deflation: 8,640 km², covers 1.2% of total desertification-prone lands;
- Water erosion: 6,900 km², represents 1.4% of total desertification-prone lands;
- Salinization of irrigated land: 40,310 km², occupies 8.3% of total desertification-prone lands;
- Salinization of soils, caused by a drop of the Aral Sea level: 15,010 km², covers 3.0% of total desertification-prone lands;
- Water-logging of pasture land: 6,930 km², covers 1.5% of total desertification-prone lands;
- Technogenic desertification: 920 km² represents 0.2% of total desertification-prone lands.

Uzbekistan

Uzbekistan is also a landlocked country. Its climate is classified as continental, with hot summers and cool winters. Uzbekistan is an agricultural country with a large and wide distribution of cropping land and grazing lands, including steppe and pasture lands (Table 9.1). Deserts and semi-deserts occupy 80% of the territory of Uzbekistan. Intensive land usage leads to degradation and unproductive irrigated lands. Secondary salinization affects more than 50% of irrigated farmlands (Khasankhanova and Tarayanikova, 2011). Overgrazing by cattle and related erosion processes, together with other anthropogenic impacts, have resulted in loss of productivity in rangelands, with 16 million hectares (out of 23 million hectares) degraded. Of these, 7 million hectares are severely degraded, where losses in forage capacity amount to 30–40% and more. Drifting sands occupy about 1 million hectares, of which 200,000 hectares have been damaged during the past years along the boundaries of irrigated lands, which has also caused the further spread of desertification and its processes (Mukimov, 2017).

Ongoing DLDD mitigation efforts and approaches taken in GCA countries

In the GCA countries, significant efforts and visible achievements have been made in line with the strategic objectives and operational objectives of the UNCCD through multilateral and bilateral channels at international, national and local levels. The following are brief summaries of the efforts made in each country.

Efforts made by Afghanistan in combating desertification

In response to the vast recovery and reconstruction needs, a large number of international organizations (UN agencies, World Bank, ADB, NGOs and development partners) have paid great attention to Afghanistan. The first significant effort made by Afghanistan was the creation and use of a National Development Framework (NDF) to set the country on the track to prosperity. Specific targets were set to measure performance. Irrigation improvement was to reach 1 million hectares by 2010 and 2.2 million hectares in another five years. By 2015, multipurpose water schemes were to generate at least 180 MW of power generation. Targets were also set for wheat production and livestock development. The area of orchards was projected to increase by over 70% from 80,000 hectares in 2003 to 138,000 hectares in 2015. The policy and strategy for the rehabilitation and development of agriculture and the natural resource sector were approved in February 2004. Afghanistan ratified the UNCCD in November 1995 and it was effective in December 1996. In Afghanistan, national institutions and facilities have been created to operationalize and implement the National Actions Program to Combat Desertification, to reduce poverty, and to practice sustainable development, including:

- Department of Forestry and Rangeland (DFR);
- National Solidarity Programs (NSPs); and
- Ministry of Energy and Water.

International agencies are also working with the Government of Afghanistan toward SLM:

- focus on forestry;
- joint project of the Ministry of Agriculture and Irrigation (MAI) and the Wildlife Conservation Society, with a focus on the rangeland component;
- Renewable National Resource Management/Green Afghanistan Initiative (GAIN) project; and
- building capacity to address land-related conflict vulnerability in Afghanistan – e.g. mine clearance.

Efforts made for mitigating DLDD in western China

The government of China has always given great importance to combating desertification and is paying more attention and implementing approaches to neutralize land degradation (Wang et al., 2018). China incorporates ecological improvement into the overall strategy of national economic and social development, including combating DLDD (Jia et al., 2011). Consequently, several significant actions have been taken, including promulgation and execution of the State Law for Combating Desertification, and implementation of a series of integrated ecological improvement programs/projects. The pace of prevention and control of desertification is speeding up, with historic breakthroughs being made. The tendency for desertification and sand encroachment/sand movement to expand has been arrested. The process of desertification has been reversed decade by decade. The National Action Plan to Combat Desertification (NAP; 2005–2010) has been developed based on the present situation and the achievements and experiences gained in combating desertification over the past decades, in accordance with the State Eco-environment Improvement Program. China has set its focus on the major challenges to be faced in combating desertification and maps out the guiding thought and management principles for arresting and reversing desertification and neutralizing land degradation. A three-step strategic objective has been defined and measures to be taken up to 2050 have been put into the work plan. After signing the UNCCD, China developed its National Action Plan (NAP) in 1996. China aligned its NAP with the 10-year strategic plan of the UNCCD in 2005 for 2005–2010 and 2011–2015. These plans are currently under implementation with a huge budget from both central and local government levels under the framework of public–private partnership and voluntary cooperation.

China has initiated the following projects/initiatives to mitigate DLDD, to prevent and control dust and sandstorms, to conserve soil, to revegetate degraded steppe and to manage resources sustainably:

- Converting Degraded Grazing Land to Grassland Projects;
- National Soil and Water Conservation Projects;
- National Integrated Desertification Control Demonstration Area Projects;
- Integrated Agriculture Development Program on Combating Desertification;
- Aksu-Kashgar Railway Line Mobile Sand Stabilization Project in Xinjiang;
- Desertification Control Project in Southeast Tibet;
- Tibet Ecological Security Barrier Establishment Project;
- Ecological Restoration Technology Research and Demonstration on Desertification Control Initiatives;
- Sustainable Development in Poor Rural Areas Initiatives;
- Sustainable Management and Biodiversity Conservation of the Lake Aibi Basin Project;
- Training and Support Measures for the Desertification Control Program;
- Cooperative Research on Desertification Control and Water-saving Agriculture;
- Early Warning System on Drought and Adaptive Technology in Desert Areas;
- Integrated Ecology Improvement Project in Qinghai Lake Watershed Area.

Efforts made by Iran to fight against DLDD

The particular climatic conditions of Iran contribute to the formation of more than 3,000 years of indigenous knowledge. Furthermore, institutional frameworks have been created and concerted efforts undertaken during the last four decades. On that basis, a broad range of strategies has been developed and a series of programs/projects implemented to address the problems associated with desertification, land degradation and drought before the endorsement of the UNCCD in 1994. In addition, the regular five-year Social and Economic Development Plans have paid particular attention to soil and the sustainable management of water, rangelands and forests. The Forest and Rangeland Organization (FRO) of Iran is an important organization with branches in all provinces of the country. This organization with 40 years history in combating desertification efforts is in charge of the development of policies and project management on combating desertification and sand dune fixation as well as on the sustainable management of forests, reforestation and afforestation activities.

Based on a resolution adopted at a ministerial meeting of the United Nations Economic and Social Commission for Asia Pacific (ESCAP), a Desertification Control in Asia and the Pacific (DESCONAP) Program Office (DPO) was established in April 1994 in Tehran. During its existence, the DPO has undertaken activities at the national and sub-regional levels, including several workshops and seminars for the NAP and the Regional Action Plan (RAP) in Asia. The Government of Iran is committed to the concept of sustainable development by linking the environment to the overall development process. In line with its obligations under the UNCCD and as a national priority, the Government of Iran has launched an elaboration of its NAP to Combat Desertification and has adopted the framework and guidelines of the NAP. The formulation of the NAP will supplement past and existing initiatives in combating desertification and mitigating land degradation in Iran. The UNCCD principles and NAP guidelines, based on past experiences, have been integrated into the national five-year development plan under implementation.

The UNCCD calls for implementing the strategy through concerted efforts and the effective participation of key actors involved in aligning NAP process and targeting priorities. The National Environment Action Plan of Iran adopted in 1999 reflects the consensus within the government that land degradation is a priority area in the country. The relationship between combating desertification and poverty reduction is well addressed in the context of the National Program for Poverty Reduction (NPPR). Once ratified by parliament, the provisions of the UNCCD are evidently considered now as a law and a national commitment. The UNCCD principles have also been integrated into the Comprehensive National Sustainable Development Plan, which includes the sectors of human resources development and national energy strategy.

Efforts made by Kazakhstan to combat DLDD

In 2005, the National Coordination Body for Implementing the UNCCD of Kazakhstan aligned with the National Action Program to Combat Desertification in Kazakhstan for 2005–2015, which is implemented under the principle of Kazakhstan's commitment to the UNCCD and based on its commitment to fulfill its Activities Plan for 2010–2016 on implementation of the Concept on Ecology Security of the Republic of Kazakhstan for 2004–2016, approved by the

Government of Kazakhstan in February 2004. Today, with a growing population, intense management of fertile cropping lands and an increase in anthropogenic pressure on the environment desertification and land degradation present the main threats to successful social-economic development. For Kazakhstan, whose main territory is located in an area of water deficiency, the desertification problem is very serious. At present, two-thirds of the country's territory suffers from desertification, land degradation and drought to various extents. Kazakhstan's NAP is the starting point for combating desertification, rehabilitating land degradation, mitigating drought effects, and halting their consequences. The NAP of Kazakhstan contains an analysis of the causes of desertification, land degradation and drought (DLDD) issues and priority activities to mitigate DLDD, which include a range of immediate and preventive actions, policy and finance inputs and approaches to fight DLDD issues and challenges. The Kazakhstan NAP (aligned to the UNCCD 10-year strategic plan) was approved in January 2005, and it contains additional information, data priority targets and directions for action to combat desertification.

The Kazakhstan NAP has set targets on: a) a sustainable natural resources management policy; b) the design of social-economic aspects of natural resources conservation and the arrest and reversal of desertification; c) research and information/data support; d) the acceleration of technology to combat desertification; e) international cooperation in the area of combating desertification and synergies among the Rio conventions; and f) coordinating activities of local authorities, land users, farmers and non-governmental organizations. The key means of achieving the goal of the Kazakhstan NAP is to improve the effectiveness of natural resources management, based on balancing ecology and economic aspects. In the course of combating desertification, Kazakhstan took the necessary steps to undertake measures to reduce the impact of anthropogenic pressure such as exhaustion of the soil, extensive pressure on rangelands, deforestation, inappropriate irrigation methods and mismanagement of water resources. At the same time, the main social-economic causes of DLDD were addressed seriously at both governmental and local community levels. The national strategy directed toward the sustainable social-economic development of society is based on the state's ability to consider ecological priorities while taking agricultural or other decisions. Kazakhstan also strengthened the integration of activities on poverty reduction, combating desertification and mitigating drought effects into its policy of the sustainable development of the country. Kazakhstan's NAP is designed to create an enabling environment for sustainable natural resource development and management and to enhance economic activities, to facilitate conservation and rehabilitate affected lands.

Efforts made by Kyrgyzstan to control DLDD

The National Action Plan to Combat Desertification and Implement the UNCCD (NAP-CCD) of Kyrgyzstan was approved in December 2000. Kyrgyzstan has actively pursued a number of activities to combat desertification, to rehabilitate land degradation and to mitigate drought effects. There are a few concrete investment projects to address the priority areas highlighted in the NAP. While a number of projects are financed as a part of government agencies' routine plans/initiatives, the most visible responses to environmental protection and management needs are those involving international cooperation. In 2004, Kyrgyzstan set up a Working Group on Development Partnership for UNCCD Implementation (PWG-CCD). The National Program Framework on Land Management during 2006–2016 was adopted with donors' assistance for NAP development in February 2006. The National Program Framework on Land Resources Management (2006–2016) in the framework of the Central Asian Countries Initiative for Land Management (CACILM) has approval at governmental level. The Kyrgyzstan National Focal Point (NPF) to the UNCCD overseas/coordinates a comprehensive program to address the serious challenge of land degradation and desertification and the need for sustainable management of land, water and natural resources in the country. To date, Kyrgyzstan has issued more than 20 laws concerning UNCCD implementation and realizing the country's improvement of degraded land and deteriorated environment.

Efforts made by Mongolia to mitigate DLDD

Mongolia became a signatory party to the UNCCD in October 1994, ratified the Convention in September 1996, and became a party to the UNCCD family in December 1996. The Mongolian National Action Plan to Combat Desertification (NAP) was initially approved by Mongolian Government Decision 169 of July 1996. Since then, the NAP was revised and aligned in 2003 and endorsed by government. The objectives of the Mongolian NAP prioritize those activities that combat desertification and involve the participation of citizens, companies and enterprises, organizations and stakeholders. Consequently, the NAP contains a set of activities and initiatives, such as establishing the Desertification Combat Center, accelerating capacity building for local organizations, education and public awareness raising, providing assistance to improve and support the livelihoods of those residing in the affected areas and localities, and involving the public, enabled through promotional mechanisms as well. Implementation of the NAP was targeted to be completed in three phases. In the short term (2003–2007), Mongolia conducted a primary investigation/survey of current desertification. The mid-term (2008–2011) was focused on desertification monitoring and evaluation, and the long term was to be prioritized on mitigation beyond 2012, using the full capacity of the mechanism to decrease disasters of desertification and evaluate the preliminary results of the adaptation policy. Mongolia also undertakes consultation activities in collaboration with the Republic of Korea, France, Germany, the USA and other development partners. Furthermore, the government carried out 12 projects nationally and locally that were connected to the UNCCD and SLM.

The Mongolian NAP was aligned and approved by government in 2010 in line with the UNCCD 10-year strategy. In general, the Mongolian NAP is aimed mainly at:

- strengthening national and institutional capacity for combating desertification;
- improving the legal and policy framework;
- enhancing science, technology and knowledge management;
- increasing advocacy, awareness rising and education; and
- supporting concrete actions at the local level and increasing investment.

The NAP is composed of two phases:

- First phase (2010–2015): Problem scoping, capacity building, and laying the foundation for Phase 2; and
- Second phase (2016–2020): Actions to cope with the intensity of desertification by rehabilitating degraded land and restoring vulnerable areas adversely affected by desertification and land degradation are being and will be continuously taken within this phase (http://www.neaspec.org/sites/default/files/Mongolia_NCCD.pdf).

Mongolia's reforestation activities were financed from the state budget every year. Some are funded by central government, and some have contributions from development donors. Also, the Government of Mongolia launched implementation of the "Greenbelt" national program from 2004. About 353.5 hectares of vegetative belts with fence protection and irrigation means were planted annually following the NAP scheme.

Efforts made by Tajikistan to tackle DLDD

The National Action Program to Combat Desertification (NAP) of Tajikistan was developed in 2000 and a series of national seminars were organized in various districts of the country on different hot issues and prevailing problems of desertification. The National Forum on NAP Implementation, with the participation of the state, public and non-governmental organizations, was sponsored by a national focal point agency. Tajikistan practiced the following measures to combat desertification, to rehabilitate land degradation and to mitigate drought effects:

- a program of economic transformations of the agricultural complex for 1995–2003;
- the preservation and rational use of lands, rehabilitation of land degradation and combating desertification;
- Tajikistan Governmental Order No. 294 on "State Control over the Use and Protection of Lands in Tajikistan" was endorsed in 1997;
- the Program of Ecology Education of the Population of Tajikistan was approved in 1996 and was extended from 2000 to 2010;

- the Program on the Prevention of Irrigated Lands from Salinization for 1998–2003 was approved;
- Tajikistan Governmental Order No. 598 on "Endorsement of the National Action Program (NAP) to Combat Desertification" was adopted on December 30, 2001.

Efforts made by Turkmenistan to rehabilitate land degradation

Turkmenistan joined the UNCCD in 1996. The Turkmenistan National Action Program to Combat Desertification (NAP) was developed in 1997. In order to integrate UNCCD implementation into national development, the priorities of the NAP were integrated into national and bilateral development plans and programs, such as the National Environmental Action Plan (NEAP) of the President of Turkmenistan (2002), Strategy of Economic, Political and Cultural Development of Turkmenistan 2003–2020, UN Development Assistance Framework (2004) and Central Asian Countries Initiative for Land Management (CACILM, 2004). At the national level, implementation of the UNCCD is closely linked to implementation of the NEAP. After submission of the Second National Report, significant progress was made in the environmental legislation process. New Land and Water Codes were adopted in 2004. There are numbers of state sectoral funds that direct their resources to implementation of sectoral investment projects to combat desertification, to rehabilitate land degradation and to control the salinization of soil and water management.

Despite the governmental and international interventions to combat desertification and rehabilitate land degradation in Turkmenistan, coherent and coordinated efforts under the UNCCD are hindered by the programmatic and methodological shortcomings of Turkmenistan's NAP.

Efforts made by Uzbekistan to reverse desertification

The Republic of Uzbekistan was the first of the Central Asian States to ratify the UNCCD and has taken an active part in all phases of its preparation. The National Action Plan to Combat Desertification (NAP) of Uzbekistan was developed in 1999. A series of pilot projects, which are a component of the Uzbekistan NAP, was created to mitigate DLDD in Uzbekistan. With the UNDP/UNSO[1] and the Government of Finland's financial and technical support, Uzbekistan has organized regional seminars and workshops. Uzbekistan implemented a series of projects aiming at combating desertification and land degradation rehabilitation. These projects are related to providing the rural population with drinking water and gas, and developing small-scale power engineering using alternative energy sources. Much has been done to change the crop pattern, put an end to the cotton monoculture and increase the area of land under cereals, vegetables and forage grasses. For the best use of the scientific-technical potential in the field of combating desertification, Uzbekistan, in the framework of the strategy of scientific-technical activity improvement, has established a special division targeting soil fertility and highly technological methods to combat desertification. Uzbekistan launched a legislative reform reviewing old and outdated Acts and adopted new regulations to address environmental protection.

Current collaboration partnerships at multilateral, bilateral and south–south cooperation levels

At the global level, the Global Environment Facility (GEF) provides incremental financing for eligible countries to invest in SLM interventions that generate multiple environmental and development benefits. GEF financing under the focal area encompasses a diversified portfolio of interventions from farm-level to wider landscapes, with emphasis on those interventions designed to control soil erosion, loss of vegetative cover, and depletion of water.

In most affected developing countries, SLM represents a major opportunity for sustainable intensification of existing farmlands through integrated management of land and water resources and diversification of mixed farming systems. SLM thus improves soil health and quality, and enables sustained productivity of farmlands. The approach also ensures improved management of agro-ecosystem services across production systems and reduces pressure on natural lands. Integrated management also helps improve and sustain the economic productivity and environmental sustainability of rangeland and agro-pastoral systems.

The GEF officially recognized land degradation as a focal area in 2002, and its investment in projects that relate to land degradation has increased steadily over time. Between 1991 and 2000, the GEF addressed land degradation through cross-cutting investments. Although GEF resources invested in such cross-cutting activities to control land degradation were relatively modest, the projects helped lay the foundation for two important operational programs (OPs): Integrated Ecosystem Management (OP12) and Sustainable Land Management (OP15). The demand for GEF resources through these OPs ultimately crystalized land degradation as a worldwide challenge.

The *third replenishment phase* (2002–2006) was a major turning point for GEF investment in mitigating land degradation, with two significant milestones. At the close of GEF-3, the total GEF investment in SLM was US$396.16 million, generating US$1.08 billion in co-financing. The *fourth replenishment phase* (2006–2010) represented another important milestone in GEF investment to mitigate land degradation. The innovative investment programs initiated in GEF-3 were expanded from national to regional and multi-country during this phase. By end of the GEF-4 replenishment phase, a total of US$340 million was invested in projects addressing SLM, with an additional US$2.3 billion generated as co-financing. The *fifth replenishment phase* (2010–2014) introduced several important reforms that further enhanced opportunities for increased GEF financing under the focal area, including support for UNCCD-enabling activities. By the close of this replenishment phase, GEF-5 has financed US$66,981,786 for

39 projects on combating land degradation, SLM, national reporting on UNCCD implementation and reforestation in 22 countries of the Asia-Pacific region, with approximately US$2 billion in co-financing.

GEF-6 has been launched and focuses on the land degradation results framework and is targeted at supporting efforts to arrest and reverse current global trends in land degradation, specifically desertification and deforestation, and to accelerate the sustained productivity of agro-ecosystems and forest landscapes in support of human livelihoods. It is also targeted to manage 120 million hectares of land under SLM regimes.

The objectives of GEF-6 are:

- LD-1: Agriculture and Rangeland Systems: Maintain or improve the flow of agro-ecosystem services to sustain food production and livelihoods;
- LD-2: Forest Landscapes: Generate sustainable flows of forest ecosystem services, including sustaining livelihoods of forest-dependent people;
- LD-3: Integrated Landscapes: Reduce pressures on natural resources from competing land uses in the wider landscape;
- LD-4: Maximizing Transformational Impact: Maintain land resources and agro-ecosystem services through mainstreaming at scale;
- LD-EA: Adaptive Management and Learning: Increase capacity to implement interventions;
- Adaptive management tools in sustainable land management, sustainable forest management and integrated resource management (SLM/SFM/INRM) by GEF and UNCCD parties.

The GEF's mandate to invest in global environmental benefits from production landscapes relates directly to its role as a financial mechanism of the UNCCD. A Memorandum of Understanding between the UNCCD Conference of Parties (COP) and the GEF Council (Decision 6/COP.7) has since paved the way for direct support to those affected countries eligible for GEF financing through enabling activities. The amendment of the GEF instrument in 2010 formally designated the GEF as a financial mechanism of the UNCCD.

The Land Degradation Focal Area mandate is directly linked to the 10-year (2008–2018) strategy of the UNCCD, which aims to forge a global partnership to combat desertification and rehabilitate degraded lands and to mitigate the effects of drought in affected areas in order to support poverty reduction and environmental sustainability. Because the GEF-6 replenishment phase (2014–2018) coincides with the final four years of the UNCCD 10-year strategy, links with the focal area strategy will ensure that countries appropriately channel GEF resources toward implementing the strategy. In this regard, the focal area resources will contribute directly toward strategic objectives to improve the livelihoods of affected populations, to improve affected ecosystems and to mobilize resources at a global level. At the same time, the GEF also welcomes the participation of the private sector, non-governmental organizations (NGOs) and civil society organizations (CSOs) as partners for co-financing the initiatives/efforts to mitigate DLDD at various levels.

Table 9.2 GEF-6 STAR allocations for the GCA countries (US$ million)

Country	Climate change	Biodiversity	Land degradation	Total	Fully flexible
Afghanistan	3.00	3.91	4.39	11.30	no
China	126.00	58.55	9.95	194.50	no
Iran	9.76	4.79	2.66	17.21	no
Kazakhstan	11.81	5.04	5.13	21.98	no
Kyrgyzstan	2.00	1.56	3.04	6.60	yes
Mongolia	3.02	5.09	3.65	11.76	no
Tajikistan	2.00	1.50	2.78	6.28	yes
Turkmenistan	4.99	1.81	3.29	10.09	no
Uzbekistan	11.46	1.78	5.12	18.36	no
TOTAL	**174.04**	**84.03**	**40.01**	**298.08**	
Percentage	58.39%	28.19%	13.42%		

Table 9.2. is indicative of the GEF's support for SLM-related projects/programs with examples from several GCA countries.

Table 9.2 provides the indicative STAR allocations for Biodiversity, Climate Change, and Land Degradation for the GCA countries during GEF-6 (1 July 2014–30 June 2018), according to the formula and indices agreed by the GEF Council in May 2014, based on a replenishment level of US$4.433 billion (GEF/C.47/Inf.08, 1 July 2014).

National-level initiatives

At the national level, the GCA countries have made significant achievements in combating desertification, rehabilitating land degradation and mitigating drought effects through multilateral, bilateral and south–south cooperation among countries.

Afghanistan

Government departments and institutions have been authorized to implement multilateral and bilateral cooperative projects on DLDD mitigation, which were initiated by development partners for carrying out the National Action Plan to Combat Desertification (NAP) and to Implement the UNCCD, in collaboration with UN agencies, international institutions and NGOs. The authors summarized some cases of multilateral and bilateral cooperative projects jointly conducted in Afghanistan from the *National Report of Afghanistan on the Implementation of United Nations Convention to Combat Desertification (UNCCD)*, Ministry of Agriculture and Irrigation, Kabul, Afghanistan, July 2006.

- The Asian Development Bank (ADB), in collaboration with the Global Environment Facility (GEF) and Department for International Development (DFID), assisted Afghanistan to protect biodiversity and develop a participatory management plan for wildlife conservation in some protected areas.
- The FAO provided Afghanistan with 21 weather stations, 88 rain gauges and 72 crop monitoring points to collect data on weather conditions for food security and agriculture analysis. The UN Food and Agriculture Organization (FAO) also completed the cleaning of irrigation canals throughout the country, which are aimed to improve irrigation water delivery to an estimated 12,000 farm families. The FAO helped, in particular, communities in Farah Province to prevent and control sandstorms.
- The United Nations Office for Project Services (UNOPS) supported Afghanistan to establish Forestry Demonstration and Training Centers in Takhar, Herat and Ghazni Provinces. The project was aimed to provide alternative sustainable livelihoods through the creation of home and community-based tree nurseries, to increase natural vegetation through a community-supported labor-intensive approach and to improve technical and institutional capacity at community and government levels.
- A Wildlife Conservation Society joint project, which was initiated on wildlife, rangeland, livelihoods and forest protection in Wakhan, Hazarajat and the eastern forest complex, was targeted on: a) identifying threats and designing initiatives for their eradication; b) environmental education and the establishment of community resources committees; and c) establishing a four-county trans-boundary peace park in Pamir.
- The UNEP National Capacity Self-Assessment (NCSA) and National Adoption Program of Action (NAPA) for Afghanistan were aimed to develop a program of action to climate change through assessing vulnerability and identifying priority adoption measures and to assess Afghanistan's needs to meet the obligation of the Rio Conventions.
- The Green Afghanistan Initiative (GAIN) was jointly initiated by UN country teams such as UNICEF, UNAMA, UNEP, WFP and UNDP, in close collaboration with the Ministry of Agriculture and Irrigation and National Environment Protection Agency (NEPA) of Afghanistan. The GAIN was targeted on reforestation, environment protection and development of a green environment in the country.
- The Program on Building Capacity to Address Land-related Conflict and Vulnerability in Afghanistan was focused on building the capacity of the Land Commission and communities to formulate policies and adopt practices that address land related to conflict and vulnerability. The program was targeted on analytic work and pilot initiatives which serve to inform subsequent policy dialogue.

The above-mentioned UN agencies, international institutions, governmental departments and projects had positive and fruitful outcomes, although the country faces challenges in the long-term course to mitigate DLDD, improve ecosystems and the livelihoods of the population affected by land degradation, desertification, drought and man-made disasters.

China

As governmental authorities, the State Forestry Administration (SFA) and National Bureau to Combat Desertification (NBCD) have launched a group of multilateral and bilateral cooperative projects to combat desertification, to neutralize land degradation and to mitigate drought effects. Under the agreement of bilateral cooperation between China and development partners, the following projects have been initiated in affected provinces and regions of China to mitigate DLDD and prevent and control dust and sandstorms:

- PRC-GEF Management and Policy Support to Combat Land Degradation Project;
- FAO Project on Land Degradation Assessment in Dryland (LADA);
- Sino-German Desertification Control Program in Ningxia;
- Sino-Japanese Forestry Cooperation on Popularization and Training Project of New Technology and Management Methods;
- Sino-Korean Cooperation on Sandstorm Monitoring Inspection.

Asia-Pacific Forest Network

As a semi-international institution on sustainable forest management, the Asia-Pacific Network for Sustainable Forest Management and Rehabilitation (APFNet) was established in 2009 in Beijing, and is focused on institutional development, demonstration projects, capacity building, policy dialogue and partnership development. APFNet was officially registered as an independent international organization under Chinese regulations in 2011. The working mechanism of APFNet focal points was expanded and the Interim Steering Committee was set up. The Strategic Plan 2011–2015 was adopted by all APFNet stakeholders after rounds of consultation.

The framework of the regional cooperation strategy in the Central Asia Region has been developed in collaboration between China's State Forest Administration (SFA) and APFNet and is composed of several elements:

- The regional forestry ministerial meeting (FMM), which is organized every two years with the assistance of APFNet, to exchange views on major issues of forestry development of the individual economy, regional forestry cooperation and essential international forestry issues.
- The regional workshop on Strategic Forestry Cooperation in Central Asia (WSFC), which is organized in conjunction with FMM. The purpose of the

workshop is to implement the decisions made by the regional forestry ministerial meeting with the support of the Technical Group of Central Asia (TGCA).

- The Technical Group of Central Asia (TGCA), which is responsible for designing project activities regarding policy dialogue, pilot demonstration, capacity building and information sharing, developing the Overall Work Plan and budget for these projects, and breaking them down at operational level under the direction of the WSFC.
- The Asia-Pacific Network for Sustainable Forest Management and Rehabilitation (APFNet Secretariat), which is responsible for daily communication with the TGCA and logistics issues for FMM and WSFC. It is suggested that the forestry agencies of the Central Asia economies and Mongolia become APFNet's partners or membership economies and officially nominate a focal point to maintain communication with APFNet on a regular basis. It has been suggested that the operational fund for the aforementioned activities will be provided by the APFNet Secretariat and its partners and membership economies, and probably the Silk Road Fund (SRF, at 40 billion CNY) and Asia Infrastructure Investment Bank (AIIB).

The priorities within the operational framework for the regional cooperation strategy in Central Asia are identified as follows:

- forestry policy dialogue, to assist the regional economies in developing forestry long-term planning, policy and legislation, and assessment of the outcomes of implementation of these planning and policy approaches;
- assessment of forest resource (*sensu lato*) and status of desertification, to provide critical information for forestry policy makers of the regional economies in GCA;
- forest land restoration and desertification control, to demonstrate the combination of advanced restoration techniques and local practices through holistic land use planning;
- trans-boundary biodiversity conservation to protect precious flora and fauna species in the GCA region;
- joint law enforcement and forest fire prevention and control;
- multifunctional utilization of sand-inhabited bio-resource, to increase local income;
- involvement of interested economies and NGOs with APFNet's partners and nomination of APFNet's focal point in each economy; and
- development of a website and electronic newsletter on Central Asia.

Iran

For implementing the NAP and UNCCD, the Government of Iran has initiated a series of closely related programs/projects on DLDD mitigation, SLM, soil and water management in the country at both national and international levels (with major support from UN agencies, ADB, and a number of bilateral donors), including:

- an integrated plan to combat desertification through sand dune fixation;
- establishment of windbreaks around farms;
- forests and rangelands conservation;
- rangeland improvement and development;
- sustainable management of rangelands;
- watershed management and aquifer recharge;
- rehabilitation of forest landscapes and degraded land with particular attention to saline soils and areas prone to wind erosion;
- specific project of sand dune fixation around oil infrastructures;
- development of agriculture and orchards in arid lands;
- official higher education in DLDD-related sciences;
- livestock improvement;
- grazing management on rangelands;
- designing desertification control projects;
- investigation of the source of dust in Iran;
- investigation of mineral exploration in the deserts of Iran;
- identification and control of dust storm hotspots;
- renewable energies (research, training and executive projects); and
- geographical organization DLDD-related projects.

The main outputs made to date can be summarized as:

- numerous outcomes and achievements in terms of sustainability of planted stands and socio-economic growth;
- careful management of the dependency between local communities and their environment, which has created more sustainable mechanisms on arid-land forest management;
- totally, 14,000 hectares of degraded lands changed to vigorous stands consisting of various species including saxaul (*Haloxylon* species);
- considerable reduction of dust storms;
- rehabilitation cost decreased – people have even used their own resources for rehabilitation;
- local communities benefit from the rehabilitation contract;
- creation of a successful model for the participatory rehabilitation of desertified rangelands in arid and hyper-arid areas addressing human development and the promotion of alternative livelihoods;
- local communities have been benefiting from different advantages of rehabilitated stands such as by-products and controlled livestock grazing permission; and
- value added products have been supported in marketing.

Kazakhstan

International cooperation in combating desertification, rehabilitating land degradation and mitigating drought effects in Kazakhstan provides a valuable opportunity for obtaining methodological, technical and financial assistance from international development partners and gives an impetus to combating desertification through that assistance. The following activities were implemented to strengthen the international cooperation:

- Phase I (2005–2007): preparation and implementation of inter-state activities directed to the conservation of transboundary ecosystems balance;
- Phase II (2008–2010): analysis and adaptation of best practices and technologies in combating desertification, strengthening cooperation in combating desertification within the Framework of the Sub-regional Program of Activities on Combating Desertification in Central Asia, Regional Plan of Actions to Promote Sustainable Development of Highlands Territories, and Regional Plan of Actions on Environment Protection; and
- Phase III (2011–2015): implementation of pilot projects, seeking practical, experimental goals.

A certain amount of progress has been achieved in the area of regional cooperation among Kazakhstan and its neighboring states. To date, the Ministry of Environmental Protection (MEP) has completed sound work on the Central Asian Countries Initiative for Land Management (CACILM) project, the primary task of which is the development of a well-coordinated integrated and complete approach for assistance to the Central Asian countries (CACs) for implementation of the UNCCD and development of the all-round National Program on SLM. The CACILM represents innovative international cooperation by donors to support the development and performance of the Framework Program at the National Level (NPF). It is aimed at the development of all-round and complex approaches to combat desertification through sustainable management of land and water resources, as stated in the GEF Operational Program on SLM. Investment and technical assistance to CACs simultaneously serve as important parts of the Multi-country Framework Partnership (MCFP). In 2006, the first phase of the CACILM/ NPF to CACs was planned and carried out, and it continues to the present day.

The CACILM project is a 10-year, multi-country, multi-donor program promoting SLM to restore, maintain and enhance the productivity of drylands (https://www.adb.org/sites/default/files/page/149300/gef-projects-focal-areas-june2013.pdf). This project aims to work toward implementation of SLM, rehabilitation of land degradation, and promotion of adaptation to climate change (http://www.adb.org/projects/38464-012/main). At present, an integrated and comprehensive project on CACILM is conducted regionally in the five "stans" in line with the partnership of Kazakhstan, Kyrgyzstan, Tajikistan, Turkmenistan and Uzbekistan, brought together by international donor partners to combat desertification, rehabilitate land degradation and improve rural livelihoods (http://www.adb.org/projects/38464-012/main). The projects are also aimed to increase capacity building for mainstreaming SLM and ensuring integrated SLM planning and management; accelerate the development of an SLM information system; reinforce research on SLM; widen information dissemination;

and improve knowledge management at both national and local levels. The goals of CACILM are: restoration, maintenance, and enhancement of the productive functions of the land in CACs, leading to improved economic and social well-being of those who depend on these resources while preserving the ecological functions of the land in the spirit of the UNCCD (http://www.adb.org/ projects/38464-012/main).

By analyzing data and information collected from the five countries to implement the CACILM project, it is recognized that the following global environmental benefits have been made as significant outcomes of the CACILM project:

- protecting threatened ecosystems of global significance;
- decreasing the basis for transboundary dust and sandstorm events;
- decreasing greenhouse gas emissions and enhancing carbon sequestration;
- improving the management of transboundary water;
- reducing the loss of soil in dust storms, which creates regional and intercontinental hazards;
- reducing soil and pesticide runoff into rivers, which causes downstream and transboundary water quality deterioration;
- improving water availability, which will help moderate the harsh climate associated with desertification;
- reversing the loss of carbon stocks sequestered in soils or forests, and hence reducing greenhouse gas emissions, as agriculture becomes sustainable and forests regenerate; and
- reversing the loss of biodiversity that is inevitable if the present trend of habitat loss continues (https://www.adb.org/sites/default/files/page/149300/ gef-projects-focal-areas-june2013.pdf).

CALCILM is being implemented in a multi-country framework which includes a 10-year program of activities in each participating country based on NAP frameworks (http://www.adb.org/projects/38464-012/main) and follow-up of the Country Pilot Partnership promoted by the GEF-3 OP15 and Strategic Partnership of the UNCCD (http://www.zoinet.org/web/sites/default/files/publications/CACILM.pdf).

The anticipated outcomes of the 10-year program in each CA country can be divided into six groups:

- a favorable environment for SLM investments in CACs, supported by SLM mainstreaming and improvements in policies, regulations and land administration;
- improved capacity of the institutions in the CACs to adopt integrated land use planning and management;
- rehabilitation and improved productivity of selected lands, thereby leading to improved livelihoods, foreign exchange earnings and food security, and providing indirect protection to threatened ecosystems;
- enhanced protection of ecosystem integrity and landscapes;

- broader involvement of civil society and other stakeholders in SLM in the CACs; and
- long-term harmonized commitments of financial and human resources through mainstreaming of SLM in donor programs for Central Asia. (https://www.adb.org/sites/default/files/page/149300/gef-projects-focal-areas-june2013.pdf).

CACILM is being implemented in three phases: Phase 1 (Inception), July 2006–December 2008; Phase 2 (Implementation), January 2009 (2010)–December 2013 (2014); Phase 3 (Consolidation), January 2014–June 2016 (http://www.zoinet.org/web/sites/default/files/publications/CACILM.pdf).

The main outcomes of CACILM include:

- improved capacity of institutions in the CACs to adopt integrated land use planning and management;
- long-term, sustained and harmonized commitments of financial and human resources through mainstreaming of SLM in donor programs for the five "stans" in Central Asia;
- efficient and effective coordination of implementation of the CACILM Multi-country Framework Partnership;
- efficient and effective coordination of the implementation of NPFs in the five "stans";
- SLM is designed, developed and operated:

    – SLM research is designed and implemented;
    – a knowledge management system is established.

Mongolia

Mongolia is one of the leading countries in the GCA region and it has created several national key projects for combating desertification, in collaboration with inline ministries and governmental agencies. These projects are:

- Government Policy on Ecology;
- Comprehensive National Development Strategy;
- Mongolian Action Program for the 21st Century;
- National Water Program;
- National Forest Program;
- National Climate Change Program;
- National Greenbelt Program; and
- National Action Program to Combat Desertification.

Mongolia has conducted international cooperative projects for mitigating DLDD, including:

- Coping with Desertification Project, financed by the Swiss Development Cooperation Agency;
- Sustainable Land Management Project for Combating Desertification, financed by UNDP and the Government of the Netherlands;
- Greenbelt Project, financed by the Government of the Republic of Korea; and
- Implementing the Regional Master Plan for Prevention and Control of Dust and Sand Storms in Northeast Asia Project, financed by ADB, UNCCD, UNEP and UNESCAP (http://www.neaspec.org/sites/default/files/Mongolia_NCCD.pdf).

The Korea–Mongolia Greenbelt Project is a 30-year replantation project financed by the Korean Forest Service and implemented by governmental departments and civil society organizations of both Mongolia and Republic of Korea in seriously affected areas of Mongolia. The Mongolian Greenbelt Project aims to plant some 3,000 hectares of trees to mitigate drought effects and combat desertification. It is estimated that about US$7 million was provided to revegetate 1,498 hectares of trees, including *Haloxylon* spp., in arid areas. With financial support and technical assistance, it is recorded that a total of 879 hectares of degraded lands were revegetated with trees and shrubs by Korean volunteers and Mongolian residents from 2000 to 2010.

The Greenbelt Project of Mongolia is one of the largest bilateral rehabilitation projects in Mongolia and the Republic of Korea. The project was launched in 2007 and is to be completed in 2016, with a total fund of US$10 million. The objectives of the project are: a) combat desertification and control dust and sandstorms through the planting of 3,000 hectares of greenbelt; and b) enhancement of the contribution to international society by sharing successful experiences and rehabilitation technologies from the Republic of Korea. The project is also aimed at building up a foundation and establishing tree nurseries to provide seedlings for the Greenbelt Project since 2007, and about 650 hectares of degraded areas were revegetated from 2008 to 2010. The main activities of the Greenbelt Project are targeted at: a) revegetation of deforested lands in desert area; b) increasing capacity building among both Mongolian technicians and local citizens through Korean expertise; c) launching joint research; and d) supporting local communities and NGOs to improve the livelihood of the population and the ecosystem of the affected areas of Mongolia (http://www.neaspec.org/sites/default/files/Korea_Forest_Service.pdf).

Future outlook for improved regional cooperation in GCA

The Chinese Government has recently accorded the rebuilding and development of *"ecology civilization"* the same strategic importance as economic, political, cultural and social development for the country's well-being, and has mobilized the whole country to greatly promote it.

Through "ecology civilization", China is striving to build a green China and realize lasting and sustainable development of the nation. It would be impossible to mobilize the nationwide campaign without public participation!

"We call for holistic and integrated approaches to sustainable development which will guide humanity to live in harmony with nature and lead to efforts to restore the health and integrity of the Earth's ecosystem" (Rio+20 outcomes document *The Future We Want*, paragraph 40). This great goal won't be achieved without public participation!

Therefore, public awareness, political willingness, effective private–public partnership and active involvement of people from urban to rural areas are the key to achieving the targets of the Sustainable Development Goal (SDG) and land degradation neutrality (Wang et al., 2018).

To realize the above-mentioned ecology civilization and global SDG, the following approaches are essential.

In strengthening the Global Partnership, the UNCCD seeks and builds a multinational approach to address DLDD based on a number of principles, as follows:

- Urgency: We share a commitment to urgent action to address critical bottlenecks supported by an increase in investment in SLM.
- Equity: We share a commitment to reducing the social and economic inequities associated with life in drought-prone areas and will strive to ensure equity of voice and representation.
- Shared responsibility: SLM is a global public good and the shared responsibility of all members of the global community. All partners and nations therefore have a responsibility to make efficient, effective and equitable use of the resources available to them, and are individually accountable for their actions.

The UNCCD Global Partnership contains the following aspects:

- Inclusiveness: We welcome all those who share the vision, mission and values of a partnership with the UNCCD for SLM and a Land Degradation Neutral World (LDNW).
- Consensus: Recognizing the diversity of mandates and priorities of each individual partner, we function through a process of consensus and act in a coordinated manner based on the comparative strengths of individual partners.
- Sustainability: We share a commitment to effective and sustained action, and emphasize strengthening national/local capacity.
- Dynamism: The global DLDD crisis continually brings about new challenges. A partnership for SLM and LDNW must be dynamic and evolving, seeking to develop innovative opportunities that facilitate effective and concerted action.

Meeting the challenges that affected country parties face

It is necessary to encourage stakeholders/governments to formulate, design and implement private–public partnership (PPP) to address DLDD through:

- identifying priority areas and overcoming barriers;
- innovative approaches;
- providing an enabling and conducive environment for PPP by:
  - reinforcing a policy, regulatory, and economic incentive framework;
  - accelerating institutional capacity and human resource development;
  - accelerating extension services at the local level;
  - strengthening infrastructure to facilitate private investments; and
  - enhancing on-the-ground investments for improvements in the economic productivity of land through sustainable management and restoration of the structural and functional integrity of dryland ecosystems;

- accelerating PPP in addressing DLDD, which is essential to ensure the interests of people at the grassroots, and governments should offer policy guarantees and economic incentives through contracts and employment arrangements to local stakeholders;
- operationalizing PPP in internationally financed projects – this is an effective approach and therefore international finance institutions and private sectors are encouraged to be involved in land degradation neutrality and DLDD mitigation and other development projects;
- mobilizing financial resources for DLDD mitigation under the UNCCD – this needs to be emphasized and operationalized through the Global Resource Mobilization Mechanism (GM) of the UNCCD, which is established "to promote actions leading to the mobilization and channeling of substantial financial resources, including for the transfer of technology, on a grant basis, and/or on concessional or other terms, to affected developing country Parties" (Article 21 (4) of the UNCCD); and
- mobilizing resources under the three Rio Conventions from the GEF-6 to implement the Conventions. The GEF has opened a window for the UNCCD during the GEF-5 circle and offers another opportunity for funding the UNCCD and SLM issues in the affected countries through the GEF-6 circle.

Therefore, the private sector, NGOs and CSOs and other stakeholders are encouraged to join international efforts in both the financial and technological aspects to support the population and communities in the affected countries and regions to mitigate DLDD, alleviate poverty and operationalize SLM and sound resource development within the spirit of south–south cooperation.

China has accumulated great experience in combating desertification, restoring degraded lands, stabilizing sand movement and mitigating the effects of dust and sandstorms (Heshmati and Squires, 2013). A large body of well-trained technical experts is available to assist throughout GCA if called upon to do so. As part of the "Belt and Road" initiative, large sums will be allocated for "greening the Silk Road" and helping to repair the degraded lands of the countries north and west of China (Squires, Shang and Ariapour, 2017).

Note

1 The United Nations Development Program/United Nations Sudano-Sahelian Office.

References and further reading

Akiyanova, F.Z., Temirbayeva, R.K., Karynabaye, A.K. and Yegemberdiyev, K.B. 2017. Assessment of the scale, geographical distribution and diversity of pastures along Kazakhstan's part of the route of the Great Silk road. In: V. Squires, Z. Shang and A. Ariapour (eds) *Rangelands along the Silk Road: Transformative Adaptation under Climatic and Global Change*. Nova Science Publishers, New York, pp. 257–298.

Annagylyjova, J. and Squires, V.R. 2017. Turkmenistan: Rangelands and Pasturelands: Problems and Prospects. In: V.R. Squires, Z.H. Shang and A. Ariapour (eds) *Rangelands along the Silk Road: Transformative Adaptation under Climate and Global Change*. Nova Science Publishers, New York, pp. 323–329.

Ariapour, A., Badripour, H. and Jouri, M.H. 2017. Rangeland and Pastureland in Iran: Potentials, Problems and Prospects. In: V.R. Squires, Z.H. Shang and A. Ariapour (eds) *Rangelands along the Silk Road: Transformative adaptations under climate and global change*. Nova Science Publishers, New York, pp. 121–48.

Bekchanov, M., Djanibekov, N. and Lamers, J.P.A. 2018. Water in Central Asia: a cross-cutting management issue. In this volume, pp. 211–36.

Bekniyaz, B. 2011. Kazakhstan: Overcoming Constraints and Barriers to Implementation of Sustainable Land Management. In: Y. Yang, L.S. Jin, V. Squires, K.S. Kim and H.M. Park (eds) *Combating Desertification & Land Degradation: Proven Practices from Asia and the Pacific*. UNCCD/Korea Forest Service, Daejon, pp. 83–6.

China National Action Programme to Combat Desertification (Abstract), China National Committee for the Implementation of the UN Convention to Combat Desertification (CCICCD), August 1996.

Emadi, M.H. 2012. Better Land Stewardship to Avert Poverty and Land Degradation: A Viewpoint from Afghanistan. In: V. Squires (ed.) *Rangeland Stewardship in Central Asia: Balancing Improved Livelihoods, Biodiversity Conservation and Land Protection*. Springer, Dordrecht, pp. 91–108.

Heshmati, G.A. and Squires, V.R. (eds). 2013. Combating Desertification in, Asia, Africa and the Middle East: Proven Practices. Springer, Dordrecht, 538 pp.

Jia, X., Ya, J., Yuan, X., Li, J. and Wang, W. 2011. China: Combating Desertification on a Grand Scale. In: Y. Yang, L.S. Jin, V. Squires, K.S. Kim and H.M. Park (eds) *Combating Desertification & Land Degradation: Proven Practices from Asia and the Pacific*. UNCCD/Korea Forest Service, Daejon, pp. 41–67.

Khasankhanova, G. and Taryanikova, R. 2011. Uzbekistan: Sustainable Land Management in Practice. In: Y. Yang, L.S. Jin, V. Squires, K.S. Kim and H.M. Park (eds) *Combating Desertification & Land Degradation: Proven Practices from Asia and the Pacific*. UNCCD/Korea Forest Service, Daejon, pp. 223–37.

Kurbanova, B. and Squires, V.R. 2017. Rangelands and pasturelands of Tajikistan: Problems and Prospects. In: V.R. Squires, Z. Shang and A. Ariapour (eds) *Rangelands along the Silk Road: Transformative Adaptation under Climate and Global Change*. Nova Science Publishers, New York, pp. 231–256.

Mukimov, T. 2017. Rangelands and pasturelands of Uzbekistan: Problems and Prospects. In: V.R. Squires, Z. Shang and A. Ariapour (eds) *Rangelands along the Silk Road:*

Transformative Adaptation under Climate and Global Change. Nova Science Publishers, New York, pp. 217–230.

National Action Programme to Combat Desertification and Mitigate the Effects of Drought of Islamic Republic of Iran Compiled by Forest, Range and Watershed Management Organization, Tehran 2004. http://www.unccd.int/ActionProgrammes/ iran-eng2004.pdf

National Report of Islamic Republic of Afghanistan on the Implementation of United Nations Convention to Combat Desertification (UNCCD); Ministry of Agriculture and Irrigation, Kabul, Afghanistan, July 2006. http://www.unccd-prais.com/Data/Reports

Program on Combating Desertification in the Republic of Kazakhstan 2005–2015, Approved by the Resolution of the Government of Republic of Kazakhstan as of 24 January 2005 No. 49, Astana, 2005. http://www.unccd.int/ActionProgrammes/ kazakstan-eng2005.pdf

Ridder, R., Isakov, A. and Ulan, K. 2017. Transformation in pasture use in Kyrgyzstan. What are the costs of pasture degradation? In: V. Squires, Z. Shang and A. Ariapour (eds) *Rangelands along the Silk Road: Transformative Adaptation under Climatic and Global Change*. Nova Science Publishers, New York, pp. 299–322.

Safar, Y. and Squires, V.R. 2017. Rangeland and grassland in Afghanistan: its current status and future prospects. In: V.R. Squires, Z.H. Shang and A. Ariapour (eds) *Transformative Adaptation under Climate and Global Change along the Silk Road: Challenges and potentials*. Nova Science Publishers, New York, pp. 149–164.

Squires, V.R., Kurbanova, B. and Madaminov, A. 2013. Pastureland management in Tajikistan's rural regions for ecological sustainability, economic profit and social equity. In: V.R. Squires (ed.) *Rangeland Ecology, Management and Conservation Benefits*. Nova Science Publishers, New York, pp. 135–46.

Squires, V.R., Shang, Z. and Ariapour, A. (eds). 2017. *Rangelands along the Silk Road: Transformative Adaptation under Climatic and Global Change*. Nova Science Publishers, New York.

Sustainable Land Management Financing in the GEF: A primer for the sixth GEF replenishment phase (GEF-6). 2015. GEF Secretariat, Washington DC, 48 pp.

Tsed, B. 2011. Mongolia: Working with Land Users to Arrest and Reverse Desertification. In: Y. Yang, L.S. Jin, V. Squires, K.S. Kim and H.M. Park (eds) *Combating Desertification & Land Degradation: Proven Practices from Asia and the Pacific*. UNCCD/Korea Forest Service, Daejon, pp. 113–26.

Turkmenistan – 20 years along the way of Independence, Photo Album.-A: Turkmen State Publishing Service, 2012; TSBCh No. 332, 2011, BBK 663.3 (2 Tu) + 66.4 (2 Tu).

United Nations Convention to Combat Desertification Performance Review and Assessment of Implementation System. Fourth UNCCD Reporting Cycle, 2010–2011 Leg Report for China. http://www.unccd-prais.com/Data/Reports

United Nations Convention to Combat Desertification Performance Review and Assessment of Implementation System Fourth UNCCD Reporting Cycle, 2010–2011 Leg Report for Iran (Islamic Republic of). http://www.unccd-prais.com/Data/Reports

United Nations Convention to Combat Desertification Performance Review and Assessment of Implementation System Fourth UNCCD Reporting Cycle, 2010–2011 Leg Report for Kyrgyzstan. http://www.unccd-prais.com/Data/Reports

United Nations Convention to Combat Desertification Performance Review and Assessment of Implementation System Fourth UNCCD Reporting Cycle, 2010–2011 Leg Report for Mongolia. http://www.unccd-prais.com/Data/Reports

United Nations Convention to Combat Desertification Performance Review and Assessment of Implementation System Fourth Reporting and Review Cycle – 2010 Report for Tajikistan. http://www.unccd-prais.com/Data/Reports

United Nations Convention to Combat Desertification Performance Review and Assessment of Implementation System Fourth UNCCD Reporting Cycle, 2010–2011 Leg Report for Turkmenistan. http://www.unccd-prais.com/Data/Reports

United Nations Convention to Combat Desertification Performance Review and Assessment of Implementation System Fourth UNCCD Reporting Cycle, 2010–2011 Leg Report for Uzbekistan. http://www.unccd-prais.com/Data/Reports

United Nations Convention to Combat Desertification Performance Review and Assessment of Implementation System Fifth Reporting Cycle, 2014–2015 Leg Report from Tajikistan as Affected Country Party. prais2.unccd-prais.com/pdfs2014/ACP2014_Tajikistan_515.pdf

Wang, F., Squires, V.R. and Lu, Q. 2018. The future we want: putting aspirations for a land degradation neutral world into practice in the GCA region. In this volume, pp. 99–112.

Part IV

Thematic issues of SLM in Greater Central Asia

Sustainable land management (SLM) in Greater Central Asia is water-dependent. As a dryland region, water is a limiting factor. The two chapters here focus on water and its role.

Bekchanov, Djanibekov and Lamers discuss water in the context of its role in achieving SLM, especially in the irrigated oases (both natural and artificially created). Special consideration is given to the Aral Sea Basin and to transboundary rivers and their management

Hagg focuses on water from the mountains of Greater Central Asia – water that drives the ecological systems and provides motive power for hydroelectric power generation. In Hagg's view, this water is a resource under threat. Melting glaciers are a major source of concern, with serious implications for water users (agriculture, hydropower, industry and urban dwellers), not only within the country but also affecting millions of people in the downstream reaches of transboundary rivers.

10 Water in Central Asia

A cross-cutting management issue

Maksud Bekchanov

INTERNATIONAL WATER MANAGEMENT INSTITUTE (IWMI), COLOMBO, SRI LANKA AND CENTER FOR DEVELOPMENT RESEARCH (ZEF), BONN, GERMANY

Nodir Djanibekov

LEIBNIZ INSTITUTE OF AGRICULTURAL DEVELOPMENT IN TRANSITION ECONOMIES, GERMANY.

J.P.A. Lamers

CENTER FOR DEVELOPMENT RESEARCH (ZEF), BONN, GERMANY

Introduction

Cropland salinization, waterlogging, and land abandonment in the five former Soviet republics ("stans") in Central Asia (see Squires and Lu, 2018) are reportedly triggered by water scarcity, flooding or low water quality for irrigation (Abdullaev et al., 2005). Although these drivers and consequences are common in this region, they still belong to the series of least understood challenges. The ongoing, different forms of land degradation reduce the area of productive irrigated land, decrease crop yields and threaten ecosystems sustainability, consequently causing enormous economic losses (Nkonya et al., 2015). For instance, in Uzbekistan, one of the five "stans", more than 60% of the arable land reportedly is saline and about 885,000 ha are considered marginal, with low or even no profits during annual crop cultivation (Ministry of Agriculture and Water Resources of Uzbekistan, 2010). Concurrently, financial losses due to the forced retirement of highly saline croplands have been estimated at 12 million USD annually (World Bank, 2002). Similarly, in Tajikistan, more than 30% of the arable croplands are salt-affected whereas in Kyrgyzstan this amounts to 40% and even to 90% in Turkmenistan (Ahrorov et al., 2012). Reduced agricultural production and income impact human health and welfare, especially in the rural areas that heavily depend on irrigated agriculture (Nkonya et al., 2015). However, only a handful of studies have addressed the complex interactions among water quantity and quality, the degradation processes and their socio-economic impacts. With this overview it is intended to focus on the ongoing discussions on anthropogenic and natural causes of land degradation in Central Asia, which hitherto considered issues of water scarcity and low water quality. Because interventions to prevent land degradation can return 5 USD for each 1 USD of investment (Nkonya et al, 2015), it is often suggested that direct (land-related) and indirect (water-related) options are the priority. Reforms of institutional policies and governance are also important if land degradation is to be arrested and reversed (Lamers et al., 2014).

In the next section, the characteristics of water availability and distribution in the five "stans" are outlined. This is followed by an assessment of the consequences of the tremendous irrigation expansion that was pushed during the Soviet Union (SU) era and that has triggered, among other problems, the desiccation of the Aral Sea Basin (ASB), which is considered one of the worst environmental disasters worldwide. Next, water management institutions in the five "stans" that deal with water distribution and environmental protection issues in the ASB are briefly described. Then, the interlinkages between water availability, quality and land degradation are highlighted before discussing the options of coping with the dominating cause of cropland degradation in the region – soil salinity. A summary and conclusion section closes this synthesis.

Water availability and use in the five former Soviet republics of Central Asia

The two main rivers of the Aral Sea Basin (ASB) – the Amu Darya and Syr Darya – are the lifelines of the economies of at least four out of the five "stans". Here, irrigated agriculture plays a pivotal role in providing food security, hard currencies, and employment for a large share of the (rural) population (Dukhovny and de Schutter, 2011). The Amu Darya has a catchment area of 309,000 km^2 and is the largest river in Central Asia. With a length of 2,574 km, it connects the headwaters of the Pyanj River on the Afghan–Tajik border with the Aral Sea. The Syr Darya, with an overall catchment area of 219,000 km^2, is the second largest river in the region and stretches approximately 2,337 km, spanning from the Naryn River headwaters in Kyrgyzstan to the Aral Sea, sequentially crossing the Ferghana Valley, the Hunger Steppe and the Kyzylkum desert. Both main rivers are fed predominantly by snowmelt and glaciers located in the upstream countries Kyrgyzstan and Tajikistan. Characteristically, water discharges are maximal in summer and minimal in winter, creating favorable pre-conditions for the use of river flow for irrigated agriculture that predominates in the summer periods. The combined annual flow of both rivers is estimated at ca. 110 km^3, of which about 73 km^3 comes through the Amu Darya and ca. 37 km^3 through the Syr Darya (Figures 10.1 and 10.2).

A closer look at the historical discharges reveals that the annual flows varied substantially for both rivers, increasing in turn the risks for all water users, such as those involved in irrigated agricultural production, industrial development, and drinking water provision. In particular, the downstream countries Kazakhstan, Uzbekistan and Turkmenistan, but specifically the downstream regions in these countries, depend for their survival on a close coordination and regulation of the water flows (Dukhovny and de Schutter, 2011). To cushion temporal and spatial water deficits and to ensure hydropower generation, water reservoirs and dams have been constructed particularly in the mountainous zones of the upstream countries. During the SU era, the management of water distribution and flows was guided

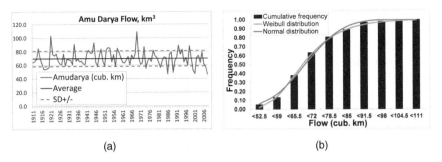

Figure 10.1 Annual amount (a) and distribution function (b) of the Amu Darya flow

Source: Based on Dukhovny et al. (2008).

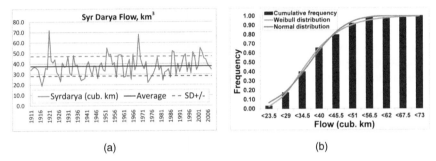

Figure 10.2 Annual amount (a) and distribution function (b) of the Syr Darya flow

Source: Based on Dukhovny et al. (2008).

by the urge to obtain an even distribution and flow along the rivers over the years and according to the necessity during the seasons, which meant providing a stable water supply for irrigating croplands, especially in midstream and downstream areas. Following independence from the SU in 1991, many reservoir operations upstream have prioritized energy production, often at the expense of securing downstream irrigation needs, thus endangering irrigated production and the environment. Although average annual water resources are numerically assessed as being more than sufficient to meet the needs (Varis, 2014), since Independence in 1991, redistribution of water resources across the basin regularly occurs, both in space and time (Dukhovny and de Schutter, 2011). This in turn results in an increasing number of conflicts among the riparian states over their common water resources. Although conflicts over water sharing among the countries in Central Asia existed even before the 1990s, their escalation was restricted by the unified coordination and inter-country "water use-energy compensation" schemes, which, however, have gradually ceased (Krutov et al., 2014; Libert et al., 2008).

Almost 90% of the basin water resources originates in the mountains of the upstream countries such as Tajikistan and Kyrgyzstan, compared to a share of

10% in Uzbekistan, and about 1% each from Kazakhstan and Turkmenistan (Figure 10.3). However, given the favorable soil and climate conditions in the midstream and downstream river areas as well as the availability of labor and arable land resources downstream, irrigated agriculture has developed and intensified over centuries on the territories of Uzbekistan and Turkmenistan (Wegerich, 2010). During the SU epoch, more than 50% of the total water withdrawals in the ASB was allocated to Uzbekistan and more than 20% to Turkmenistan. Due to the dominance of high water-demanding crops, such as cotton, in the cropping portfolio across the entire territory, and rice in the downstream regions, irrigation water used almost 90% of the total water withdrawals in the ASB (Figure 10.3).

Following the disintegration of the unified water management system after 1991, the absence of a mutual, beneficial coordination and distribution of water triggered the "use it or lose it" management approach in the water sector causing a "tragedy of commons" (Hardin, 1968). Upstream countries and regions have access to abundant water, in contrast to the downstream countries and regions. According to water consumption statistics covering 1980 to 2008, the gap between water use levels in the upper, middle and lower reaches of the rivers has even widened (Figures 10.4 and 10.5): e.g., the level of variation in water use satisfaction among the different river zones has become more pronounced after 1990, especially increasing in general the marginal value of water for downstream regions but particularly in dry years. Furthermore, inefficient water distribution strategies have generally grown more common, but again particularly in the Amu Darya Basin and during dry years (Bekchanov et al., 2010). For instance, in 2000,

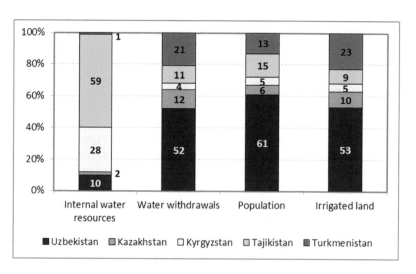

Figure 10.3 Distribution of runoff, water use, land, and population in the Aral Sea Basin by five Central Asian countries

Source: Based on McKinney (2004).

Note: Northern parts of Afghanistan and Iran are not included.

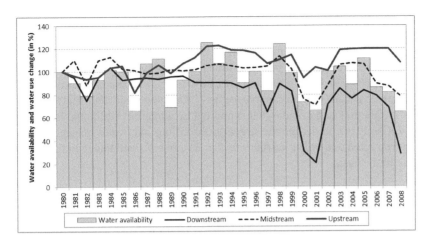

Figure 10.4 Water withdrawals in upstream, midstream and downstream reaches, and water availability during the growing season in the Amu Darya Basin, 1980–2008

Source: Data on water availability (river runoff) was taken from Dukhovny et al. (2008); water supply (source) for 2008 was extrapolated based on Dukhovny et al. (2008); water flow observations at Kerki station were selected from secondary sources (UzHydromet 2009); water withdrawals upstream (represented by the regions of Tajikistan), midstream (represented by the Kashkadarya region of Uzbekistan) and downstream (represented by the Autonomous Republic of Karakalpakstan) are based on SIC-ICWC (2009) and SIC-ICWC (2011).

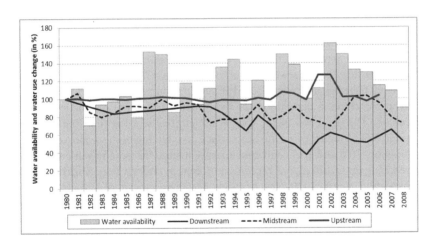

Figure 10.5 Water withdrawals in upstream, midstream and downstream reaches, and water availability during the growing season in the Syr Darya Basin, 1980–2008

Source: Data on water availability (river runoff) was taken from Dukhovny et al. (2008); water supply (source) for 2008 was based on Dukhovny et al. (2008) and water flow observations at Kal station were selected from secondary sources (UzHydromet 2009); water withdrawals upstream (represented by the regions of Kyrgyzstan), midstream (represented by Jizzakh region of Uzbekistan) and downstream (represented by the Kyzylorda region of Kazakhstan (1980–1990) and Kyzylkum canal (1991–2006)) are based on SIC-ICWC (2009) and SIC-ICWC (2011).

2001 and 2008, years during which overall water supply dramatically decreased in Central Asia, the upstream regions experienced much less water deficiency than the downstream regions, which experienced even severe periods of drought. Water abundance, defined as the ratio of total water withdrawal to the total amount of required water, was 90% in the upstream regions such as Tajikistan, but not more than 40% and 45% in the downstream regions – Dashauz (Turkmenistan) and Karakalpakstan (Uzbekistan) respectively (Dukhovny and de Schutter, 2011). These inequalities in particular, but supported by numerous additional indicators (e.g. Varis, 2014), underline the urgent need to improve water management in the entire region and create institutions that ensure efficient and equitable water sharing in the basin.

The Aral Sea desiccation

The established irrigation and drainage network during the SU era was enormous and became one of the largest in the world. The irrigated area in the ASB grew from ca. 4.5 million ha (Mha) in the 1960s to 7.9 Mha in the 1990s. Since then, about 96–100 km^3 or 90% of the total water consumption in ASB is used for irrigated agriculture, mostly for irrigated cotton production. These impressive endings contrast, however, with the perception that the ASB has become the epi-center of a large, human-made environmental disaster that has reached global dimensions: the Aral Sea Syndrome (WBGU, 1998). The excessive diversion of river water for irrigation has led to the gradual desiccation of the Aral Sea (Figures 10.4 and Figure 10.5). In 1960, the Aral Sea was the fourth largest freshwater lake in the world, with a depth of 53 m, a volume of 1,064 km^3, and a surface area of 66,000 km^2 (Mirzaev and Khamraev, 2000), which is equivalent to the approximate area of the territories of the Netherlands and Belgium together.

Inflows to the Aral Sea were about 50 km^3 annually and the Sea was vital for regional fisheries and water-based transportation, and integral to maintaining favorable climatic conditions for living and agricultural production in the Aral region. The diversion of the river waters throughout the first four decades resulted in a 60% reduction of the Sea surface, a reduction of the water volume of 80% (Figures 10.6 and 10.7) and concurrently a large increase in salt concentration, reaching up to 100 gL^{-1}. According to recent overviews (e.g. Micklin, 2007; Rudenko and Lamers, 2010), the Aral Sea became a mosaic of various smaller, mainly disconnected water bodies known as the "Small" Aral in the north in the territory of Kazakhstan and the "Large" Aral in Uzbekistan. Due to the continuous drying of the Large Aral, it turned into the "West" and "East" Aral. These water bodies are now surrounded by the newly created Aralkum desert, which refers to the area in the epicenter of the Aral Sea crisis region (Breckle et al., 2012; Orlovsky and Orlovsky, 2018).

The desiccation of the Aral Sea was expected by SU planners and policy-makers even from the beginning while realizing certain environmental and socio-economic costs (Micklin, 2010). These estimates, however, turned out to be far below actual costs. Because of the worsening environmental and health conditions, the circum-Aral Sea region was declared an ecological disaster zone

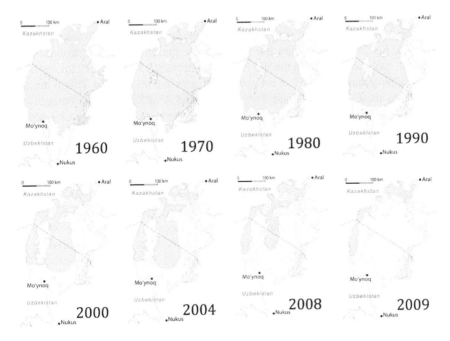

Figure 10.6 The desiccation of the Aral Sea over time (1960–2009)

Source: Based on Kartenwerkstatt (2008).

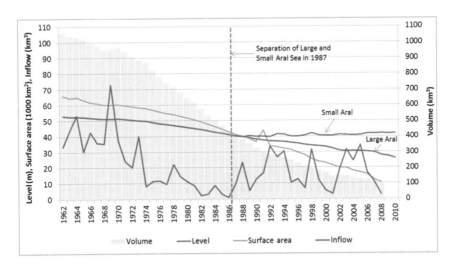

Figure 10.7 Annual inflows to the Aral Sea over time (1962–2010)

Source: Based on data for 1962–2000 from INTAS (2006), data on the sea level for 2002–2007 from Zonn et al. (2009), and data for the Small and Large Aral Sea levels for 1987–2010 from Micklin (2010). Data on annual inflows and surface area was extrapolated based on river flow observations at the hydroposts (Farkhad GES in the Syr Darya and Chatli in the Amu Darya).

even during the late Soviet era, on October 27, 1989 (Levintanus, 1992; Mirzaev and Valiev, 2000).

The rapid expansion of the irrigated croplands and the concurrent shrinking of the Aral Sea have nevertheless been a creeping issue, and gradually various hypothetical "solutions" to refill the Sea have been debated. Championed, for instance, was tapping water from Siberian streams, channeling this via the Turgai depression into the Caspian Sea and from there to the Aral Sea. But the compulsory lift of at least 80 meters of massive amounts of water would require a mammoth amount of energy. Another idea was to speed up the ice melting process from the glaciers in the Pamir Mountains (the origins of the Amu Darya and Syr Darya), or spraying sodium-iodine crystals to produce rain clouds, or damming the two rivers and channeling the water flow directly to the Aral Sea. All these ambitious options could not be realized owing to huge costs and adverse impacts on the environment. Despite a wealth of archived documents underscoring the drying out of the Aral Sea, the latter is still shrinking, since large amounts of water are being extracted from the rivers for irrigation, drinking or industrial purposes for millions of people.

Water management institutions in Central Asia

The emergence of the five new, independent states in Central Asia resulted in new issues over water management that consequently required new institutional settings to solve the challenges that used to be associated with sharing water resources and the use of infrastructure. From the outset, a need for cooperation on water management among the five Central Asian states was recognized. This led to the establishment of the Interstate Commission for Water Management Coordination (ICWC) in 1992. The ICWC was mandated in particular to: (i) determine and approve annual water withdrawal limits for each state; (ii) approve the reservoir operation regimes; and (iii) regulate the rational use and protection of the water resources in the ASB. The two Basin Water Management Organizations (BWMOs), one for Amu Darya and the other for Syr Darya, were made subordinate to the ICWC. Together, these two organizations were given the responsibility for water allocation plans, water quality control, and environmental protection throughout the entire ASB (Vinogradov and Langford, 2001).

According to the agreements, the rules of water allocation adopted during the Soviet period remained temporarily valid despite the fact that the five new republics had different interests regarding water resource use (Dukhovny and de Schutter, 2011; Mirzaev, 2000). Once the upstream states claimed that the historical water distribution rights did not consider the current needs for irrigation development and energy generation, it was decided to increase control over their own water resources, meaning those formed within their respective territories, rather than releasing these water resources aiming at the benefits of the other downstream countries of the ASB. This decision was also fueled by payment delays and disagreements in energy trade among these countries. Consequently, a new round of agreements took place between 1993 and 1995, resulting in the establishment of four additional intergovernmental water management organizations for maintaining and enhancing regional cooperation (Vinogradov and Langford, 2001; Cruz Del-Rosario, 2009):

1   The Interstate Council on the ASB (ICAS), with the task of establishing water management policies and providing inter-sectoral water use coordination;
2   The Executive Committee of the ICAS (EC-ICAS), with the purpose of implementing the Aral Sea Program financed by the World Bank, the UNDP, and the UNEP;
3   The International Fund for Saving the Aral Sea (IFAS), with the purpose of coordinating the financial resources provided by member countries and donors;
4   The Sustainable Development Commission (SDC), established to ensure an equal importance of economic, social and environmental factors during decision making.

The five "stans" also adopted an agreement on March 26, 1993 that committed them to maintaining cooperative management of the water resources in the basin and combining their efforts to solve the Aral Sea crisis. The agreement included rules to ensure a minimal flow of the key rivers into the Aral Sea and the river deltas, and to prevent the discharge of municipal and industrial wastewater, agricultural return flows, and other pollution sources into the rivers (Nanni, 1996).

In 1997, IFAS was established as the successor to the former ICAS and the previous structure of IFAS was reformed. IFAS thus became the highest political authority and was led by a board composed of the deputy prime ministers from each of the five member states, with portfolios comprising agriculture, industry and the environment (Vinogradov and Langford, 2001). Yet, all decisions by IFAS need approval from the heads of state. The board makes the final recommendations on policies and proposals suggested by a permanent working body of the fund called the Executive Committee of IFAS. The board usually meets three times a year. While IFAS is responsible for policy and financial decisions, ICWC and its subordinates are the implementing agencies (Vinogradov and Langford, 2001). Specifically, ICWC controls compliance with the interstate agreements on water distribution, distributes annual water quotas (limits) to the countries and the Aral Sea, and develops measures to maintain water supply and distribution regimes.

One of the main activities of IFAS was to implement the Aral Sea Basin Program (ASBP; 1993–2002) as a joint action plan with the World Bank, UNDP and UNEP. This plan aimed at maintaining a sustainable environment in the ASB, restoring the environmentally devastated zones adjacent to the sea, promoting improved water management on the transboundary rivers, and maintaining the capacity of the regional and local water management organizations responsible for the implementation of the plan (Sehring and Diebold, 2012). In 2003, IFAS implemented the next phase of the ASBP (ASBP-II; 2011–2015) with a focus on economic development and environmental protection plans, including the development of comprehensive water management mechanisms in the ASB, rehabilitating hydro-economic and irrigation facilities, improving environmental monitoring and flood management systems, combating desertification, promoting the rational use of return flows, and maintaining cooperation among the riparian states to effectively implement the action plan (IFAS, 2003). Nevertheless, the envisaged cooperation among the states to share common water, environmental and infrastructural resources could be only partially reached, and that only in some years. Furthermore, the high dependence of IFAS on funds from member states with a low payment mentality prevented IFAS from fulfilling most of its initially planned tasks concerning improvements of the environmental and economic situations in the circum-Aral Sea region (Weinthal, 2002). A third phase of the program (ASBP-III; 2011–2015) was approved in 2010 with the objectives of implementing IWRM principles and developing mutually beneficial agreements on water use among the basin states to maintain sustainable socio-economic development in the ASB (Dukhovny and de Schutter, 2011). It is too soon for an in-depth assessment, but in general the progress made can be considered modest at best, and that only in some countries, or better than modest in some regions only. In addition, special agreements were adopted concerning the Syr Darya Basin by the governments of Kazakhstan, Kyrgyzstan and Uzbekistan in 1998. These agreements provided the legal basis for the cooperation of the riparian states on water and energy resources in the ASB.

Nevertheless, despite their intentions, none of the agreements addressed the maintenance of the hydrological management infrastructures, nor the common view to introduce innovative water-saving technologies, nor the exchange of information or the coordination of joint activities in response to extreme events (Vinogradov and Langford, 2001). Water problems were viewed from the perspective of water consumption only (i.e., each member state attempted to increase its water use share, neglecting efficient water use). Despite an agreement among all relevant governments to consider the Aral Sea as an independent user and to guarantee at least some minimum inflow to the sea, annual river discharges have never again met the stipulated volumes (UNEP, 2005). In the context of the predicted increase of drought risks for the region, the governments are challenged to develop more effective institutions, change the organizational structure of water management, and develop innovative water allocation mechanisms that provide incentives to water users for cooperation with one another for more efficient water use, ecosystem protection, and sustainable economic development (Linn, 2008).

Key water and land degradation problems

Conceptual framework

Reductions in water availability and quality reportedly have strong implications for the ongoing land degradation processes in the region (Figure 10.6), which is alarming since irrigated land and environmental degradation processes in Central Asia threaten environmental and agricultural sustainability (World Bank, 2016). The desiccation of the Aral Sea and its consequences are the most cited example of such degradation in the region, although they are just a part of the entire ongoing environmental degradation. As the water line receded, the desiccated floor exposed the previously deposited salts, pesticides and toxic substances, which were picked up by swirling continental winds (Orlovsky and Orlovsky, 2001; Micklin 2010). Hence the desiccated area has become, according to some, a source of toxic-laden dust storms that are extremely harmful for humans and animals, as substantiated by a dramatic boost in respiratory diseases, hepatitis and anemia among the people in the vicinity of the d(r)ying Aral Sea (Rudenko and Lamers, 2010).

Agricultural land degradation and abandonments have become common in GCA (Löw et al., 2015), also because of reduced water supply and worsening water quality. Increased temperatures, also during the irrigation season, may have been contributing to water scarcity as well as consequently leading to the abandonment of irrigated and pasture lands. The lack of water supply during the hot summer months increases groundwater capillary rise, which in turn contributes to soil salinization. In downstream regions, surface water supply is challenged to face increasing pollution owing to deposits from upstream return flows. Nowadays, even surface water applications contribute more and more to soil salinization. For preventing soil salinity, historically drainage systems have been constructed, but their efficiency is highly debatable (Akramkhanov et al., 2010) whilst the clearing and maintenance costs have become immense. Due to an irregular and highly uncertain irrigation water supply and a lack of adherence to agreed delivery schedules, farmers often react by blocking drainage systems (Forkutsa et al., 2009), which obviously ensures some level of water supply, but concurrently causes water logging and soil salinization. Enormous amounts of (fresh) water

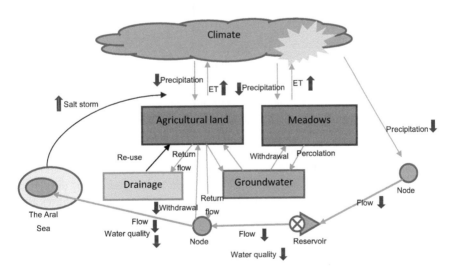

Figure 10.8 Simplified representation of key factors contributing to the ongoing land degradation within the irrigated areas of Central Asia

are being used for leaching salts out of the rooting zone and thus attempting to reduce soil salinity, frequently with limited results (Akramkhanov et al., 2010; Forkutsa et al., 2009; Qadir et al., 2014). In addition, reduced water availability is triggered not only by reduced natural river runoff, but also by increased upstream water uses for hydropower generation, which subsequently restricts water availability for irrigation and leaching. In fact, the complex interlinkages form a major challenge (Figure 10.6).

Climate change-induced water scarcity and its effect on irrigation, pastures and ecosystems

Expected temperature rises for the regions in GCA are set to become above global averages (Lioubimtseva and Henebry, 2009) and predicted to decrease upstream river runoff, increase crop evaporation requirements and increase water losses through transpiration of the water in the irrigation and drainage network, although some studies have shown that these impacts are likely to be modest (Chub, 2007). Reduced river runoff decreases irrigation water availability, which consequently may trigger cropland abandonment (Bekchanov and Lamers, 2016). Increased capillary rise of the groundwater, especially on those areas where the groundwater has high salt contents, will increase soil salinity, consequently reducing crop yields (Schieder, 2011; Dubovyk et al., 2013). Pastures also will be influenced by reduced precipitation and irrigation water availability, consequently causing lower grazing opportunities, which can be detrimental owing to a lack of alternatives (e.g., Siegmund-Schultze et al., 2013). Ecosystem services as provided, for instance, by the small-scale riparian forests in downstream regions or the numerous small water bodies in the region (Klein et al., 2014), in particular those at the tail ends of irrigation zones, will be reduced owing to water deficits (Oberkircher et al., 2011).

Hydropower generation and water availability for downstream irrigation and ecosystems

Under the centralized management system implemented during the SU epoch, a comprehensive energy production and delivery grid was developed to ensure a stable energy supply throughout the ASB (Dukhovny and de Schutter, 2011). This energy production and supply system included mutual compensation mechanisms to balance out deficiencies in hydroelectric power production by the stations of the upstream countries, and the fossil fuel-driven power stations of the downstream countries. The emergence of the five "stans" as independent countries after the disintegration of the SU in 1991 has generated different interests in issues related to sharing common water resources in the basin (Eshchanov et al., 2011). The coordination of the basin resources, which was previously conducted by a unified water management system, has become extremely challenging for water managers and policymakers alike. The dual nature of the water resources – being abundant and concurrently not always accessible and available – has hampered the further development of the local economies and is a major driver of the competition among these countries over the common water resources. Conflicts of interest between the irrigated agriculture and hydroelectric power generation sectors, together with the change in the reservoir operation mode after choosing to prioritize energy production, endanger water availability to downstream countries. Consequently, cropland degradation and abandonment in these regions are growing and are predicted to continue, unless a common understanding is reached.

The reservoirs in the ASB have initially been pivotal in ensuring timely irrigation needs in the downstream reaches. But the changes introduced in reservoir storage regimes aimed at prioritizing winter water releases for hydroelectric power generation, with the result that downstream flooding risks increased as a consequence (Mirzaev and Khamraev, 2000). For instance, between November and January 2003–2004, winter floods destroyed the water control structures and damaged settlements in the Kyzylorda region of Kazakhstan (UNEP, 2005) and more than 2,000 people needed to be evacuated and more than 55,000 ha were flooded, causing ca. 2.4 million USD worth of damage (Dukhovny and de Schutter, 2011). Flooding occurred in 2005 as well, causing ca. 7.2 million USD of damages (Dukhovny and de Schutter, 2011). Only after 2005, and with the support of the World Bank, the Kazakhstan government initiated a series of complex measures to improve the conveyance capacity of the Syr Darya riverbed located between the Aral Sea and the Shardara reservoir, and to restore partially the northern part of the Aral Sea ("Small Aral Sea").

Environmental effects of irrigation

At the start of the 19th century, the political and economic costs of maintaining imports of cotton from the US to the SU were extremely high and hence alternatives were urgently looked for. Although land, water and labor resources were abundant in the eastern parts of Russia, the cold temperatures prevailing in these regions were unsuitable for cotton cultivation. On the other hand, Central Asia had appropriate climatic conditions, sufficient land and water resources, as well as cheap labor for cotton cultivation, and hence Central Asia was turned into the "cotton basket" for the SU. The early irrigation expansion in Central Asia had mainly been driven by the increasing needs of the SU and its satellite states for securing irrigated production, in particular of the "white gold" cotton (Field, 1954).

Diverting excessive amounts of water to irrigate croplands in the upstream and midstream sections of Central Asia decreased the water flow to the downstream reaches. As a response to this, some sections of the river in the downstream areas were blocked by (small) dams to supply water to the numerous small lakes in the delta. For instance, by the 1990s, the Syr Darya had practically lost its capacity to channel excess water annually from the reservoirs to the Aral Sea. In 1995, when winter-water releases from the Tokhtogul reservoir were excessive, 19 km^3 of water (50% of the average annual flow of the Syr Darya) was wastefully discharged into the Arnasay depression because the water could no longer be directed to the Aral Sea owing to the narrow and at that time frozen downstream riverbed (Mirzaev and Khamraev, 2000: 360). The increased water levels in this depression subsequently destroyed the landscape surrounding the lake and hence aggravated the environmental problems.

The desiccation of the Aral Sea is a result of an impressive expansion of the irrigated areas and irrigation and drainage networks as well as of the unproductive use of water (Varis, 2014; Martius and Lamers, 2016). This led to the collapse of the fishing industry and water-based transportation in the lower reaches of the Amu Darya, increasing overall unemployment in the circum-Aral region as well. The reduction of the river flows into deltaic zones decreased the area of the *tugai* forests from 1,000,000 ha in 1950 to only 20,000–30,000 ha in 2000 (Micklin, 2007) and damaged habitats for a diverse array of animals, including 60 species of mammals, about 3,000 species of birds, and 20 species of amphibians (Micklin, 2010). In addition, storms blew toxic salts and dust from the dried seabed onto the surrounding irrigated areas, water bodies and pastures, increasing soil salinization and hence contributing to a further degradation of the natural ecosystems (Micklin, 2010). Airborne salts and dust are the main factors associated with the increased frequency of a host of health issues in the circum-Aral region, including illnesses of the respiratory organs, eye problems, and cancers of the throat and esophagus (Micklin, 2007). On the other hand, it was recently argued that the quantitative data on dust and salt depositions from the Aralkum on irrigated croplands is of such low quantity that they cannot explain the monitored growing salt loads in agricultural soils in Northern Uzbekistan, which are rather a result of a combination of high inputs of even modest saline irrigation water and capillary rise of the groundwater (Martius and Lamers, 2016). Hence, it was cautioned that when relying on the wrong cause of soil salinization, solutions to reduce and even arrest soil salinization can hardly be expected.

Water overuse, waterlogging, return flows and land degradation

In addition to the massive expansion of the irrigation system during the SU era, excessive water losses during the delivery of water to the fields as well as during crop irrigation also determined the still high share of the agricultural sector in total water use. Water losses are enormous also owing to the poor efficiency of the outdated irrigation and drainage (I&D) network (Kyle and Chabot, 1997; Purcell and Currey, 2003), which has been in operation without any modernization for more than three decades now. Due to the predominance of unlined, mainly earthen canals, the average efficiency of water delivery is about 60%, as estimated for the downstream areas in Uzbekistan, for example (Tischbein et al., 2012). A small proportion of water is lost through evaporation during conveyance, but most is lost by seepage and percolation along the main inter-farm and on-farm canals (Lamers et al., 2014). In addition, substantial operational water losses due to overflows from these canals into the drainage system occur regularly (Veldwisch, 2007). Because conventional irrigation methods are still common at the field level, 50% more water than the actually required volume is applied to irrigate crops (Tischbein et al., 2012).

Economic reasons for the dominance of inefficient, conventional irrigation practices were mainly related to the lack of incentives to implement modern irrigation technologies (Djanibekov et al., 2013a). At the initial stages of the irrigation and network construction, land, rather than water, was considered to be scarce (Field, 1954) and hence an efficient use of water was not prioritized. As a result of concerns about the growing environmental degradation, which became alarming after the 1970s, plans for irrigation modernization were developed. However, since water was delivered free of charge and the command-and-control-based management was interested predominantly in meeting the production targets at whatever costs, incentives for farmers to improve water use efficiency were hardly ever issued (Weinthal, 2002). Targeted investments for improving irrigation infrastructure occurred from the late 1980s onwards, but these investments discontinued following the fall of the SU and the consequent political and economic crises. In the early periods of independence, inadequate investment in the entire irrigation sector was considered the main culprit behind the accelerated deterioration of the infrastructure. Low incomes from state-controlled cotton and wheat production and insecure land rights prevented the farming population from investing in technological improvements (Djanibekov, 2008). Water was insufficiently priced for cost to be an incentive for a more efficient use of water (Djanibekov, 2008).

Irrigation expansion combined with inefficient water conveyance and application techniques resulted in huge water losses often ending up in groundwater bodies. These water losses have thus raised groundwater levels, which can be highly polluted due to the seepage of agrochemicals and salts from the cotton fields. High groundwater tables, in turn, caused waterlogging and salinization of irrigated lands and the degradation of drinking water quality, and accelerated the deterioration of the rural and urban infrastructure. Waterlogging and soil salinization gradually decreased crop yields as well: for instance, over the period 1976–1992, yields decreased in Karakalpakstan by 1.77 ton/ha for cereals, 0.98 ton/ha for cotton, 5.4 ton/ha for potatoes, 5.68 ton/ha for vegetables, 5.75 ton/ha for grapes, and 0.32 ton/ha for rice (Mirzaev and Khamraev 2000: 361). The decreased quality of drinking water, mainly in the rural areas, extracted from pipe-wells and contaminated due to the seepage of chemicals from the cotton fields, increased the number of water-borne illnesses such as typhoid, hepatitis A, and diarrheal diseases (Glantz, 1999).

Because of high percolation losses during conveyance and irrigation, the volume of return flows has increased whilst the greatest proportion of return flow ends up in desert depressions at the tails of the irrigation zones, in turn damaging the landscapes (Chembarisov, 1996; Mirzaev and Khamraev, 2000). In 1990, the total area of the tail-end desert sinks reached 6,289 km^2 with a total water volume of about 51 km^3 or virtually one-sixth of the total volume of the Aral Sea (Chembarisov, 1996). Particularly in Turkmenistan, drainage water discharge into natural depressions resulted in the formation of about 275 lakes with the total combined area of 4,286 km^2 (Mansimov, 1993), causing long-term flooding over 80,000 ha of formerly productive rangelands, periodical flooding of another

150,000 ha, and waterlogging of 2,300,000 ha (Babaev and Babaev, 1994). Some of the return flows were released back into the rivers, consequently contaminating river flows and creating external costs to the downstream water users and the environment. It was estimated that the total drainage flow in the ASB accounts for 46–47 km³, of which 25–26 km³ is discharged into rivers, 11–12 km³ into lakes, and 14–15 km³ into desert sinks (Levintanus, 1992: 63).

Another challenging issue is the high amount of small water bodies (lakes) scattered over the irrigated landscape. These small water bodies are a testimony of the deteriorated irrigation and drainage infrastructure. Many inland water bodies throughout the entire lowlands of the ASB are classified as lentic (standing-water) lakes. Owing to the endorheic character of the region, which is typified by low precipitation, and the fact that river water stems from outside the region whilst the inland water bodies are fed from underground water sources, many of these water bodies are often temporary. Combined with the poor management of irrigation water, the water losses refill these "lakes", many of them each year. On the other hand, an archive of evidence (e.g., Awan et al., 2011, 2012; Conrad et al., 2007; Ibragimov et al., 2007; Forkutsa et al., 2009; Tischbein et al., 2012) calls for the much needed increase in water use efficiency throughout GCA. But while doing so, numerous small water bodies could subsequently disappear, will reduce in size and will experience increased salinity levels. Hence, knowledge of the dynamics of the water volume of these bodies is vital to the inclusion of lake water in overall management strategies.

Direct (land-related) and indirect (water-related) measures for coping with land degradation in Greater Central Asia

The prevention of land and environmental degradation processes requires the implementation of both direct (land-related) and indirect (water-related) measures. Direct measures involve a wide range of technical options such as land reclamation, drainage improvement, soil fertility improvement, land leveling, conservation agricultural practices, afforestation and others, but also institutional measures such as improved land ownership rights or long-term land rental security, effective auxiliary services (extension, laboratories), and financial measures such as (low-interest) credits. Improved drainage system management reduces groundwater capillary rise, which allows, in turn, for effective soil leaching and thus creating favorable conditions for crop growth (Akramkhanov et al., 2010). Due to the lack of effective maintenance of the drainage system in vast areas, the reconstruction of drainage and irrigation systems is also advised (Pender et al., 2009; Toderich et al., 2008a, 2008b). Conservation agricultural practices are among the means that have been allocated a high potential in dealing with water and land issues concurrently (Kienzler et al., 2012). Integrated water and fertilizer management is an option for enhancing soil health and crop yields (Djumaniyazova et al., 2010). In marginalized lands with low water access and low soil quality, establishing vegetation by cropping drought- and salt-resistant trees is technically feasible and financially viable (Djumaeva et al., 2009; Khamzina et al., 2009,

2012; Djanibekov et al., 2013b). Alternative land uses such as crop diversification with sorghum (*Sorghum bicolor*), indigo (*Indigofera tinctoria*) or other crops such as licorice (*Glycyrrhiza glabra*) could be recommended as alternatives on saline lands due to their health and cash benefits (e.g. Bobojonov et al., 2013; Kushiev et al., 2005). A focus could also be on supporting feed demands for livestock (e.g., Siegmund-Schultze et al., 2013; Bekchanov et al., 2015d). For instance, fodder availability can be improved through establishing palatable halophytes in degraded lands (Toderich et al., 2008a, 2008b). Many more options for action have recently been assessed from different perspectives, such as with the aim of increasing water use efficiencies both at field and system level, or improving land and water use concurrently by introducing institutional and policy measures (Lamers et al., 2014; Tables 10.1 and 10.2).

Table 10.1 Technical options for action to improve irrigation water use efficiency at different levels of the irrigation and drainage systems

Options for action	*Anticipated effect on water use*
	Field level
Adjusting irrigation time and amount based on flexible irrigation scheduling at field level (using tools for modeling water fluxes and balances).	Optimizing fulfillment of time-dependent and site-specific crop water demand (in case of sufficient supply); avoiding water stress (in case of insufficient supply); minimizing impact of water stress on yield (controlled deficit irrigation).
Determining amount (and timing) of leaching (using tools for modeling water fluxes and salt dynamics/balances).	Site-specific leaching amount; improved leaching effectiveness.
Optimizing application discharge (considering irrigation method, soil characteristics, field geometry, irrigation amount).	Improving application uniformity and efficiency.
Advanced handling of furrow irrigation (surge flow, alternate furrow).	Improving application uniformity and efficiency.
Intermittent rice irrigation strategies.	Raising water productivity of rice irrigation.
Double-side irrigation.	Improving application uniformity and efficiency in case of zero-slope lands.
Laser-guided leveling.	Improving application uniformity, efficiency and leaching effectiveness.
Introduction of equipment for discharge dosage.	Achieving appropriate irrigation amount and application efficiency.
Introduction of modern irrigation techniques.	Improving irrigation efficiency and water productivity by more targeted irrigation.

(continued)

Table 10.1 (continued)

Options for action	Anticipated effect on water use
	Field level
Decentralized storage (lakes, aquifer, artificial storage).	Improving temporal matching of supply made available by the irrigation system and demand.
	System level
Linked irrigation scheduling–groundwater model.	Optimizing water distribution at WUA/WCA* level.
	Conjunctive use of surface and groundwater.
	Assessing the impact of improved efficiency on groundwater recharge and drainage.
Assessing irrigation performance.	Detecting problematic areas and reasons for problems and improving large-scale allocation.
Groundwater monitoring.	Assessing drainage performance.
	Options (regions) for conjunctive use. Improving input for balancing approaches.
Inclusion of small water bodies in water management schedules.	Increased ecosystem services benefits.
	Regional level
Basin-wide coordination.	Higher planning security for water managers and distribution managers; enhances mutual benefits and reduces water scarcity impact.
Synchronization of water needs for hydropower production and irrigation needs.	Overall increase in benefits and compromise in water-sharing disputes.
Sustainable sectoral adjustments by focusing on livestock rearing and agricultural processing.	Increase resilience to water risks and improve regional incomes.

Source: Adapted from Lamers et al. (2014).

Note: * WUA or Water Users Associations were renamed as Water Consumers Associations (WCA) in 2009.

Indirect measures to prevent land abandonment and environmental degradation are often related to improved water supply to the fields and environmental sites. The basin-wide coordination of water resources that considers interests across sectors and boundaries is advisable for preventing artificial water abnormalities caused by water reservoir mismanagement (Bekchanov et al., 2015a). The wider adoption of water-saving irrigation technologies reduces not only irrigation water demand, but also the amount of highly saline return flows (Bekchanov et al., 2015b). As a consequence of improved economic incentives for water saving, larger amounts of flow will become available for downstream environmental needs (Bekchanov et al., 2015c). The expansion of local surface

Table 10.2 Institutional and policy-oriented options for action to support and enable the use of improved technologies for increasing land and water use efficiency

Options for action	Anticipated effect on land and water use
	Institutional level
Flexible water pricing (taking into account irrigation months, crops, location) and progressive water tariffs (higher fees with increased water use).	Recognition of the (monetary) value of water leading to water saving and increased water use efficiency; conducive for adoption of innovations.
Stability of irrigation water supply.	Higher water use efficiency; reduction of water waste; lowering of groundwater level and in turn secondary soil salinization.
Farmer-supporting services (agricultural implements, closer links with research and other institutions).	Better informed farmers and land users; increased adoption of innovations; increased efficiencies of land and water; water saving.
	Policy level
Shift water use from the water-demanding agricultural sector to the less water-consuming industrial sector (cotton-processing industry).	Reducing water use and coping with water scarcity with lowest possible detriment to the regional economy.
Easing cotton policy restrictions.	Increased resource use efficiencies; increased investments and adoption of innovations; modernization of management; reduced rate of natural resource degradation.
Land tenure security.	Increased investments and adoption of innovations; modernization of management; reduced rate of resource degradation.
Increasing farmer–market links.	Higher yields; cheaper and higher quality products (fruits and vegetables); value added through processing and trade.
	International level
Coordination bodies and legislation.	Enforcing the rule of law and cooperation.
Information sharing.	Enhancing cooperation and improving decision making.

Source: Adapted from Lamers et al. (2014).

and groundwater storage also has been assessed as an effective option to cope with droughts; however, their relationship with soil salinity should be studied further (Karimov et al., 2012, 2014; Tischbein et al., 2012).

Conclusions

This overview summarizes the impacts of water scarcity and water salinity on cropland and environmental degradation as they occur in Greater Central Asia. We focus on the ASB as a specific case study but believe that the problems experienced there encapsulates the range of water- and land-related issues throughout GCA, especially in western China. The analysis of mainly secondary sources confirmed that salt and water balances at field and basin scales are highly interdependent, thus demanding concerted efforts of technological, institutional and financial policies, rather than isolated means. Direct measures should be assessed simultaneously against different perspectives. For example, land reclamation measures that would improve soil fertility and soil health could be taken jointly with indirect measures of improved water management and enhanced water supply that are effective means for the prevention of land degradation processes. It should also be noted that under the irrigation conditions that prevail in GCA, land reclamation measures influence both salinity and water balances. Hence, measures of improving water management in turn impact not only on water balances, but also on soil salt regimes. Despite the present level of understanding and knowledge, more integrated and quantification studies, especially at basin scale, are needed to increase the understanding of the complex interlinkages between water and salinity balances and the financial benefits of implementing measures of land degradation prevention at the macro-scale.

References and further reading

Abdullaev, I., Giordano, M. & Rasulov, A. 2005. Cotton in Uzbekistan: Water and welfare. In: Conference on "Cotton Sector in Central Asia: Economic Policy and Development Challenges". University of London, School of Oriental and African Studies, London, 3–4 November 2005.

Ahrorov, F., Murtazaev, O. & Abduallev, B. 2012. Pollution and salinization: compounding the Aral Sea disaster. In M.R. Edelstein, A. Cerny & A. Gadaev (eds) *Disaster by Design: The Aral Sea and its Lessons for Sustainability*. Research in Social Problems and Public Policy 20. Bingley, Emerald Group Publishing Limited, pp. 29–36.

Akramkhanov, A., Ibrakhimov, M. & Lamers, J.P.A. 2010. Managing soil salinity in the lower reaches of the Amudarya delta: How to break the vicious circle? Case Study #8-7. In: Per Pinstrup-Andersen and Fuzhi Cheng (eds.), *Food Policy for Developing Countries: Case Studies*. Cornell University Press, Ithaca NY, 13 pp.

Awan, U.K., Ibrakhimov, M., Tischbein, B., Kamalov, P., Martius, C. & Lamers, J.P.A. 2011. Improving irrigation water operation in the lower reaches of the Amu Darya River – Current status and suggestions. *Irrigation and Drainage* 60 (5): 600–12.

Awan, U.K., Tischbein, B., Kamalov, P., Martius, C. & Hafeez, M. 2012. Modeling irrigation scheduling under shallow groundwater conditions as a tool for an integrated management of surface and groundwater resources. In: C. Martius, I. Rudenko, J.P.A. Lamers & P.L.G. Vlek (eds) *Cotton, water, salts and soums: economic and ecological restructuring in Khorezm, Uzbekistan*. Springer, Dordrecht, pp. 309–27.

Babaev, A. & Babaev, A.A. 1994 Aerospace monitoring of dynamic of desert geosystems under watering. *Problems of Desert Development* 1: 21–9.

Bekchanov, M. & Lamers, J.P.A. 2016. Economic costs of reduced irrigation water avail-ability in Uzbekistan (Central Asia). *Regional Environmental Change*. doi:10.1007/s10113-016-0961-z

Bekchanov, M., Lamers, J.P.A. & Martius, C. 2010. Pros and Cons of Adopting Water-Wise Approaches in the Lower Reaches of the Amu Darya: A Socio-Economic View. *Water* 2: 200–16. doi:10.3390/w2020200

Bekchanov, M., Ringler, C., Bhaduri, A. & Jeuland, M. 2015a. How would the Rogun dam affect water and energy scarcity in Central Asia? *Water International* 40 (5–6): 856–76. http://dx.doi.org/10.1080/02508060.2015.1051788

Bekchanov, M., Ringler, C., Bhaduri, A. & Jeuland, M. 2015b. Optimizing irrigation effi-ciency improvements in the Aral Sea Basin. *Water Resources and Economics*. http://dx.doi:10.1016/j.wre.2015.08.003

Bekchanov, M., Ringler, C. & Bhaduri, A. 2015c. A water rights trading approach to increasing inflows to the Aral Sea. *Land Degradation & Development*. http://dx.doi.org/10.1002/ldr.2394

Bekchanov, M., Lamers, J.P.A., Tischbein, B., Bhaduri, A. & Lenzen, M. 2015d. Input-output model-based water footprint indicators to support IWRM in the irrigated drylands of Uzbekistan, Central Asia. In: D. Borchardt, J.J. Bogardi & R.B. Ibisch (eds), *Integrated Water Resources Management: concept, research and implementation*. Springer, Dordrecht/Heidelberg/London/New York. doi:10.1007/978-3-319-25071-7

Bobojonov, I., Lamers, J.P.A., Bekchanov, M., Djanibekov, N., Franz-Vasdeki, J., Ruzimov, J. & Martius, C. 2013. Options and constraints for crop diversification: A case study in sustainable agriculture in Uzbekistan. *Agroecology and Sustainable Food Systems* 37 (7): 788–811.

Breckle, S.W., Wucherer, W., Dimeyeva, L.A. & Oga, N.P. (eds). 2012. *Aralkum – A Man-made Desert: Desiccated floor of the Aral Sea (Central Asia)*. Ecological Series Vol. 218. Springer, Heidelberg.

Cai, X., McKinney, D.C. & Rosegrant, M.W. 2003b. Sustainability analysis for irrigation water management in the Aral Sea region. *Agricultural Systems* 76: 1043–66.

Chembarisov, E. 1996. Return flows of the Basin. *Aral Herald* 1: 40–6 (in Russian).

Chub, V.E. 2007. *Climate change and its impact on hydrometeorological processes, agro-climatic and water resources of the Republic of Uzbekistan*. Center for Hydro-meteorological Service under the Cabinet of Ministers of the Republic of Uzbekistan (Uzhydromet)/Scientific and Research Hydro-meteorological Institute (NIGMI), Tashkent, Uzbekistan.

Conrad, C., Dech, S.W., Hafeez, M., Lamers, J.P.A., Martuis, G. & Strunz, G. 2007. Mapping and assessing water use in a Central Asian irrigation system by utilizing MODIS remote sensing products. *Irrig. Drain. Syst.*, 21: 197–218.

Cruz-Del Rosario, T. 2009. Risky riparianism: cooperative water governance in Central Asia. *Australian Journal of International Affairs* 3: 404–15. http://dx.doi.org/10.1080/10357710903104869

Deng, M., Long, A, Zhang, Y., Li, X. and Lei, Y. 2010. An analysis of the exploitation, cooperation and problems of trans-boundary water resources in the five Central Asian countries. *Advances in Earth Science* 2010–12.

Djanibekov, N. 2008. Introducing Water Pricing among Agricultural Producers in Khorezm, Uzbekistan: An Economic Analysis. In: J. Qi & K. Evered (eds) *Environmental Problems of Central Asia and their Economic, Social and Security Impacts*. NATO Science for Peace and Security Series C: Environmental Security, Springer, Dordrecht, pp. 217–40. http://dx.doi.org/10.1007/978-1-4020-8960-2_16

Djanibekov, N. & Valentinov, V. 2015. Evolutionary governance, sustainability, and systems theory: The case of Central Asia. In: R. Beunen, K. Van Assche & M. Duineveld (eds) *Evolutionary governance theory: Theory and applications*, Springer, Cham/Heidelberg/ New York/Dordrecht/London, pp. 119–34.

Djanibekov, N., van Assche, K., Bobojonov, I. & Lamers, J.P.A. 2012. Uzbekistan: New Farms with Old Barriers. *Europe-Asia Studies* 64 (6): 1101–26.

Djanibekov, N., Sommer, R. & Djanibekov, U. 2013a. Evaluation of effects of cotton policy changes on land and water use in Uzbekistan: Application of a bio-economic farm model at the level of a water users association. *Agricultural Systems* 118: 1–13.

Djanibekov, U., Djanibekov, N., Khamzina, A., Bhaduri, A., Lamers, J.P.A. & Berg, E. 2013b. Impacts of innovative forestry land use on rural livelihood in a bimodal agricultural system in irrigated drylands. *Land Use Policy* 35: 95–106. doi:10.1016/j.landusepol.2013.05.003.

Djumaeva, D., Djanibekov, N., Vlek, P.L.G., Martius, C. & Lamers, J.P.A. 2009. Options for optimizing dairy feed rations with foliage of trees grown in the irrigated drylands of Central Asia. *Research Journal of Agriculture and Biological Sciences* 5 (5): 698–708.

Djumaniyazova, Y., Sommer, R., Ibragimov, N., Ruzimov, J., Lamers, J.P.A. & Vlek, P.L.G. 2010. Simulating water use and N response of winter wheat in the irrigated floodplains of Northwest Uzbekistan. *Field Crops Research* 116 (2010): 239–51.

Dubovyk, O., Menz, G., Conrad, C., Lamers, J., Lee, A. & Khamzina, A. 2013. Spatial targeting of land rehabilitation: A relational analysis of cropland productivity decline in arid Uzbekistan. *Erdkunde* 67: 167–81.

Dukhovny, V.A. & de Schutter, J.L.G. 2011. *Water in Central Asia: Past, Present, Future.* Taylor and Francis, London.

Dukhovny, V.A., Sorokin, A.G. & Stulina, G.V. 2008. Should we think about adaptation to climate change in Central Asia? CAREWIB, Tashkent. Available online at: http://www.cawater-info.net/library/eng/adaptation_climate_en.pdf (accessed on 20.07.2011).

Eshchanov, B.R., Stultjes, M.G.P., Salaev, S.K. & Eshchanov, R.A. 2011. Rogun Dam: Path to Energy Independence or Security Threat? *Sustainability* 3 (9): 1573–92.

Field, N.C. 1954. The Amu Darya: a study in resource geography. *Geographical Review* 44: 528–42.

Forkutsa, I., Sommer, S., Shirokova, Y.I., Lamers, J.P.A., Kienzler, K., Tischbein, B., Martius, C. and Vlek, P.L.G. 2009. Modeling irrigated cotton with shallow groundwater in the Aral Sea Basin of Uzbekistan: I. Water dynamics. *Irrigation Science* 27 (4): 331–46.

Glantz, M.H. 1999. Creeping Environmental Problems and Sustainable Development in the Aral Sea Basin. Cambridge University Press, Cambridge, UK.

Ibragimov N., Evet, S.R., Esanbekov, Y., Kamilov, B., Mirzaev, L. & Lamers, J.P.A. 2007. Water use efficiency of irrigated cotton in Uzbekistan under drip and furrow irrigation. *Agricultural Water Management* 90: 112–20.

INTAS. 2006. *Final report: The rehabilitation of the ecosystem and bioproductivity of the Aral Sea under conditions of water scarcity.* INTAS Project-0511 REBASOWS, Tashkent.

International Fund for Saving the Aral Sea (IFAS). 2003. *Program of concrete actions on improvement of environmental and socio-economic situation in the Aral Sea Basin for the period of 2003–2010.* Dushanbe, Tajikistan.

Karimov, A., Mavlonov, A., Miryusupov, F., Gracheva, I., Borisov, V. & Abdurahmonov, B. 2012. Modelling policy alternatives toward managed aquifer recharge in the Ferghana Valley, Central Asia. *Water International* 37 (4): 380–94.

Karimov, A., Smakhtin, V., Mavlonov, A., Borisov, A., Gracheva, I., Miryusupov, F., Akhmedov, A., Anzelm, K., Yakubov, S. & Karimov, A. 2014. Managed aquifer recharge: a potential component of water management in the Syrdarya River basin. *Journal of Hydrologic Engineering.* doi:10.1061/(ASCE) HE.1943-5584.0001046

Kartenwerkstatt. 2008. Animated picture of the Aral Sea desiccation. Available online at: http://en.wikipedia.org/wiki/File:Aral_Sea.gif (accessed on 15.02.2012).

Khamzina, A., Sommer, R., Lamers, J.P.A. & Vlek, P.L.G. 2009. Transpiration and early growth of tree plantations established on degraded cropland over shallow saline groundwater table in northwest Uzbekistan. *Agricultural and Forest Meteorology* 149 (11): 1865–74.

Khamzina, A, Lamers, J.P.A. & Vlek, P.L.G. 2012. Conversion of degraded cropland to tree plantations for ecosystem and livelihood benefits. In: C. Martius, I. Rudenko, J.P.A. Lamers & Vlek, P.L.G. (eds) *Cotton, Water, Salts and Soums – Economic and Ecological Restructuring in Khorezm, Uzbekistan.* Springer, Dordrecht, pp. 235–48.

Kienzler, K., Lamers, J.P.A., McDonald, A., Mirzabaev, A., Ibragimov, N., Egamberdiev, O., Ruzibaev, E. & Akramkhanov, A. 2012. Conservation agriculture in Central Asia – What do we know and where do we go from here? *Field Crops Research* 132: 95–105.

Klein, I., Dietz, A.J., Gessner, U., Galayeva, A., Myrzakhmetov, A. & Kuenzer, P. 2014. Evaluation of seasonal water body extents in Central Asia over the past 27 years derived from medium-resolution remote sensing data. *Int J Appl Earth Obs Geoinf* 26: 335–49.

Krutov, A., Rahimov, S. & Kamolidinov, A. 2014. Republic of Tajikistan: Its role in the Management of Water Resources in the Aral Sea Basin. In: V.R. Squires, H.M. Milner & K.A. Daniell (eds) *River Basin Management in the Twenty-first Century: Understanding People and Place.* CRC Press, Boca Raton, pp. 325–44.

Kushiev, H., Noble, A.D., Abdullaev, I. & Toshbekov, U. 2005. Remediation of abandoned saline soils using Glycyrrhiza glabra: A study from the hungry steppes of Central Asia. *International Journal of Agricultural Sustainability* 3 (2): 102–13.

Kyle, S. & Chabot, P. 1997. Agriculture in the Republic of Karakalpakstan and Khorezm Oblast of Uzbekistan. Working Paper No 97-13. Cornell University Department of Agricultural, Resource, and Managerial Economics, Ithaca, NY.

Lamers, J.P.A., Vlek, P.L.G., Khamzina, A., Tischbein, B. & Rudenko, I. 2014. Conclusions and Options for Action: Conclusion, recommendations and outlook. In: J.P.A. Lamers, A. Khamzina, I. Rudenko & P.L.G. Vlek (eds) *Restructuring land allocation, water use and agricultural value chains. Technologies, policies and practices for the lower Amudarya region.* V & R Unipress, Bonn University Press, Göttingen, pp. 365–78.

Levintanus, A. 1992. Saving the Aral Sea. *International Journal of Water Resources Development* 8 (1): 60–4.

Libert, B., Orolbaev, E. & Steklov, Y. 2008. Water and energy crisis in Central Asia. *China Eurasia Forum Q* 6 (3): 9–20.

Linn, J. 2008. The impending water crisis in Central Asia: an immediate threat. Brookings Institution, 19 June. Available online at: https://www.brookings.edu/opinions/the-impending-water-crisis-in-central-asia-an-immediate-threat/

Lioubimtseva, E. & Henebry, G.M. 2009. Climate and environmental change in arid Central Asia: Impacts, vulnerability, and adaptations. *J Arid Environ* 73: 963–77. doi:10.1016/j.jaridenv.2009.04.022.

Löw, F., Fliemann, E., Abduallev, I, Conrad, C. & Lamers, J.P.A. 2015. Mapping abandoned agricultural land in Kyzyl-Orda, Kazakhstan using satellite remote sensing. *Applied Geography* 62: 377–90.

Mansimov, M. 1993. New data on lakes of Turkmenistan from the space surveys data. *Problems of Desert Development* 3: 64–7.

Martius, C. & Lamers, J.P.A. 2016. Let there be science: separating environmental misperceptions from reality in the Aral Sea Basin. In: E. Freedman & M. Neuzil (eds) *Environmental Crises in Central Asia. From Steppes to Seas, from Deserts to Glaciers.* Taylor & Francis Group, London and New York, pp. 67–77.

McKinney, D.C. 2004. Cooperative management of transboundary water resources in Central Asia. In: D. Burghart & T. Sabonis-Helf (eds) *In the Tracks of Tamerlane: Central Asia's Path to the 21st Century.* Washington, DC: National Defense University, pp. 187–220.

Micklin, P. 2007. The Aral Sea Disaster. *Annu. Rev. Earth Planet. Sci.* 35: 47–72. Available online at: http://www.annualreviews.org/doi/pdf/10.1146/annurev.earth.35. 031306.140120 (accessed on 28.07.2011).

Micklin, P. 2010. The past, present, and future of the Aral Sea. *Lakes & Reservoirs: Research & Management* 15 (3): 193–213. Available online at: http://onlinelibrary. wiley.com/doi/10.1111/j.1440-1770.2010.00437.x/abstract (accessed on 29.07.2011).

Ministry of Agriculture and Water Resources of Uzbekistan. 2010. *Land and Water Use. Values for Uzbekistan for 1992–2009.* Government of Uzbekistan, Tashkent.

Mirzaev, S.Sh. 2000. Main institutions of managing water resources in the Aral Sea basin (ASB). In: U. Umarov & A.H. Karimov (eds) *Water Resources, Problems of the Aral Sea and Environment.* Universitet, Tashkent, pp. 308–31 (in Russian).

Mirzaev, S.Sh. & Khamraev, N.R. 2000. The fate of the rivers and the water depressions of irrigation zones. In: U. Umarov & A.H. Karimov (eds) *Water Resources, Problems of the Aral Sea and Environment.* Universitet, Tashkent, pp. 357–63 (in Russian).

Mirzaev, S.Sh. & Valiev, H.I. 2000. Main principles of saving the Aral Sea and restoring ecological balance in its basin. In: U. Umarov & A.H. Karimov (eds) *Water Resources, Problems of the Aral Sea and Environment.* Universitet, Tashkent, pp. 99–120 (in Russian).

Nanni, M. 1996. The Aral Sea basin: legal and institutional issues. *Review of European Community and International Environmental Law* 5 (2): 130–7.

Nkonya, E., Mirzabaev, A. & von Braun, J. 2015. *Economics of land degradation and improvement: A global assessment for sustainable development.* Springer International Publishing, Cham, Switzerland.

Oberkircher, L., Shanafield, M., Ismailova, B. & Saito, L. 2011. Ecosystem and social construction: An interdisciplinary case study of the Shurkul lake landscape in Khorezm, Uzbekistan. *Ecol. Soc.* 16 (4): 20.

Orlovsky, L. & Orlovsky, N. 2018. Biogeography and natural resources of Greater Central Asia: an overview. In this volume, pp. 23–47.

Orlovsky, N. & Orlovsky L. 2001. White sandstorms in Central Asia. In: Yang Youlin, V.R. Squires & Lu Qi (eds) *Global Alarm: Desert and sandstorms from the world's drylands.* UN, Bangkok, pp. 169–201.

Pender, J., Mirzabaev, A. & Kato, E. 2009. Economic analysis of sustainable land management options in Central Asia. Washington, DC: Final report submitted to ADB.

Purcell, J. & Currey, A. 2003. *Gaining Acceptance of Water Use Efficiency: Framework, Terms and Definitions.* National Program for Sustainable Irrigation (NPSI).

Qadir, M., Quillérou, E., Nangia, V., Murtaza, G., Singh, M., Thomas, R.J., Drechsel, P. & Noble, A.D. 2014. Economics of salt-induced land degradation and restoration. *A United Nations Sustainable Development Journal* 38: 282–95.

Rudenko, I. & Lamers, J.P.A. 2010. The Aral Sea: An Ecological Disaster. Case Study #8-6. In: P. Pinstrup-Andersen and Fuzhi Cheng (eds) *Food Policy for Developing Countries: Case Studies*. Cornell University Press, Ithaca NY, 14 pp.

Schieder, T.-M. 2011. Analysis of water use and crop allocation for the Khorezm region in Uzbekistan. In: P. Wehrheim, A. Schoeller-Schletter and C. Martius (eds) *Continuity and Change: Land and water use reforms in rural Uzbekistan*. Studies on the Agricultural and Food Sector in Central and Eastern Europe 43. Leibniz-Institut für Agrarentwicklung in Mittel- und Osteuropa (IAMO), Halle, pp. 105–27.

Sehring, J. & Diebold, A. 2012. *From the Glaciers to the Aral Sea: Water Unites*. Available online at: http://www.waterunites-ca.org/

SIC-ICWC. 2009. Water use by regions in Uzbekistan in 2000–2008. Unpublished report. SIC-ICWC, Tashkent, Uzbekistan.

SIC-ICWC. 2011. CAREWIB (Central Asian Regional Water Information Base). Available online at: www.cawater-info.net (accessed on 27.01.2012).

Siegmund-Schultze, M., Rischkowsky, B., Yuldashev, I., Abdalniyazov, B. & Lamers, J.P.A. 2013. The emerging small-scale cattle farming sector in Uzbekistan: highly integrated with crop production but suffering from low productivity. *Journal of Arid Environments* 98: 93–104.

Squires, V.R. & Lu, Q. 2018. Greater Central Asia: its peoples and their history and geography. In this volume, pp. 252–72.

Tischbein, B., Awan, U.K., Abdullaev, I., Bobojonov, I., Conrad, C., Jabborov, H., Forkutsa, I., Ibrakhimov, M. & Poluasheva, G. 2012. Water management in Khorezm: current situation and options for improvement (hydrological perspective). In: C. Martius, I. Rudenko, J.P.A. Lamers & P.L.G. Vlek (eds) *Cotton, Water, Salts and Soums: Economic and Ecological Restructuring in Khorezm, Uzbekistan*. Springer, Dordrecht.

Toderich, K., Shoaib, I., Juylova, E., Rabbimov, A., Bekchanov, B. & Shuyskaya, E. 2008a. New approaches for biosaline agriculture development, management and conservation of sandy desert ecosystems. In C. Abdelly, M. Ozturk, M. Ashraf & K. Grignon (eds) *Biosaline agriculture and high salinity tolerance*. Birkhauser Verlag, Basel, Switzerland.

Toderich, K., Tsukatani, T., Shoaib, I., Massino, I., Wilhelm, M. & Yusupov, S. 2008b. Extent of salt-affected land in Central Asia: Biosaline agriculture and utilization of salt-affected resources. Discussion Paper no. 648. Kier Discussion Paper series. Kyoto Institute of Economic Research, Kyoto.

UNEP. 2005 *Aral Sea, GIWA Regional assessment 24*. Prepared by Severskiy, I., Chervanyov, V., Ponomarenko, Y., Novikova, N.M., Miagkov, S.V., Rautalahti, E. & Daler, D. University of Kalmar, Kalmar, Sweden.

UNEP and ENVSEC. 2011. *Environment and security in the Amu Darya basin*. UNEP, Bresson, France.

UzHydromet. 2009. *The State Water Cadastre: Water Flows at Hydroposts (Kerki and Kishlak Kal) of the Amu Darya and Syr Darya in 1980–2008*. Tashkent, Uzbekistan.

Varis, O. 2014. Curb vast water use in central Asia. *Nature* 514 (7520): 27–9.

Veldwisch, G.J. 2007. Changing patterns of water distribution under the influence of land reforms and simultaneous WUA establishment: Two cases from Khorezm, Uzbekistan. *Irrigation and Drainage Systems* 21 (3–4): 265–76.

Vinogradov, S. & Langford, V.P.E. 2001. Managing transboundary water resources in the Aral Sea basin: In search of a solution. *International Journal of Global Environmental Issues* 1: 345–62.

WBGU (German Advisory Council on Global Change). 1998. *Worlds in transition: Ways towards sustainable management of fresh water resources.* Springer-Verlag, Berlin.

Wegerich, K. 2010. *Handing over the sunset: External factors influencing the establishment of water user associations in Uzbekistan: Evidence from Khorezm Province.* Göttingen: Cuvillier Verlag.

Weinthal, E. 2002. *State Making and Environmental Cooperation: Linking Domestic and International Politics in Central Asia.* Cambridge, MA: MIT Press, pp. 274.

World Bank. 2016. *High and Dry: Climate Change, Water, and the Economy.* World Bank, Washington, DC.

World Bank for Reconstruction and Development. 2002. *Global Condition of Environment.* Tashkent.

Zonn, I.S., Glantz, M., Kosarev, A.N. & Kostianoy, A.G. 2009. *The Aral Sea Encyclopedia.* Springer-Verlag, Berlin-Heidelberg, VIII, pp. 290.

11 Water from the mountains of Greater Central Asia

A resource under threat

Wilfried Hagg

DIPARTIMENTO DI SCIENZE GEOLOGICHE E GEOTECNOLOGIE,
UNIVERSITÀ DEGLI STUDI DI MILANO-BICOCCA, ITALY

Introduction

The lowlands of Greater Central Asia are dry: steppe, semi-deserts and deserts. The main reasons for the aridity are predominant westerly winds and lee-side effects, which cause very low precipitation sums (less than 100 mm per year) in the desert regions of Gobi, Karakum and Taklamakan. As a consequence of this aridity, Greater Central Asia consists of several so-called endorheic basins, which means that their runoff does not reach the ocean, but evaporates in the dry environment. Mountains are the water towers in the region; due to orographic effects, they receive annual precipitation sums of 1000–2000 mm. Greater Central Asia is a well-known example of a region that strongly depends on runoff formed in mountain regions (Viviroli et al., 2007).

The largest of the closed drainages are the Aral Sea (~1.2 million km²), Tarim (530,000 km²) and Ili-Balkhash (413,000 km²) basins. The Aral Sea has two potential inflows, but they do not reach the lake every year: Amu-Darya (465,000 km²) drains the largest part of the Pamir and is the main tributary to the Aral Sea. About 68% of total runoff in the Aral Sea basin is formed in its catchment (United Nations Environment Programme, 2006). The remaining ~32% is provided by the Syr-Darya river (782,000 km²), which originates in the Central Tian Shan. The Tarim river is formed at the confluence of three rivers from Inner and Central Tian Shan, Eastern Pamir and Kunlun Shan. The Ili river, which flows into Lake Balkash, drains the northeastern part of the Tian Shan. Alcamo and Henrichs (2002) applied the global water model WaterGAP to identify critical regions with highly sensitive water resources. Their results show that the western part of Greater Central Asia is entirely under severe water stress, indicated by a withdrawal-to-availability ratio greater than 0.4. This measure illustrates that water stress is not only caused by low availability, but also by high withdrawals. In the five former Soviet republics, the high water consumption is particularly caused by cotton production, irrigated by old canals with high seepage losses and inefficient flooding methods (Orlovsky and Orlovsky, 2018; Bekchanov et al., 2018).

In this millennium, a series of droughts over large parts of Greater Central Asia have demonstrated the high human vulnerability of this region to water shortages. Water is an important limiting factor for ecosystems, food and fiber production, human settlements, and human health.

Greater Central Asia consists of several independent nations which partly have different needs and requirements for water resources. While downstream riparian nations conduct irrigation in summer, the mountainous upstream countries have an interest in hydropower generation during the cold season. This inevitably leads to water use conflicts along transboundary rivers. It is obvious that water is the most precious and conflict-prone natural resource in the region. In Kyrgyzstan, Tajikistan and western China, especially the Tibetan Plateau, large quantities of water are stored in the mountain glaciers, which again are strongly influenced by climate change. Kaser et al. (2010) found that in the Aral Sea basin, the share of glacier meltwater in total discharge is the highest among eighteen large river basins investigated around the globe. They concluded that a population of more than 10 million would suffer from water shortages if the glaciers disappeared. To assess future water availability, which is a prerequisite for water resources planning, a careful examination of the process chain climate–glaciers–water is essential.

Observed and predicted climate change

In the long-term meteorological data, which exists since the end of the 19th century, a constant increase of temperature can be observed (Lioubimtseva et al., 2005), whereas no significant precipitation change can be recognized throughout most of the region, with the exception of a slight decrease in the western part (Lioubimtseva and Henebry, 2009). Giese et al. (2007) conducted a statistical analysis of 21 stations with long-term data and discovered that from 1950 to 2000, a remarkably stronger warming occurred throughout the Central Asian region compared with the global mean. The strongest warming (always given per 100 years) was observed in the lowlands and foothills (3.1–3.7°C) and in intramontaneous basins (3.4–4.2°C). The corresponding global mean value is 1.9°C. Temperature increase is particularly pronounced in winter, probably due to a weakening of the Siberian anticyclone (Giese et al., 2007). The strongest increase in air temperature in the western regions of Greater Central Asia commenced in the beginning of the 1970s; from 1972 to 2000, the stations of the lowlands and foothills experienced warming by 4.2–4.8°C. Giese and Mossig (2004) also confirm the more differentiated picture of precipitation changes. Due to their cyclic behavior, trends are more difficult to determine and often depend more on the subjective choice of start and end years. Significant trends could only be found for some stations in the Tian Shan, where annual precipitation has been decreasing since 1950, contributing to the increasing aridity of the region (Giese and Mossig, 2004).

The analysis of observed climate change is hampered by the fact that many meteorological stations were closed after the collapse of the Soviet Union and by the general sparseness of measurements from high altitudes. The most sophisticated way to produce scenarios for future climatic conditions are so-called Atmospheric Ocean Global Climate Models (AOGCMs), which simulate the atmospheric response to increasing greenhouse gases on a physical basis. According to the Intergovernmental Panel on Climate Change (IPCC, 2013), the

Central Asian region (30°N to 50°N, 60°E to 75°E) will experience a warming in winter (December–February) of 1.6°C–5.2°C until 2081–2100. In summer (June–August), the corresponding values range from 1.2°C–5.7°C. The same study reveals small precipitation increases for the same period of approximately 5–7% (IPCC, 2013). This is also confirmed by the work of Malsy et al. (2012), who combined three General Circulation Models (GCMs) with two emission scenarios (A2 and B1). However, increased evaporation in the warmer climate will overcompensate the surplus in precipitation, and, overall, a further increase of aridity and desertification is expected (Lioubimtseva et al., 2005).

Glacier state and response to climate

Greater Central Asia contains the largest glacier area outside of polar regions, and is therefore referred to as the "third pole". According to the Randolph Glacier Inventory (Pfeffer et al., 2014), 30,200 glaciers with a total area of 64,497 km² exist in the western region. This ice mass corresponds to a sea water equivalent of 16.7 mm (Vaughan et al., 2013). Another large area of glaciers and ice fields exists in the Tibetan Plateau, the eastern Tian Shan and the Qilian mountains of China.

Glacier retreat caused by climate warming is a worldwide effect and also affects GCA glaciers. As in many other mountain systems of the mid-latitudes, the majority of Tian Shan glaciers were more or less in equilibrium from the late

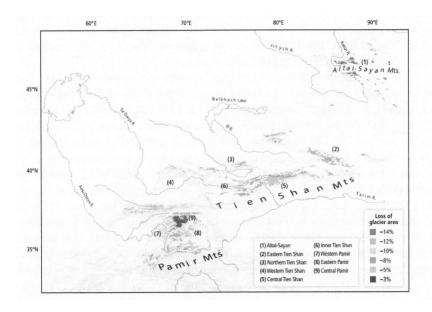

Figure 11.1 Losses of glacier area in the Altai-Sayan, Pamir and Tien Shan. Remote-sensing data analysis from the 1960s (Corona) through 2008 (Landsat, ASTER and Alos Prism)

Source: Taken from Hijioka et al. (2014).

1950s into the 1970s (Makarevich and Liu, 1995). This steady state ended in the mid-1970s as a consequence of the above-mentioned enforced warming. All recent catchment-scale studies reveal a loss of glacier area over several decades (Figure 11.1), and most of them show accelerating loss (Hijioka et al., 2014).

In the Altai-Sayan mountains, the rate of glacier area change varied between −0.13% and −0.38% per year over the past 40–50 years (Surazakov et al., 2007; Shahgedanova et al., 2010; Mayer et al., 2011). From the Qilian mountains in West China, very high retreat rates of −0.73% per year have been reported for the same period (Huai et al., 2014). The other regional studies from China report rates between −0.51% per year from the source of the Yellow River (1967–2000, Liu et al., 2002) to −0.05% per year at the source of the Yangtze River (1969–2000, Lu et al., 2002). Farinotti et al. (2015) have assessed the area retreat for the whole Tian Shan as −18% from 1961 to 2012, equaling a rate of −0.45% per year. A spatially closer look shows larger area changes in the outer ranges (−0.38 to −0.76% per year) compared with the inner (−0.15 to −0.40% per year) and eastern ranges (−0.05 to −0.31% per year) since the middle of the 20th century (Sorg et al., 2012). In general, small and low-lying glaciers in the more humid margins of the mountains suffered more than the large and high-lying ice masses in the interior. This is also the reason for the comparably small area losses from −3% to −0.30% per year in the Pamir (Konovalov and Desinov, 2007; Aizen, 2011). In the Central Pamir and the Karakoram (Gardelle et al., 2013) as well as in the Western Kunlun Shan (Neckel et al., 2014), even gains in the overall glacier mass, probably caused by increased snowfall, were observed for the most recent period. These regional variations in glacier behavior are mainly caused by differences in glacier size (large glaciers react more slowly), debris cover extent and thickness, distributions of glacier area by altitude and precipitation changes.

The estimation of future glacier extents is generally based on the output of climate scenarios and different methods to translate the climate signal into volume, area or length changes. According to Sorg et al. (2014), even in the most glacier-friendly out of four climate scenarios, the glaciers in the Chon Kemin catchment in the northern Tian Shan will lose two-thirds of their 1955 area by the end of this century. For the Aksu catchment, the most heavily glacierized part of the Tarim basin, Duethmann et al. (2016) predict an area retreat of 32–90% for the same period. Depending on the continentality of the respective location, Shi and Liu (2000) predict glacier area changes between −15% for arctic and −50% for maritime environments in West China until 2050. Based on four different global climate models, Aizen et al. (2007) make projections for the whole Tian Shan and project volume losses of up to 43% as a worst case until 2070–2099. In the best case, meaning the most glacier-friendly climate scenario, there will be almost no volume losses. This is in contrast to the findings of Farinotti et al. (2015), which are based on a combination of three independent approaches and more recent climate scenarios. They suspect that half of the current ice mass in the Tian Shan could disappear as early as the 2050s. Based on the output of 14 global climate models, Radic and Hock (2014) simulated glacier volume changes for the entire High Mountain Asia and found a mean loss of 49% until 2100.

Hydrological response

Current situation

Changes in glacier melt water amount have a huge hydrological impact in Greater Central Asia, because here the mountain ranges are surrounded by arid lowlands where water is a precious resource. This preserves the signal of snow and ice melt far downstream, because it is not superimposed by precipitation, as in more humid climates. In the Tarim basin, for example, glacier melt contributed 41.5% to total runoff (Gao 2010).

By storing solid precipitation and releasing it in summer, glaciers control the seasonal water availability. In arid regions, glacier meltwater is sometimes almost the only source of water during the main period of plant growth. As they build up their ice masses in cold-wet years and reduce them again during hot-dry years, glaciers have a compensating effect on streamflow: rivers with glaciers in their catchment have a guaranteed minimum flow, either from rainfall or from melt, whereas non-glaciated catchments completely rely on rainfall during summer. This effect reduces the year-to-year variability of runoff in moderately glacierized river basins, which is highly beneficial for drinking water security, agriculture and water resources management. Where glaciers disappear, dams can partly take over the role of inter-annual and seasonal runoff redistribution (Farinotti et al., 2016).

The responses of glacier melt water yield to gradual warming signals are outlined in Figure 11.3. The temperature increase is followed by an increase in annual streamflow as a consequence of enhanced melt (phase b). After a temporal delay which increases with glacier size, glacier tongues begin to retreat and the glacier area is reduced. This effect decelerates the increase. After a peak discharge is reached at some point in time, the area reduction overcompensates enhanced melt rates, and, as glacier degradation continues, annual glacier melt again decreases (c) and even drops below the initial level (d). When glaciers have adjusted their extent to the new climate, mean annual melt becomes stable again (e). If the warming is so severe that glaciers cannot adjust but are forced to disappear completely, glacier melt ceases towards zero (high ΔT in Figure 11.2). As glaciers are water storages and do not produce additional water, total runoff – in contrast to meltwater runoff – is not affected by the disappearance of glaciers. However, it will be lowered by the increased evapotranspiration, which is another consequence of the warming.

Currently, the mass loss of glaciers leads to increased streamflow (b and c) compared with the period before the mid-1970s, when most glaciers were in steady state. From 1961 to 2012, the Tian Shan glaciers have lost 5.4 Gt mass per year (Farinotti et al., 2015), additionally contributing to runoff. For the Syr Darya, Zeravshan, Varzob, Vanj and Gunt rivers, Finaev (2008) reports an average 13% runoff increase from 1990 to 2005. In the Tarim river, up to 20% of total runoff can be attributed to an increase in glacial melt in recent decades (Pieczonka and Bolch, 2014).

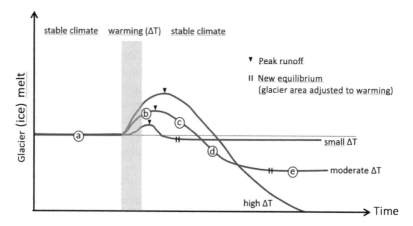

Figure 11.2 Schematic response of glacier melt water yield to a small, moderate and high signal of atmospheric warming (ΔT)

Note: Typical phases are marked in the graph for moderate warming: a – steady state in stable climate; b – increasing runoff during and after the warming; c – declining, but still increased runoff; d – declining runoff, below initial steady state; e – new equilibrium (steady state).

In heavily glacierized regions, runoff is still on the ascending part of the curve (b), whereas in little glacierized regions, peak runoff may already have passed (c) and annual flows may already be below the level of the initial steady state (d). At least for the Ile Alatau in Northern Tian Shan, Vilesov and Uvarov (2001) reported negative runoff trends caused by glacier shrinkage. Since climate warming is still ongoing today, a new equilibrium (e) cannot be reached so far.

Ice melt is concentrated in the months July to September and follows the main snow melt period. The observed warming causes an earlier snow melt, increasing runoff and flood risk in spring (Hagg et al., 2006).

Future scenarios

If observed trends are extrapolated into the near future, heavily glaciated basins are likely to still be on the ascending branch (b), because increased melt rates will overcompensate the as yet moderate area losses. This means that these catchments can still expect more water in the near future. But also here a tipping point will be reached after which water shortages can be expected. In the far future, a reduction of water availability compared with the current situation can be expected overall in Greater Central Asia. Based on an ensemble of different emission scenarios, global circulation models and regional climate models, Duethmann et al. (2016) have simulated transient runoff changes in the Aksu catchment, the main tributary of Tarim river. According to their results, discharge will be increased in the period 2010–2039 (compared with 1971–2000), but will decrease in 2070–2099. Also here, the timing of peak runoff is earlier in a sub-catchment with little glacierization (Kakshaal, around 2020) and more delayed in heavily glacierized sub-catchments (Sari-Djaz, 2030–2040).

For the whole territory of Kyrgyzstan, the United Nations Environment Programme (2009) predicts a runoff decline of 12–24% by 2050 and of 18–36% by 2100, mainly due to increasing evapotranspiration. The greatest uncertainty considering future water availability is the future precipitation climate. It is not only the annual changes that are of interest for water resources planning, but also the shifts of seasonal patterns. In a first phase, an earlier and more intense snow melt will increase runoff in spring. Summer flows are also increased due to enhanced ice melt. In a second phase, summer discharge will drop again after the glacier area goes below a certain threshold. Eventually, a thinner snow cover due to warmer winters will smooth the melt peaks in spring.

These general patterns were simulated by computer models in several catchments of the Tian Shan (Hagg et al., 2007; Sorg et al., 2012) and the Pamir (Hagg et al., 2013). The exact timing and extent of first increased and later decreased snow and ice melt depend on many factors, such as glacierization, mean snow water equivalent or the seasonality of climate change (e.g., if warming predominantly occurs in summer or winter). Figure 11.3 displays possible changes for a heavily glacierized catchment and is based on modeling attempts at Abramov glacier (Hagg et al., 2007). One can easily observe that, after glacier cover is reduced by 50% (left), runoff increases during spring and early summer, but during the main ice melt in July and August, enhanced melt rates and reduced glacier surface area perfectly balance each other in this example. If glaciers disappear completely, the whole winter snowpack is melted until June, while from July to September a massive shortage of water can be expected.

For the Pamir mountains, regional climate scenarios predict a temperature rise of 2.2°C–3.1°C until 2050, which is sufficient to reduce the glacierized area by more than 50% (Hagg et al., 2013). In the headwaters of the Amu-Darya river, only minor changes in annual runoff are expected until 2050, because here the area

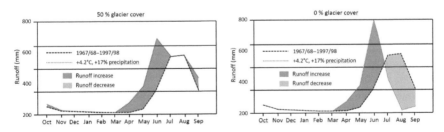

Figure 11.3   Sub-recent (1968/69–1988/89) mean monthly runoff and possible hydrographs for a future climate (+4.2°C warming, 17% precipitation increase) and two steps of deglacierization

Source: Modified after Hagg et al. (2007).

loss is also compensated by enhanced melt rates. Only the seasonal redistribution of discharge, with summer shortages of up to 30%, will unfavorably affect agriculture downstream (Hagg et al., 2013).

The Chon-Kemin river is the glacier-fed main tributary to the Chu river, which provides fresh water to Bishkek, the capital of Kyrgyzstan. Sorg et al. (2014) have applied a glacio-hydrological model to simulate the transient changes in runoff until the end of this century. They found a reduction in summer runoff by 9–66%, severely affecting the water security of Bishkek.

Another theoretical consequence of deglacierization is the loss of the compensating effect of glaciers on runoff and thus an enhanced year-to-year variability in runoff: Duethmann et al. (2016) confirm a tendency towards increased interannual variability in the Tarim basin. In other words: without glaciers, runoff depends solely on precipitation and therefore is less reliable.

Over the entire western region of Greater Central Asia, Bliss et al. (2014) found a continuous reduction in glacier runoff for the 21st century, which would mean that many glaciers and glacierized catchments have already passed peak runoff. Already now, or a bit later, "the days of plenty might soon be over in glacierized Central Asian catchments" (Sorg et al., 2014).

Conclusion

Already today, Greater Central Asia faces serious water stress, caused by increasing aridity of the climate and – so far much more important – the growing demand for water: by simulating the hydrological cycle of the catchment with and without water withdrawals, Aus der Beek et al. (2011) quantified the water use impact as high as 86%, thus only 14% is caused by climate change. According to Shiklomanov and Rodda (2001), water withdrawals in the Central Asian states of the former Soviet Union have increased by 175% over the second half of the 20th century. There was a massive increase in irrigation over recent decades: in the Aksu-Tarim region, agricultural land area more than doubled from 1989 to 2011 (Feike et al., 2015). This resulted in a decrease of downstream runoff, causing serious degradation of the riparian ecosystem (Hou et al., 2007). Also, the environmental crisis in the Aral Sea basin was mainly caused by excessive and ineffective irrigation water use (Micklin, 2007; Bekchanov et al., 2018). This irrigation mismanagement led to subsequent problems with water quality, e.g. the contamination of groundwater with salts (Saiko and Zonn, 2000).

Until the end of the 21st century, a further increase in temperature is very likely (Lioubimtseva and Henebry, 2009). Prolonged glacier retreat will transform melt-dominated runoff regimes into rain-dominated regimes. This will reduce water availability during summer months dramatically. Furthermore, evapotranspiration losses will increase in the warmer atmosphere, especially in low-lying downstream regions, further tightening water availability during the growing season.

Given the already very high level of water stress in many parts of Greater Central Asia, a further water shortage will surely increase the challenges for water management (Siegfried et al., 2012). Considering the immense dependence of

Uzbekistan, for example, on its irrigated agriculture, which consumes more than 90% of the total water resources of the Amu Darya basin, these changes will also strongly affect the economy (Schlüter et al., 2010).

Along transboundary rivers, water-related issues have already caused diplomatic tension between the independent states and may even lead to open conflicts in the future (Glantz, 2005), requiring multilateral planning of adaption and impact mitigation strategies to secure the livelihood of the population.

Acknowledgment

Wilfried Hagg's research is supported by the Heisenberg program of the Deutsche Forschungsgemeinschaft (DFG, project HA 5061/3-1).

References and further reading

Aizen, V.B. 2011. Pamirs. In: Singh, V.P., Singh, P., Haritashya, U.K. (eds) *Encyclopedia of Snow, Ice and Glaciers*. Springer, Dordrecht, pp. 813–15.

Aizen, V.B., Aizen, E.M., Kuzmichonok, V.A. 2007. Glaciers and hydrological changes in the Tien Shan: simulation and prediction. *Environ. Res. Lett.* 2 (2007): 045019. doi:10.1088/1748-9326/2/4/045019

Alcamo, J., Henrichs, T. 2002. Critical regions: A model-based estimation of world water resources sensitive to global changes. *Aquat. Sci.* 64: 352–62.

Aus der Beek, T., Voß, F., Flörke, M. 2011. Modelling the impact of global change on the hydrological system of the Aral Sea basin. *Physics and Chemistry of the Earth, Parts A/B/C* 36 (13): 684–95. ISSN 1474-7065. http://dx.doi.org/10.1016/j.pce.2011.03.004

Bekchanov, M., Djanibekov, N., Lamers, J.P.A. 2018. Water in Central Asia: a cross-cutting management issue. In this volume, pp. 211–36.

Bliss, A., Hock, R., Radic, V. 2014. Global response of glacier runoff to twenty-first century climate change. *J. Geophys. Res. – Earth Surf.* 119: 717–30.

Duethmann, D., Menz, C., Jiang, T., Vorogushyn, S. 2016. Projections for headwater catchments of the Tarim River reveal glacier retreat and decreasing surface water availability but uncertainties are large. *Environmental Research Letters* 11 (5).

Farinotti, D., Pistocchi, A., Huss, M. 2016. From dwindling ice to headwater lakes: could dams replace glaciers in the European Alps? *Environ. Res. Lett.* 11: 054022.

Farinotti, D., Longuevergne, L., Moholdt, G., Duethmann, D., Molg, T., Bolch, T., Vorogushyn, S., Güntner, A. 2015. Substantial glacier mass loss in the Tien Shan over the past 50 years. *Nat. Geosci.* 8: 716–22.

Feike, T., Mamitimin, Y., Li, L., Doluschitz, R. 2015. Development of agricultural land and water use and its driving forces along the Aksu and Tarim River, P.R. China. *Environ. Earth Sci.* 73: 517–31.

Finaev, A. 2008. Review of hydrometeorological observations in Tajikistan for the period of 1990–2005. In Braun, L., Hagg, W., Severskiy, I., Young, G. (eds) *Assessment of Snow, Glacier and Water Resources in Asia*. IHP/HWRP-Berichte 8, Koblenz, pp. 73–88.

Gao, X., Ye, B.S., Zhang, S.Q., Qiao, C.J., Zhang, X.W. 2010. Glacier runoff variation and its influence on river runoff during 1961–2006 in the Tarim River Basin, China. *Science China Earth Sciences* 53: 880–91.

Gardelle, J., Berthier, E., Arnaud, Y., Kääb, A. 2013. Region-wide glacier mass balances over the Pamir–Karakoram–Himalaya during 1999–2011. *Cryosphere* 7 (4): 1263–86. doi:10.5194/ tc-7-1263-2013.

Giese, E., Mossig, I. 2004. Klimawandel in Zentralasien. Diskussionsbeiträge. Zentrum für internationale Entwicklungs- und Umweltforschung No. 17, 70 pp. https://www.econstor.eu/bitstream/10419/50489/1/382885805.pdf

Giese, E., Mossig, I., Rybski, D., Bunde, A. 2007. Long-term analysis of air temperature trends in central Asia. *Erdkunde* 61: 186–202.

Glantz, M.H. 2005. Water, climate, and development issues in the Amu Darya basin. *Mitigation and Adaptation Strategies for Global Change* 10 (1): 23–50.

Hagg, W., Braun, L.N., Weber, M., Becht, M. 2006. Runoff modelling in glacierized Central Asian catchments for present-day and future climate. *Nordic Hydrology* 37 (2): 93–105.

Hagg, W., Braun, L.N., Kuhn, M., Nesgaard, T.I. 2007. Modelling of hydrological response to climate change in glacierized Central Asian catchments. *J. Hydrol.* 332: 40–53.

Hagg, W., Hoelzle, M., Wagner, S., Mayr, E., Klose, Z. 2013. Glacier and runoff changes in the Rukhk catchment, upper Amu-Darya basin until 2050. *Glob. Planet. Change* 110: 62–73.

Hijioka, Y., Lin, E., Pereira, J.J., Corlett, R.T., Cui, X., Insarov, G.E., Lasco, R.D., Lindgren, E., Surjan, A. 2014 Asia. In: Barros, V.R., Field, C.B., Dokken, D.J., Mastrandrea, M.D., Mach, K.J., Bilir, T.E., Chatterjee, M., Ebi, K.L., Estrada, Y.O., Genova, R.C., Girma, B., Kissel, E.S., Levy, A.N., MacCracken, S., Mastrandrea, P.R., White, L.L. (eds.) *Climate Change 2014: Impacts, Adaptation, and Vulnerability. Part B: Regional Aspects. Contribution of Working Group II to the Fifth Assessment Report of the Intergovernmental Panel on Climate Change.* Cambridge University Press, Cambridge, UK and New York, pp. 1327–70.

Hou, P., Beeton, R.J.S., Carter, R.W., Dong, X.G., Li, X. 2007. Response to environmental flows in the lower Tarim River, Xinjiang, China: an ecological interpretation of water table dynamics. *Journal of Environmental Management* 83 (4): 383–91.

Huai, B., Li, Z., Wang, S., Sun, M., Zhou, P., Xiao, Y. 2014. RS analysis of glaciers change in the Heihe River Basin, Northwest China, during the recent decades. *J. Geogr. Sci.* 24 (6): 993–1008.

IPCC. 2013. Annex I: Atlas of Global and Regional Climate Projections (eds van Oldenborgh, G.J., Collins, M., Arblaster, J., Christensen, J.H., Marotzke, J., Power, S.B., Rummukainen, M., Zhou, T.). In: Stocker, T.F., Qin, D., Plattner, G.-K., Tignor, M., Allen, S.K., Boschung, J., Nauels, A., Xia, Y., Bex, V., Midgley, P.M. (eds) *Climate Change 2013: The Physical Science Basis. Contribution of Working Group I to the Fifth Assessment Report of the Intergovernmental Panel on Climate Change.* Cambridge University Press, Cambridge, UK and New York.

Kaser, G., Großhauser, M., Marzeion, B. 2010. Contribution potential of glaciers to water availability in different climate regimes. *Proceedings of the National Academy of Sciences of the United States of America* 107 (47): 20223–7.

Konovalov, V., Desinov, L. 2007. Remote sensing monitoring of the long-term regime of the Pamirs glaciers. *International Association of Hydrological Sciences Publications* 316: 149–56.

Lioubimtseva, E., Henebry, G.M. 2009. Climate and environmental change in arid Central Asia: Impacts, vulnerability, and adaptations. *Journal of Arid Environments* 73 (11): 963–77. ISSN 0140-1963. http://dx.doi.org/10.1016/j.jaridenv.2009.04.022

Lioubimtseva, E., Cole, R., Adams, J.M., Kapustin, G. 2005. Impacts of climate and land-cover changes in arid lands of Central Asia. *Journal of Arid Environments* 62 (2): 285–308.

Liu, S., Lu, A., Ding, Y. 2002. Glacier fluctuations and the inferred climate changes in the A'nyemaqen Mountains in the source area of the Yellow River. *Journal of Glaciology and Geocryology* 24: 701–6 (in Chinese with English abstract).

Lu, A., Yao, T., Liu, S. 2002. Glacier change in the Geladanong area of the Tibetan Plateau monitored by remote sensing. *Journal of Glaciology and Geocryology* 45: 559–62 (in Chinese with English abstract).

Makarevich, K.G., Liu, C. 1995. *Glacierization of the Tian Shan* (eds Dyurgerov, M., Liu, C., Zichu, X.). VINITI, Moscow, pp. 189–213 (in Russian).

Malsy, M., Aus der Beek, T., Eisner, E., Flörke, M. 2012. Climate Change impacts on Central Asian water resources. *Adv. Geosci.* 32: 77–83.

Mayer, C., Lambrecht, A., Hagg W., Narozhny, Y. 2011. Glacial debris cover and melt water production for glaciers in the Altay, Russia. *The Cryosphere Discuss.* 5: 401–30.

Micklin, P. 2007. The Aral Sea disaster. *Annu. Rev. Earth Planet. Sci.* 35: 47–72.

Neckel, N., Kropaček, J., Bolch, T., Hochschild, V. 2014. Glacier mass changes on the Tibetan Plateau 2003–2009 derived from ICES at laser altimetry measurements. *Environmental Research Letters* 9 (1): 14009–15.

Orlovsky, L., Orlovsky, N. 2018. Biogeography and natural resources of Greater Central Asia: an overview. In this volume, pp. 23–47.

Pfeffer W.T., Arendt, A.A., Bliss, A., Bolch, T., Cogley, J.G., Gardner, A.S., Hagen, J.-O., Hock, R., Kaser, G., Kienholz, C., Miles, E.S., Moholdt, G., Moelg, N., Paul, F., Radic, V., Rastner, P., Raup, B.H., Rich, J., Sharp, M.J., Andreassen, L.M., Bajracharya, S., Barrand, N.E., Beedle, M.J., Berthier, E., Bhambri, R., Brown, I., Burgess, D.O., Burgess, E.W., Cawkwell, F., Chinn, T., Copland, L., Cullen, N.J., Davies, B., De Angelis, H., Fountain, A.G., Frey, H., Giffen, B.A., Glasser, N.F., Gurney, S.D., Hagg, W., Hall, D.K., Haritashya, U.K., Hartmann, G., Herreid, S., Howat, I., Jiskoot, H., Khromova, T.E., Klein, A., Kohler, J., Koenig, M., Kriegel, D., Kutuzov, S., Lavrentiev, I., Le Bris, R., Li, X., Manley, W.F., Mayer, C., Menounos, B., Mercer, A., Mool, P., Negrete, A., Nosenko, G., Nuth, C., Osmonov, A., Pettersson, R., Racoviteanu, A., Ranzi, R., Sarikaya, M.A., Schneider, C., Sigurdsson, O., Sirguey, P., Stokes, C.R., Wheate, R., Wolken, G.J., Wu, L.Z., Wyatt, F.R. 2014. The Randolph glacier inventory: A globally complete inventory of glaciers. *Journal of Glaciology* 60 (221): 537–52.

Pieczonka, T., Bolch, T. 2014. Region-wide glacier mass budgets and area changes for the Central Tian Shan between 1975 and 1999 using Hexagon KH-9 imagery. *Global and Planetary Change* 128: 1–13.

Radic, V., Hock, R. 2014. Glaciers in the Earth's Hydrological Cycle: Assessments of Glacier Mass and Runoff Changes on Global and Regional Scales. *Surv Geophys* 35: 813–37. doi:10.1007/s10712-013-9262-y.

Saiko, T.A., Zonn, I.S. 2000. Irrigation expansion and dynamics of desertification in Circum-Aral region of Central Asia. *Applied Geography* 20: 349–67.

Schlüter, M., Hirsch, D., Pahl-Wostl, C. 2010. Coping with change: responses of the Uzbek water management regime to socio-economic transition and global change. *Environmental Science & Policy* 13 (7): 620–36.

Shahgedanova, M., Nosenko, G., Khromova, T., Muraveyev, A. 2010. Glacier shrinkage and climatic change in the Russian Altai from the mid-20th century: an assessment using remote sensing and PRECIS regional climate model. *Journal of Geophysical Research: Atmospheres* 115: D16107. doi:10.1029/ 2009JD012976.

Shi, Y., Liu, S. 2000. Estimation of the response of the glaciers in China to the global warming of the 21st century. *Chinese Science Bulletin* 45: 668–72.

Shiklomanov, I.A., Rodda, J.C. (eds). 2001. *World Water Resources at the Beginning of the Twenty-first Century*. Cambridge University Press, Cambridge, UK, 449 pp.

Siegfried, T., Bernauer, T., Guiennet, R., Sellars, S., Robertson, A.W., Mankin, J., Bauer-Gottwein, P., Yakovlev, A. 2012. Will climate change exacerbate water stress in Central Asia? *Clim. Change* 112: 881–99.

Sorg, A., Bolch, T., Stoffel, M., Solomina, O., Beniston, M. 2012. Climate change impacts on glaciers and runoff in Tien Shan (Central Asia). *Nat. Clim. Change* 2: 725–31.

Sorg, A., Huss, M., Rohrer, M., Stoffel, M. 2014. The days of plenty might soon be over in glacierized Central Asian catchments. *Environ. Res. Lett.* 9: 104018 (8 pp). doi:10.1088/1748-9326/9/10/104018.

Surazakov, A.B., Aizen, V.B., Aizen, E.M., Nikitin, S.A. 2007. Glacier changes in the Siberian Altai Mountains, Ob river basin (1952–2006) estimated with high resolution imagery. *Environmental Research Letters* 2: 045017. doi:10.1088/ 1748-9326/2/4/045017.

United Nations Environment Programme. 2006. *Tajikistan: State of the Environment 2005*. http://ekh.unep.org/files/State%20of%20the%20Environment_1.pdf (accessed 13 April 2015).

United Nations Environment Programme. 2009. Second National Communication to the UN Convention on Climate Change. UNDP, Bishkek, 37 pp.

Vaughan, D.G., Comiso, J.C., Allison, I., Carrasco, J., Kaser, G., Kwok, R., Mote, P., Murray, T., Paul, F., Ren, J., Rignot, E., Solomina, O., Steffen, K., Zhang, T. 2013. Observations: Cryosphere. In: Stocker, T.F., Qin, D., Plattner, G.-K., Tignor, M., Allen, S.K., Boschung, J., Nauels, A., Xia, Y., Bex, V., Midgley, P.M. (eds) *Climate Change 2013: The Physical Science Basis. Contribution of Working Group I to the Fifth Assessment Report of the Intergovernmental Panel on Climate Change*. Cambridge University Press, Cambridge, UK and New York.

Vilesov, E.N., Uvarov, V.N. 2001. *Evolution of the Recent Glaciation in the Zailyskiy Alatau in the 20th Century*. Kazakh State University, Almaty (in Russian).

Viviroli, D., Durr, H.H., Messerli, B., Meybeck, M., Weingartner, R. 2007. Mountains of the world, water towers for humanity: typology, mapping, and global significance. *Water Resour. Res.* 43: W07447.

Part V

Consolidating and summarizing findings

The way forward

Plotting a path that will achieve SLM and ecological security so that the continued flow of ecological goods and services will be ensured is a major task. The three chapters here explore different aspects of the path-seeking exercise.

Squires makes the case for regarding Greater Central Asia (GCA) as the 'new frontier' in the twenty-first century as a direct consequence of China's 'Belt and Road' initiative to rejuvenate the Silk Road and the Maritime Silk Road. He argues that economic and geopolitical forces are acting to promote the GCA region to close to the top of the agenda for China, Russia, India, Pakistan, Iran and other interested parties, as vast amounts of money will continue to be invested.

Swanström explores several scenarios related to the development path of countries in GCA. He evaluates the prospects for a clear domination in the GCA region of China's or Russia's economic and geopolitical interests, or whether multilateralism will be the principal outcome.

Squires and Lu, as co-editors, seek to sum up the key points raised in the book and to provide unifying perspectives on land, water, people and development (economic, political and social). An attempt is made to put forward an agenda for future social-ecological research.

12 Greater Central Asia as the new frontier in the twenty-first century

Victor R. Squires

INSTITUTE OF DESERTIFICATION STUDIES, BEIJING, CHINA

Greater Central Asia: the region and beyond

Greater Central Asia (GCA) has been variously defined as the five post-Soviet Central Asian republics (the five 'stans'), plus Mongolia, Iran, Pakistan and Afghanistan. Throughout this book we have restricted our focus to the five 'stans', plus Mongolia and the far western parts of China, but we are mindful of the significance of the neighboring countries such as Pakistan (one of the keys to the proposed New Maritime Silk Road) and Iran and Afghanistan. At one time the five 'stans' plus Mongolia, Iran, Pakistan and Afghanistan used to form a coherent and inter-connected sub-region for millennia until the impenetrable Soviet border sliced through the continent in the early 20th century and destroyed the old trade and cultural routes. Impressive levels of trade and cultural exchanges, but also military and political competition that had effects far beyond the region itself, characterized GCA (*sensu lato*) in earlier times. Today, something similar to the old GCA region is reforming after a brief interlude with the Soviet Union (Russia), China (Mongolia) and Great Britain as lords and occupiers, even if the Central Asian states view their international relations with great suspicion and tension and regionalism is still a far-fetched concept in GCA (*sensu lato*) or in post-Soviet Central Asia. Historically, Central Asia, a confluence of three major civilizations, sits on the hub of the Silk Road, but it has failed to evolve into an independent area or region. In the second half of the 19th century, this region was a scene for struggles and rivalry between Russia and Britain, and ended up with its annexation by Tsarist Russia.

This has not hampered the importance of GCA; on the contrary, its importance has been accelerated with the Russian retreat (and partial return) from Central Asia (and Afghanistan and Mongolia), combined with a rising China that has begun to challenge Russian pre-eminence on several accounts, but also with an increased interest from India, the EU, the United States and other actors in the region and beyond. This has begun to redefine the strategic landscape in GCA, with profound implications for strategic thinking within and outside the region. This is not to say that any external power views the region as its most important focal point.

GCA (however defined) is important for many reasons, not only in terms of energy security and the combating of terrorism and fundamentalism that tend to dictate the news today. These are all important, particularly for the bordering states, but, moreover, we have seen that GCA has increasingly positioned itself as a nexus for inter-regional trade between many of the concerned states, such as China, Iran, the EU and Russia. This has significantly increased the importance of the region in a very positive way. On the negative side, GCA has become a transit route as well as the origin of some forms of organized crime, particularly heroin production, weapons and human trafficking. The creation and sustainability of more stable state structures in the region, particularly in the so-called weak states/ economies, are increasingly important, both in the context of the organized crime that thrives in weak states, and in terms of sustaining the positive political and economic developments that the region exhibits. Much of the current development in GCA has been a cause for concern on the part of the Chinese government in terms of security and also among private companies, which are plagued by the increased corruption and unpredictability of the economic system. Despite this, it is evident that the Chinese influence in the region has increased dramatically since 1991 (post-independence for the five 'stans') and China has emerged as one of the most important players in the region.

There is no doubt that the chief engine of growth in the region has been the abundant natural resources of Kazakhstan, Uzbekistan and Turkmenistan (Squires and Lu, 2018a). By 2030, the entire territory of GCA will be covered with a great infrastructure of highways, railways, airports, and logistics centers that will handle goods and passengers moving between Europe and Asia thanks to the New Silk Road initiative (see below). For many GCA countries, however, envisioning such a future is a complex matter. Political, economic, and social crises caused by the sudden collapse of the Soviet Union have dominated the relatively short history of independence enjoyed by these states. In 2011, they celebrated only the twentieth anniversary of the end of Soviet rule. Memories of wars, unresolved conflicts, economic hardships, and coups still haunt the generation old enough to remember the days of communist control. Fortunately, the most difficult times have been left behind, though a few crucial challenges persist. The countries of GCA are now at the stage of development where they must complete their political and economic transitions and choose a path that would lead them into the ranks of prosperous developed nations.

Euro-Asian trade: the big picture

The countries of GCA have always acted as a land bridge along the major commercial routes between Europe and Asia. The Silk Road trade (see below) brought wealth and prosperity to the region's inhabitants at different stages in history. The exchange of goods introduced new ideas and technologies, enriching and advancing the development of these societies. The disruption of the ancient trade routes, however, brought suffering and hardship to the region with long-lasting impacts. Some regions were gradually able to recover, while others never did. Over time, a number of commercial cities faded away as they lost the prominence they once held in the Silk Road trade, and new vibrant megacities emerged in their places. Euro-Asian trade was the economic backbone of most GCA countries for centuries. Today, the majority of this trade bypasses the region, and so do the attendant benefits. Large ships that can carry thousands of containers at a time have replaced the ancient caravans of the Silk Road. Most of the trade between Europe and Asia is conducted by maritime transportation via the Suez Canal, which makes up more than 90% of total cargo exchanged between the two continents. The success of any Central Eurasian 'hub strategy' largely depends on the ability of the regional states to attract some of this Euro-Asian continental container trade by creating integrated and competitive intermodal transportation and logistics networks across Eurasia. China is the EU's major trading partner, and the most important country for the potential land- and air-based Euro-Asian trade via Central Eurasia. Currently, most of China's industrial output comes from its eastern and south-eastern provinces. As the Chinese economy continues to grow and expand westwards, its north-western Xinjiang, which borders Central Asia, will start to generate significant extra volumes of trade.

The westward expansion of the Chinese economy will create new opportunities for the countries of Central Eurasia, aiming to increase commercial ties with neighboring Chinese provinces or attract land-based transit traffic to/from Europe. The Transport Corridor Europe–Caucasus–Asia (TRACECA) program is an EU-led international intermodal transport initiative (see above). It dates back to the May 1993 Brussels conference between three South Caucasus and five Central Asian countries. The key concerns with the TRACECA route, especially with regard to the shipment of non-oil cargo, are cost and predictability. There are significant delays caused by loading/unloading operations, border crossings, customs clearance, police checkpoints and queues along this route. The route crosses a number of countries, and therefore a number of different border and customs checkpoints. Thus there is a chain of dependency in terms of timing. In many ways, this interdependency is a positive development, particularly for the land-locked countries in the region. However, if the strategies, priorities and transport policies of the bordering states are not synchronized, this interdependence could become an impediment. This problem requires a 'bird's eye view' – not only of the national and regional sections of the TRACECA network, but also of the whole supply chain, from Europe to the Caucasus, and from GCA to China (see also above and below).

China's development initiatives in the GCA region

Much as North America built its strength from the opening of the west in the 19th century, an analogous strategy is assumed to undergird China's westward engagement. To promote cross-border trade, China has also sponsored a number of upgrades of already existing railroads and highways that link up with Central Asian, Pakistani and Iranian infrastructure. China has for long been implementing integration policies in the GCA region; the future of Xinjiang (China's westernmost region that borders four countries) is to a significant degree linked to the economic and political development in GCA.

Many issues in the region are of a multilateral rather than purely bilateral nature, but GCA's relations with China are crucial. After the collapse in 1991 of the Soviet Union and the rise of independent states in the former Soviet republics, it came as no shock to the leaders in China, both in Beijing and Xinjiang, that they would have to fight an uphill battle to gain influence, acquire economic advantages and tackle the joint security challenges across the Central Asian borders. On the contrary, the initial Russian failure to engage Central Asia was more of a shock, and China was not ready to step into the power vacuum after the Russian withdrawal; therefore, China's first few years in Central Asia were rather clumsy and without clear direction. Initially, China was not primarily interested in energy – which today has become one of the most important issues (Swanström, 2018) – but rather in border security issues and in stabilizing its own minority provinces, e.g. Xinjiang. Resolving the outstanding border issues, as well as increasing military and security cooperation with the bordering states, thus became the foremost concerns in the first years for New China, whose focus is on its economic and energy interests in the region (Squires and Lu, 2018b) but also increasingly on military security. China has enhanced its cooperation with Central Asia both bilaterally and multilaterally. Domestic security concerns have been the most important motivator behind China's strategy towards GCA, due, in large part, to the continued insecurity along the Chinese borders, namely with Tajikistan, Kyrgyzstan and Pakistan.

Regardless of intentions or lack thereof, trade, infrastructure and investments will unavoidably bring a great amount of influence that could be used in whatever way China considers to be in its national interests. The issue of national interest is clearly an issue in the GCA region for China, but this cannot simply be argued in terms of national or regional security. The five 'stans' and Afghanistan are also areas that could either function as a tool in a containment strategy from the West or Russia or as a window to Europe, Iran and the coastal regions (and sea ports) in the South. There is a real fear of containment among the Chinese elite, even if it is not necessarily seen as a likely outcome. This is not to say that China views GCA as its *Lebensraum* (定位), but more importantly as a strategic region for trade and security over the long term. This, however, has resulted in China trying to create an irrevocable presence in the region both through bilateral relations as well as through multilateral institutions and its strategy of multilateral diplomacy. China has been relatively successful in expanding its operations space within GCA, despite some noticeable drawbacks. It is apparent that China has increased its influence in the region and is expected to grow much stronger over time. Considering the impact of modern transport technologies (see below), this influence is also bound to extend further and further into the Eurasian interiors (Azerbaijan, Moldova, etc.) and to raise the stakes for passivity from other powers.

It has become apparent that a failure to understand the *practical and political* implications of China's engagement with GCA will unavoidably lead to an inadequate understanding of both the opportunities that these investments open up and the challenges they present. China has not been silent about its intention to integrate the GCA region into its fold, but on the other side they have not been explicit about it either. What needs to be understood are the silent but aggressive infrastructural investments in the region in collaboration with political cooperation and their impact (Starr, 2004).

As of today there is no clear-cut consensus on the main drivers of China's westerly expansion and their respective explanatory strengths. Most commentators consider economic concerns, especially energy resources, and food security as being China's main motivations today in GCA (Squires and Lu, 2018b; Swanström, 2018; Hagg, 2018). The reality, as is often the case, is a very mixed picture in which all factors (border security, investment opportunity) are closely connected, and it depends on of whom and when the question is asked. Arguably, there was more of a security concern in the early days of the post-Soviet Central Asian states' existence, as the borders were fragile and what China defined as Central Asian fundamentalists and extremists were perceived as supporting separatists in China. This has transferred to a much stronger economic interest and more long-term political goals. Mongolia on its side has never (in modern times) been perceived as a security threat, something that can be seen in the small interest the China's People's Liberation Army (PLA) has in Mongolia, but Mongolia has been much more of economic and strategic interest for China. Afghanistan, for its part, is the single region that has been seen as a security threat rather than an economic opportunity. The economic interest in Afghanistan is still substantial, especially in natural resources, as China tends to look at security and economic development as being in a 'Ying and Yang' relationship to each other. Without economic development in Afghanistan, there will be no security on China's south-western borders. The logic is identical to the internal argumentation on how to stabilize its own minority regions and agricultural areas, a policy which has been very successful in most cases.

Frederick Starr's edited volume *Xinjiang: China's Moslem Borderland* is an elaborate study of how these centripetal and centrifugal forces affect Xinjiang's role at the crossroads of China–GCA engagements. The Chinese strategy of expanding its influence and territorial control through trade and investment also has its historical precedents. As has been noted by other scholars, particularly Perdue (2010), there are noteworthy similarities between the Chinese Qing Dynasty's conquering of the eastern half of GCA from 1600 to 1800 with today's Chinese expansionist policies in the wider region.

Geopolitics of the Silk Road: from ancient to modern times

Lessons from history

From about 1600 to 1800, the Qing empire of China expanded to unprecedented size. Through astute diplomacy, economic investment, and a series of ambitious military campaigns into the heart of Central Eurasia, the Manchu rulers defeated the Zunghar Mongols, and brought all of modern Xinjiang and Mongolia under their control, while gaining dominant influence in Tibet. The China we know is a product of these vast conquests. Perdue (2010) chronicles this little-known story of China's expansion into the north-western frontier. Unlike previous Chinese dynasties, the Qing achieved lasting domination over the eastern half of the Eurasian continent. Rulers used forcible repression when faced with resistance, but also aimed to win over subject peoples by peaceful means. They invested heavily in the economic and administrative development of the frontier, promoted trade networks, and adapted ceremonies to the distinct regional cultures. Perdue thus illuminates how China came to rule Central Eurasia and how it justifies that control, what holds the Chinese nation together, and how its relations with the Islamic world and Mongolia developed.

China remains a multi-national and multi-ethnic state with diverse relations across its southern, northern and western borders. From the 3rd century B.C. onwards, trade contacts were made westwards along the ancient Silk Road, while by the Tang Dynasty China had established strong influence in Central Asia. In the Tang and Sung periods, China was able to enter into dialogue and synthesis among diverse religious and cultural traditions (Perdue, 2010).

The several linked trade routes that crossed Central Asia, along with related routes across southern Russia and branching lines, can thus be called the Silk Roads and Steppe Roads. The long route between East and West was in sections extremely difficult: the Gobi and Taklamakan Deserts, the Karakum and Kyzyl Kum (Uzbekistan) Deserts and the mountain heights of the Tian-shan and Pamirs make the route both difficult and dangerous. However, the route itself was not a continuous one, but was established through flows of silk from China to Rome via different groups that controlled sections of the road. The Huns, followed by the Turks and then the Persians, controlled key sectors of the trade route, followed by the Mongols and the Mongol-Turkish empire of Timur (Puri, 1987: 11–16). China reasserted strong control of the eastern section of the route between A.D. 630 and A.D. 658 with the defeat of various Turkish tribes and renewed control over the Tarim Basin and Eastern Turkestan (Puri, 1987: 48). This, of course, later on would become in part the strategic region of Xinjiang, which today remains a problematic area for Chinese political and economic governance. From the 7th century onwards, the Chinese had to face two powerful enemies in this region: the Tibetans from the south and the Persians from the west, and for a time in the late 9th century a locally powerful Uighur kingdom.

Only high prestige physical goods would be traded at great distances between East and West. The most important product was silk from China, which was exported through two routes in Central Asia – the northern one passing through Turfan, Karashahr (old Agnidesa) and Kucha, and the southern one through Miran, Niya, Khotan and Yarkand. The terminal points of the two routes at the eastern end were Tunhuang (now Dunhuang in Gansu Province) and Kashgar in Xinjing. Trade provided stimulus and incentive to the merchants of different per- sonalities for participation in it and for settling down at various points on the trade routes (Puri, 1987: 226). Silk, though the most valuable of items, especially when it reached the West (Rome and Constantinople), was in fact one among many items. Silk actually composed a relatively small portion of the trade along the

Silk Road: eastbound caravans brought gold, precious metals and stones, textiles, ivory and coral, while westbound caravans transported furs, ceramics, cinnamon bark and rhubarb as well as bronze weapons. The oasis cities of Central Asia flourished when engaged heavily in this local and international trade. One of the centers that flourished from the 16th century onwards was Bukhara in Uzbekistan.

The Silk Road was never fully destroyed but came under specific, limiting pressures. The route around the south of the Tarim Basin was eventually partially lost due to shifting rivers that led to the abandonment of centers such as Miran, Endere, Niya and areas around Khotan. Economic forces would also weaken long-distance trade as the Ottoman Turks took control of the western end of the route and as Portuguese and then the Spanish began extending ocean trade routes between Europe and Asia. After World War II, of course, the region was largely divided under the fracture lines created by the Cold War and by tensions between China and the Soviet Union, with armed borders restricting trade and influence along both east–west and north–south axes. Today, however, it is possible that new initiatives will begin to reintegrate these regions again. Dialogue and economic coordination along these 'new Silk Roads' are crucial for the future of Eurasia. If a stable government can be established in Afghanistan, and a viable state rebuilt, then this will remove a major blockage and irritant within the trade and cultural flows of Eurasia as a whole, especially along the north–south aspect (see further below). This will depend upon the stability of a new Central Asian order as it is being reshaped now.

As Ferguson (2001: 49–50) states:

> The prospective integration of Eurasia in the 21st century may have an equally significant impact on the current world system and on wider globalisation processes. China thereby has an opportunity to enrich itself (economically and culturally) through a reinvigorated 'Eurasian process' that can also help shape a more just and diverse global system. A stable Eurasian region would go some way to mitigating these dangers and reduce the attendant loss of confidence in the international financial system.

Today, China needs to be able to engage in local and global dialogue that will stabilize its prospects internally and externally. Sensitivities to such diverse legacies remain important in ensuring a stable Eurasian region. China, with its economic and cultural resources, represents one key player within Eurasia and an Asian power that can play a constructive role in the future global order. Changes to the Eurasian political system due to developments (political and military) in Afghanistan have intensified China's need to take a proactive role if its western regions and borders are to remain a zone for positive development and peaceful diplomacy.

Even if the current expansion goes far beyond modern Tibet, Xinjiang and Mongolia, which have now been brought under Chinese control, there are striking similarities in the use of economic incentives, promotion of trade networks and investments to control and subdue these territories, even if this did not always mean occupying them. Contrary to many perceptions, the Chinese did not rely exclusively on military means, but to bring external territories under its economic and cultural influence was (and still is) often a primary concern. Xinjiang is today a springboard for economic cooperation with the GCA region that lies to China's west, something that is seen in the extensive improvements in border trade and rapidly increasing trade volumes from Xinjiang to the five 'stans' and beyond.

Today, Chinese capital and trade interest are penetrating an area stretching from Azerbaijan and Iran in the west, to Pakistan in the south, and Mongolia and the five 'stans' in the north. This landmass is tied together by a number of Chinese-sponsored infrastructural interconnections, a number that is rapidly increasing, creating a Chinese economic presence as far away as Moldova. These include the arterial railroad between Kazakhstan and China at Druzbaala Pass, railway modernizations between Rawalpindi and Karachi in Pakistan, and a railway between Turkmenistan and Iran. Other proposed railway projects include a North Xinjiang railway between Urumqi and Kashgar, as well as a railroad from Tajikistan to Pakistan to support the Chinese extraction of Afghanistan's Aynak copper field. The new railway from Chendgu in Sichuan province to Eastern Europe is also a significant development.

Highways have also been built or upgraded between Xinjiang and the three neighboring 'stans' as well as the Karakorum highway linking Pakistan with Xinjiang. The Chinese government has actively promoted trade and has invested heavily in ways to improve the conditions for trade and transport. This being said, there are still major difficulties in handling the Sino–GCA trade. In the case of Afghanistan, the investments were modest up to the late 2000s, but this has changed and now China has, to give just two examples, invested US$3 billion in the Aynak copper deposits in Logar province and, as of 2012, had some 43 infrastructural projects valued at US$70 million underway in Afghanistan. Tajikistan has contracted with China to build roads and tunnels and engage in other infrastructure developments. Kazakhstan is also a major trading partner and there are many joint venture arrangements with China.

China's increasing role in the region is made easier by the fact that its engagement 'comes with none of the pesky human rights conditions, good governance requirements, approved-project restrictions and environmental quality regulations'[1] that characterize Western engagement. This has been crucial, as China has not always been seen in a positive light in the GCA region. This is especially true in both Mongolia and Kyrgyzstan, which perceive China as a potential threat, not only for historical reasons, but also due to their increased dependence politically and, more importantly, economically on China.

According to Swanström (2011):

> It was, and has often been, a direct strategy for China to compromise and give each of the bordering states enough in terms of economic and territorial gains to accept Chinese demands to support and prevent *terrorism, separatism* and *extremism*, the so-called 'three evils.' In doing this, Beijing has been a loyal supporter of the sitting governments in Central Asia and has no interest

in assisting in overthrowing the current secular governments in favor of potentially more religiously oriented regimes. Military cooperation between China and the Central Asian states [five 'stans'] has as a result increased significantly, with China focusing on anti-terrorist activities. China's military sales to Central Asia are very low and do not threaten the Russian arms sales to the region. China has refrained from openly expressing further interest in establishing military bases in any of the states. It is not a proponent of regime change in the countries in the region. China does not rule out a democratic development in the region but, according to its view, the focus should be on creating an economic base and a stronger institutional base for the governments before embarking on the political adventures of democratization. China pursues a direct strategy to field its soft power in GCA, using tools such as cultural exchanges, environmental cooperation, education and trade. The result has been favorable for China, which is increasingly perceived as a good neighbor by countries in the region, even if many difficult scars, imagined or real, from the past history of occupation, expansion and cultural hegemony are still visible. China is trying to create an irrevocable presence in the region both through bilateral relations as well as through multilateral institutions and its strategy of multilateral diplomacy.

China's economic concerns and resource hunger

Economic concerns, especially energy resources, are considered by many as China's main motivation today in GCA, something that is not entirely true. It is true that China has adopted an 'open door policy' to increase its economic benefit, which has led to an inflow of capital, energy and traded goods as well as a pooling of resources from eastern to western China. Millions of Han Chinese have migrated to the western borderlands, and China has invested massively in roads, railroads, and energy infrastructure. China has sponsored a number of upgrades of already existing railroads and highways that link China with Central Asia, Pakistan and Iran. Chinese influence (capital and trade) has spread over a vast area stretching from Azerbaijan, Moldova and Iran in the west, to Pakistan in the south, and Mongolia/Central Asia in the north. The five 'stans' have become the focal point of Chinese infrastructure investments, especially related to the energy sector and trade.

The energy resources of GCA are important sources of diversification for China's energy needs (Swanström, 2018). There has been a drive from China for alternative energy supplies, but also diversification of energy corridors bypassing the Malacca Straits. China is now a serious competitor in the previously Russian and US-dominated oil fields in Western Kazakhstan and is exploring oil and gas fields as far away as the Caspian Sea. Most of GCA is subtly but directly being pulled into China's orbit by transport and energy infrastructure investments. Even Turkey has begun to flirt with China and has initiated military cooperation with China and engaged in both political and economic dialogue and cooperation.

The reason behind China's expansive strategy (as evidenced by the New Silk Road, the 'Belt and Road' initiative, and the New Maritime Silk Road initiative that is elaborated below) is not only to secure natural resources, increase trade-related economic gains and establish trans-regional trading links (primarily with the Middle East and Europe), but also to influence and secure friendly governments in the region. This is not without problems for China, as Central Asia is one of the most corrupt regions in the world and one in which organized crime has the greatest political leverage. Few economic or political changes in GCA can be made without involving organized crime and illegal transactions (Swanström, 2011: 8).

China's land acquisitions in GCA

In recent years, China's rapid economic growth has been coupled with a rising demand for natural resources. Great international concern has arisen over China's land acquisitions for agricultural and biofuel production, pejoratively called 'land grabbing'. The significant rise in China's global activities in agriculture with particular reference to its alleged 'land grabbing' should not be seen as separate from the country's global expansion in other sectors (Squires and Lu, 2018b).

Over the past decades, China's sustained economic growth has put a rising pressure on the country's domestic natural resources. The oft cited numbers portraying the country's dire situation are that China boasts 21% of the world's population, while the country possesses only 8.5% of the world's available arable land, and 6.5% of the world's water reserves. To complicate matters, China lost 8.2 million hectares of arable land between 1997 and 2010, due to urbanization and environmental degradation (UNOHCHR, 2010).

The pressure on the country's land and water resources is unquestionable. It is manifested in the different strategies that the authorities undertake to increase domestic food production. For the government, affordable food prices are perceived as being crucial to maintain social stability and guaranteed supplies are of utmost importance. To fuel its economic development, China increasingly projects its domestic shortages to other countries and regions abroad. The stimulus for this development has become even more pressing since the country's growing middle class pursues more luxurious lifestyles and consumption patterns. An increase in a range of particular food products, such as coffee, cacao and wine, but also animal feed, are more efficiently produced overseas, and thus imply new grounds for Chinese investments. In the past five years, the country has become a major player in the global land market. New, unexpected agreements have emerged under which the Chinese government seeks to acquire large tracts of land and to access overseas resources. Global land acquisitions are high on the socio-political agenda today. The recent developments have resulted in numerous research initiatives and reports in the last five years, with fierce debates about the impacts of the investments on local livelihoods and the environment. A frequently mentioned issue by critics is that the socio-political processes through which the land use changes are implemented are undemocratic and a testimony of 'bad land governance'. The recent land acquisitions regularly contain formally arranged lease or concessionary rights, ranging from 30 to 99 years. Due to shortages in food and fuel, rapidly emerging economies have begun to outsource agricultural production by leasing or buying rural land in developing countries, including some in GCA. The diversity of Chinese investments involves multiple Chinese actors that may have distinct interests to operate overseas and expand their endeavors. Land is now acquired for industrial farming (in order to produce biofuels), timber extraction/logging, tourism, aquaculture, the establishment of special economic zones (SEZs) and industrial centers.

Acquisition of minerals, oil and gas

China's huge population, which represents a third of the world, and its unprecedented economic growth have resulted in the country's supplies of food, energy and metals and minerals falling far short of demand. This has prompted China to step up efforts to acquire farmlands, oil fields and mining assets abroad (Squires and Lu, 2018b). China has been viewed as the 'dragon' that strives to swallow mineral commodities and assets around the world. It is one of the largest consumers and producers of ores and metals. Even though China is the largest consumer of certain minerals such as copper, its domestic production accounts for less than 6% of global production. There is a widening gap in China between its mined production and consumption demand, which it is trying to fill through imports and promoting companies for overseas acquisitions. The government of China is encouraging consolidation in its domestic mining sector through mergers and acquisitions in order to reduce in-house competition and shore up finances for overseas projects.

The contexts in which Chinese (and other) actors acquire land (including mining leases) differ widely in socio-economic, political, cultural and environmental conditions, as do the particular resources that Chinese companies aim at and the approach they pursue. This influences the way in which Chinese companies approach the host society. Several studies have been conducted on China's expansion in resource extractive industries and other sectors (Swanström, 2018). Yet, China's global expansion in general is still poorly understood (Squires and Lu, 2018b; Hofman and Ho, 2012).

The New Silk Road: dream or an economic and political imperative?

From the point of view of the regions of Europe, South Asia and North-East Asia, Central Asia is a crucial linkage area of interregional contact.

China has developed a 'strategic partnership' with Russia aimed at establishing a multi-polar world, while from 1994 extensive negotiations have led to strong diplomatic ties among China, Russia, Kazakhstan, Kyrgyzstan and Tajikistan (the 'Shanghai Five', sometimes dubbed the G-5, which has since developed into the Shanghai Cooperation Organization[2]). China not only seeks to engage in trade and get better access to Central Asian energy reserves, but also has sought to create a zone of stability to its west that in some way mirrors ancient efforts to ensure peaceful frontiers. This has led to an intense interest not just in the bordering states of Central Asia, but also in the wider issue of the new geopolitics of the region. In large measure, access to this new Silk Road is a major strategy in stabilizing western borders as well as aiding an economic development strategy for the multi-ethnic region of Xinjiang. There are also key benefits for the post-Soviet states of the region in cooperation with China, especially in balancing Russian influence along a new 'silk route'. China, moreover, has also made sure that strong cooperation with Russia is sustained, and has reduced any sense of overt geopolitical contest that could lead to renewed tensions with the states of the region.

Transport links: keys to future integration

China's ambitions to improve transport routes to Europe are driven in part by a desire to improve inland and particularly western China's connections with Europe. Taken at face value, they amount to what could be the biggest shake-up in global trade routes since China began its emergence as a major player in international trade two decades ago. The vast majority of the trade between China and the EU still moves by sea. Each of the four to five trains from China that arrive in Duisburg (Germany) weekly carries 40–50 shipping containers. Modern cargo ships carry thousands. Currently, Sino–European trade is confined to the sea route or, to a minor extent, the Trans-Siberian railway or air transport, but there are prospects for new economic linkages between Europe and China. Central Asia has increasingly become a transport hub. The current revitalization of the Silk Road and development of a continental transport corridor running from China's east coast to Europe are expected to bring massive gains to the landlocked countries of GCA as well as to China and Europe. At present, the sea journey from China to Europe takes 20 to 40 days, while the new links could cut transport time to 11 days. The trans-regional trade and transport system could in fact be one of the most positive factors for the post-Soviet GCA states and one in which cooperation between China, GCA and Europe is very likely and could be profitable for all actors. Of course, this would need a more stable and economically sound GCA that has been able to control the rampant corruption and narco-trade.

The key to the success of the concept is the development of an unblocked transport network. Just like the opening up of China's eastern coastal regions, the focus should be on infrastructure construction to better realize this concept. If the infrastructure in the country's eastern regions is concentrated on developing the traffic and transport systems, such as ports, highways and high-speed railways, then infrastructure construction in its western regions needs to be focused on the establishment of an all-dimensional system, such as modern railways, expressways, energy pipelines, the power grid, telecommunications and a modern capital circulation system that operates throughout the Eurasian continent. Such development will help realize the flow of people, goods and capital on an unprecedented scale on the Eurasian continent, and this will provide unprecedented opportunities for the development of China's western regions, its overall development and even the development of all countries along the Silk Road Economic Belt.

Central Asia is situated in the heartland of Asia, and its railway network is the wide-gauge railway of Russia, with its three artery railway lines running from Almaty and Tashkent northward or north-westward to Russia, joined by the Siberian railway (the First Continental Bridge). The Second Continental Bridge, which was opened to traffic in recent years and goes westward from Alataw Pass, is linked with the old Almaty–Petropavlovsk artery line near Astana, which later joins with the First Continental Bridge. But this line is not that convenient for the following three reasons: 1) Russia is not willing to see it replace the First Continental Bridge of the Siberian railways; 2) the railway going through Kazakhstan and Russia is outdated and involves several customs formalities; and 3) both China and Europe use standard-gauge railways, while Russia mainly uses a wide-gauge one, so trains have to change rails twice when travelling from one end of Eurasia to the other.

Despite being a vast country of 3,000 kilometers from east to west, Kazakhstan has no railway going from its east to west, so it has to rely on Russia to link up with Europe. In 2004, Kazakhstan put forward a plan for a standard-gauge Pan-Eurasia rail artery, proposing a rail line that would go through Kazakhstan to the Caspian Sea, then move southward to Turkmenistan, Iran and Turkey and integrate with the European rail network.

Just as Central Asia was once connected via trade routes into the mainstream of world history and linked great civilizations of the past, today the future development of Central Asia depends on its ability to deepen infrastructure and communications linking it into east–west and center–south flows of goods and information. Some positive steps have already been taken in this direction. One of the most important of these was the completion of the railway line between China's Xinjiang and Almaty during 1991–1992, which helped facilitate trade and contact between Kazakhstan and Xinjiang. More recently, efforts have been made by Kazakhstan to improve the operations of the Druzhna rail transport center near the Chinese border, while regional railway and highway links among Kazakhstan, Uzbekistan, Tajikistan and China are being improved. Other trends are increased rail and road links through Iran to the Persian Gulf and Indian Ocean, with new rail links from northern Iran into Turkmenistan, and the extension of the existing rail system to go westwards from Almaty to join other Central Asian states to Turkey and then to Europe (see maps in Razumov, 2001). At present, it is possible to speak of two major Asia–Europe linked railway systems. The first of these is the Trans-Siberian railway (see above), with the associated Baikal–Amur (BAM) that passes north of Lake Baikal and then runs down near the coast to Vladivostok, the little BAM branch line, and earlier western branches. The Second Asia–Europe network of rail links connects 10 provinces and autonomous regions in China with countries in Central Asia and western Europe, with a new southern branch linking China, Kyrgyzstan and Uzbekistan, and then westward into Europe.

Along with the Qinghai–Tibet railway, this is part of China's effort to reduce developmental disparities between the eastern and western zones of China.

In parallel, a major transport project is also being developed to connect Europe, the Caucasus region, and Asia. Called the Transport Corridor Europe-Caucasus-Asia (TRACECA), it has received serious support from the EU in an effort to rebuild sea, road and railway links. Traffic across this 'new' Silk Road grew by 60% between 1996 and 1998, with up to US$1 billion of infrastructure investment and loans eventually being needed (for the western areas, this was largely drawn from the EU and from the European Bank for Reconstruction and Development). Problems have emerged with TRACECA (see above).

The paucity of communications, links and pipelines had greatly slowed the development of Central Asia in the post-1992 period. This is evident in the fact that, in the past, most oil and gas pipelines had been directed into the old Soviet Union. In spite of Western deals for access to Azerbaijan's three largest fields, and to the huge Tengiz field in Kazakhstan, the volume of exports through alternative routes has been very limited until recent times. There have been proposals for major east–west pipelines to cross the Caspian Sea, and then go either overland through Azerbaijan and Georgia, or via longer routes through Turkey. Other pipelines routed through Iran have also been slow to develop.

Since 1993, China has accepted that its rapid industrial development means that it will need to be a net importer of oil, with China currently importing up to 25% of its current oil needs. China is therefore emerging as one of the major petroleum importers in the Asian region and as Asia's largest refiner of oil. In the long term, aside from developing its own oil resources, China is also deeply interested in the possibility of accessing oil from Kazakhstan's western oil fields via connections into its Xinjiang area, with large funds committed already to improving regional cooperation and transport corridors. Xinjiang itself has sizeable energy deposits of oil and gas in the Tarim Basin, the Junggar Basin, and the Turfan-Hami area, which are currently being developed, though more slowly and with fewer proven reserves than had at first been hoped. Oil is being imported from Siberia in Russia's far eastern regions.

In the long run, development strategies in Xinjiang seem to parallel a strong effort to develop trade and cooperative diplomacy with Central Asia: a kind of 'Open Door Policy towards Central Asia'. One of the crucial moves in this 'oil diplomacy' has been the Chinese signing of a US$4.4 billion memorandum of understanding with Kazakhstan to build pipelines to China and Iran in exchange for oil and gas concessions and a 51% stake in Kazakhstan's state-controlled oil production company. It had been hoped, perhaps optimistically, that with adequate investment such a pipeline between Kazakhstan and Xinjiang (some 3000 km long) could be developed by 2010, but the revisions to the project (now costed at over US$3 billion), based on oil prices, development costs and other obstacles, seriously slowed the project. Even more ambitious projects have been considered, e.g. a major gas pipeline from Turkmenistan into Kazakhstan and thence eastwards to the Tarim Basin. In the long term, this could form part of a proposed 'Energy Silk Route' which could connect the gas fields of Central Asia with end-users in northeast Asia, including China and Japan, a project taken quite seriously by Japanese planners. Other important projects include a gas pipeline from the Tarim Basin to refineries in Sichuan, aimed at boosting continued industrial development in the western provinces. These programs also dovetail into China's planned partial shift from coal-based to gas-based energy infrastructure, which began during the 2000–2005 period, a huge project which requires a large degree of supply, technical and pricing adaptation, forming part of China's East–West Gas Transmission project.

There are two other issues that need to be addressed.

First, some people equate energy and resource cooperation with the plundering of resources. In the face of the rich energy reserves in Russia and Central Asia, some Chinese enterprises hold that cooperation with them should focus on seizing oil and gas resources from them. Many years ago, some Russian scholars asked the questions: What else is China interested in other than oil and gas? And China took away our resources, but what could China do in return to help us?

Second, economic and trade cooperation with relevant countries is regarded as the dumping of China's goods. It is truly gratifying that trade in Xinjiang has increased 800-fold since the adoption of the reform and opening-up policy. Xinjiang's opening up to the outside world started from border tourism and small-volume border trade in the 1980s, and this played an important role in boosting the region's opening up to the outside in subsequent years. However, the cheap and low-quality products making their way into Central Asian markets have had a negative influence on China's image there, causing local people to regard Chinese products as synonymous with fake and shoddy products. As a matter of fact, up until today, the majority of Chinese products in Central Asia are cheap and low-end products and thus it is imperative for China to expand its industrial, capital and technological input into this region. China should conduct cooperation with neighboring countries and make great efforts to build a network of close-knit common interests and raise the scale of converging interests to a higher level on the basis of the principles of mutual benefit and reciprocity. China has also stressed that it is essential that neighboring countries benefit from the growth of China, and that China can also benefit from their development. In other words, China and Central Asian countries should become 'a community of shared destiny and a community of shared interests'.[3]

It is incumbent on China to consider three questions. *First*, what can China offer to these countries in return for exploiting their resources? *Second*, while conducting oil and gas cooperation with these countries on a large scale, what can be done to compensate these countries through cooperation on non-resources projects and on particular assistance with ecological restoration? *Third*, economic and trade cooperation is not simply selling Chinese products, so how can capital and technology benefit local residents?

Inasmuch as cooperation in Central Asia is concerned, the countries involved are mainly underdeveloped ones. The conventional wisdom of developing border trade at the beginning is necessary and plausible, but it is futile to expect to attract capital and technology in this fashion in the long run. The right modality should be as follows: trade from the 'gray area' moves to the regular and standard; economic and trade cooperation gives way to investment; private enterprises give way to cooperation involving both private enterprises and state-owned enterprises, with the state-owned enterprises playing a major role; singular energy and resource cooperation is replaced by the combination of energy and resource cooperation and non-energy and non-resource cooperation (see Yang et al., 2018).

If one wishes to retain the notion of separate regions (Europe, South Asia, and North-East Asia), then Central Asia becomes a crucial linkage area of inter-regional contact, which can either result in division and conflict, as in the very hot conflicts in Afghanistan, or in a new series of connections which allow more positive relationships. Here, the long, 1,700 km eastern frontier of Kazakhstan with China is a case in point – if viewed in a negative sense, the frontier is wide, porous, and an extremely expensive defense liability. It is also important to note that China's Xinjiang region consists of 60% ethnic groups who have major con-nections with cross-border populations, including Uighurs, Kazakhs and Uzbeks. Periodic harsh security crackdowns by China have made Xinjiang once again an area of instability. Likewise, China has pressured Kyrgyzstan to take a stronger line in controlling Uighur minorities within their territory. However, taken as a whole from 1996 onwards, the borders between China and its eastern neighbors are now much more stable and open than before.

One of the most important shifts in Central Asia has been *Kazakhstan's improved relationship with China*, with the trade between these two nations being the largest for Kazakhstan. There has also been a joint project between Xinjiang and Kazakhstan to develop a large electrical power plant in western China. Likewise, the fact that the standard of living in Xinjiang is at present higher than in most Central Asian states helps reduce serious calls for independence within groups in Chinese territory. In fact, a vigorous border trade has flourished based on visa-free short visit regulations for Xinjiang. It is estimated that more than 60% of Kazakhstan's foreign trade is now from China, while there are at least 300,000 ethnic Chinese within Kazakhstan. Indeed, in the long term these two nations are likely to deepen their economic and political cooperation, a policy consciously followed by Kazakhstan. Major projects for infrastructure and resource develop-ments have been jointly considered between China and Kazakhstan, with China thereby eventually accessing Central Asian oil resources.

Justification for the proposed New Maritime Silk Road

With the development of sea lanes and the Suez and Panama canals, the Eurasian continental zone lost some of its strategic significance, except for those states con-tiguous to Europe or its sea routes, e.g. powerful regional states such as Russia, Iran, Turkey and China (all of whom have at times placed major military forces along their internal land borders). The significance of this Eurasia has once again increased, if placed in the wider social, economic and cultural aspects of strategic thought. Although a potential area of instability, Central Asia also represents a major economic and cultural opportunity for new, positive relations to extend beyond Europe's border. Once divided by the Cold War and by Soviet–Chinese confrontations, Eurasia now has the potential to become more interconnected in the future as a zone of relative integration for security and economic affairs. Whether it does depends on a number of major factors, including the impact of Russian, Chinese and US policies, as well as the way in which regional and cultural interests develop.

It is evident that China harbors a strong sense of non-intervention and respect for sovereignty but is nevertheless fully integrated into international and regional multilateral organizations. To a large extent, the Shanghai Cooperation Organization (SCO), along with the ASEAN Regional Forum (ARF), has functioned as a testing ground for its multilateral commitments. China needs to engage the region in a more effective and coherent way, without being seen as too great a threat in Moscow and the regional capitals. After the demise of the Soviet Union, China's preferred strategy was to pursue an increased but weak multilateralism in the region, including in Russia (Swanström, 2018). China does not consider Russia as a long-term threat in GCA due its relative decline, but views the possible increased role of the United States and the EU as much more threatening. The more strident assertions from India (in particular) and Pakistan that they have a vested interest in the 'Belt and Road' initiative need to be taken into account.

In the 'New Maritime Silk Road' plan, China stresses win–win cooperation. The Maritime Silk Road is one of the ways to realize the 'China Dream' as 'the great rejuvenation of the Chinese nation'.[4] The Maritime Silk Road from China to its west will touch important Indian cities on its way to West Asia and involve Pakistan (see below), as is currently being envisaged. The Maritime Silk Road already represents China's most vital sea lines of communication because it is the route for many of China's strategic materials, including oil, iron ore and copper ore imports. Moreover, it offers an opportunity for active efforts to develop strategic and economic relationships along the Maritime Silk Road.

China–Pakistan Economic Corridor (CPEC)

This is a development megaproject that aims to connect Gwadar Port in southwestern Pakistan to China's north-western autonomous region of Xinjiang via a network of highways, railways and pipelines to transport oil and gas. The economic corridor is considered central to China–Pakistan relations and will run about 3000 km from Gwadar to Kashgar. Overall construction costs are estimated at US\$46 billion, with the entire project expected to be completed by 2020. The corridor is an extension of China's proposed 21st-century Silk Road initiative. Other than transport infrastructure, the economic corridor will provide Pakistan with telecommunications and energy infrastructure. The project also aims to improve intelligence sharing between the countries. China and Pakistan hope the massive investment plan will transform Pakistan into a regional economic hub as well as further boost the growing ties between Pakistan and China. The Pakistani media and government called the investments a 'game and fate changer' for the region. The project will also open trade routes for western China and provide

China direct access to the resource-rich Middle East region, bypassing longer logistical routes currently through the Strait of Malacca. The Karakoram Highway connecting the two countries will also be widened, while the rail network between Karachi in southern Pakistan and Peshawar in the north will be upgraded. The two countries also plan a fiber-optic communications link between them. These closer links between Pakistan and China have given way to concerns from India, which has long-term historical interaction with the five 'stans' (Puri, 1997). India has no other option than to take visible interest in Central Asia: to take engaging, enduring, and enlightened interest in the unfolding dynamics there. *Geopolitics*, rather than anything else, defines the parameters of Indian perception and policy. A basic geopolitical premise, and reality, that flows from this is that any meaningful analysis of the dynamics of these states can result only if they are studied *as a region* with considerable cohesion and commonalities rather than as national states. Their geographic contiguity ensures that developments in one of them will impinge upon others, whether in the form of refugee spill-overs, the occurrence and management of ecological disasters, establishing access to scarce water resources, oil and gas pipelines, laying down routes to ports out of these landlocked states, proliferating communication links among themselves, and so on. A somewhat coordinated developmental effort among them would evidently yield more satisfactory results to their mutual benefit than otherwise; appreciation of this fact necessitates a regional rather than a unit-wise view of these republics (Maini, 2012).

The Asian Development Bank (Brunner, 2013) describes the project thus: 'CPEC will connect economic agents along a defined geography. It will provide connection between economic nodes or hubs, centered on urban landscapes, in which large amounts of economic resources and actors are concentrated. They link the supply and demand sides of markets.'

CPEC is a vital part of the Silk Road Economic Belt in that it gives China access to a deep warm-water port at Gwada via a rail and road link from Kashgar in Xinjiang to Gwada in Pakistan. If there was a blockade in the South China Sea or the Straits of Malacca (through which 80% of China's oil imports pass) then China has access to a port in Gwada far to the west of any conflict.

India has the view that it can be key player in the new Silk Road Economic Belt (Maini, 2012) and is not pleased with being excluded. India's recent (June 2017) admission (along with Pakistan) to membership of the SCO may provide a forum for negotiation and an opportunity for India to be a partner in the 'Belt and Road' initiative.

Remaining challenges to closer cooperation in GCA

Major issues of concern for the future of GCA include:

- The ability to restrain ethnic violence and prevent narrow forms of ethnic or religious nationalism, a problem for the region as a whole and the Russian Federation in particular.

- The need to create viable states which have legitimate, democratic governments that can undertake economic reform and avoid corruption and the
  manipulation of power.
- Relationships with a potentially assertive Russia with a strong foreign policy
  under President Putin.
- The types of Islam that will penetrate the region. Fortunately, the mystical
  and individualistic trends of the Sufism common to the region will tend to
  counterbalance various forms of 'fundamentalism'. At the same time, mili-
  tant Islamic groups have forced a stronger security clampdown in the region
  from time to time.
- Access to the sea ports via improved road, rail and air links (something that
  is being addressed by attempts to develop the New Maritime Silk Road – see
  below).
- Relationships with China and other regional traders, especially India.
- Stabilization of local currencies and inflation rates, as well as avoiding loan
  defaults that might reduce foreign investment.
- The issue of the continued treatment of Russian minorities, who form a powerful but resented group with needed technical skills. This has complicated
  the politics of Kazakhstan, Kyrgyzstan and Turkmenistan.
- The problem of illegal drug flows, smuggling (arms and people), and
  misdirected efforts to control drug production, including research for bio-
  engineered fungi designed to attack opium plants.

However, certain challenges remain for China and GCA in the creation of a stable Eurasia. The ability to restrain ethnic violence and prevent negative forms of
ethnic or religious nationalism is important, but must be balanced by the effort to
support the genuine needs of different nationalities and ethnic groups. Likewise,
religious issues will require careful consideration. The possible dilemma of development that does not balance regional and inter-ethnic needs can lead to further
instability along China's borders with its Central Asian neighbors.

The need for an integrated development strategy

The emergence of this Greater Central Asia will roll back the forces that give rise
to extremism and enhance continental security. It will bring enormous benefit to
all the countries and peoples of the region, and, significantly, also to major powers nearby, notably Russia, China, and India. Close examination of the ongoing
and planned infrastructure and transportation projects in the region, particularly
in the resource-rich states, would reveal a lack of coherence with regard to a
non-oil economy strategy. Important and useful projects are being planned and
initiated independently of one another, without the necessary cross-sector and
intra-sector coordination. In other words, these projects do not seem to be guided
by a unified objective or directed by a cohesive state policy. Unless a clear, integrated 'big picture' strategy is set forth today, the development trajectory of any
country in the GCA region is likely to be halting and subject to change. Most of

the GCA states are landlocked, and they depend on each other's transportation infrastructure. Building highways, railways, ports, and airports is a necessary part of a strategy to develop transport hubs, but it is not a sufficient one. Without a bird's eye approach and a coherent policy, which will view all these projects as components of a single strategy, the transportation and infrastructure projects are likely to have outcomes that will be insufficiently efficacious, because they will lack complementarity. Hence, the compartmentalized mindset has to give way to an integrated vision that will direct each project towards a common goal that includes, *inter alia*, more sustainable land management to be enjoyed by present and future generations.

Conclusions

It must not be forgotten that the concept of a 'Silk Road Economic Belt', first 'floated' by the USA,[5] as envisioned by China could provide an unprecedented opportunity for China's overall development and its relatively backward western regions. But, of course, it has other value to those countries to China's west. Today, one end of the Eurasian continent is the highly developed European economy and the other the fast-growing Asia-Pacific economy, with the vast, less developed, middle region in between. The notion of a Silk Road Economic Belt stands for great cooperation in this broad region aimed at connecting the Asia-Pacific and European economies.

To sum up, Central Asia (*sensu lato*) was once the hub of the Silk Road, and if a Silk Road Economic Belt is realized, it would be so again. The development of this economic area hinges on the development of Central Asia. In this sense, the Silk Road Economic Belt would reach the Atlantic and Indian Oceans. Although the Central Asian area is rich in mineral resources, it remains an underdeveloped area. And the total population of Central Asia is only 60 million, almost equivalent to a middle-sized province of China. So, in China's push for westward opening up, it is incumbent on it, on the basis of a good Central Asia development strategy, to include the Atlantic as well as the Indian Ocean into a broader cooperation of 3 billion population involving Central Asia, West Asia and South Asia, and even dozens of European countries. The significance of the concept lies in the creation of a Silk Road Economic Belt that links China's most vigorous Yangtze River Delta, Pearl River Delta and Bohai Sea economic zones to the European economy.

China should combine the construction of the Silk Road Economic Belt with a new diplomatic approach to its neighboring countries and actively involve them in geopolitical game playing, though this kind of game playing differs from that of Russia and the US in that China does not seek to create a sphere of influence, and instead strives to construct with the Central Asian countries a community of shared destiny and shared interests. Only in this way can we win the hearts and minds of the peoples of the Central Asian countries and so create a new geopolitical situation in Central Asia by jointly building a Silk Road Economic Belt.

Acknowledgments

I am grateful to Dr. Swanström of the Central Asia-Caucasus Institute & Silk Road Studies Program – A Joint Transatlantic Research and Policy Center, Johns Hopkins University-SAIS USA and Dr. James Ferguson of the Centre for West–West Cultural and Economic Studies, Bond University, Australia for sharing their insights and allowing reproduction here.

Notes

1 *Far East Economic Review*, 12 February 2011.
2 In 2017, India and Pakistan became members, and Iran and Turkey are potential members.
3 Speech by President Xi Jinping on the CCTV channel and used on billboards along the route from Beijing airport to the city during the recent 'Belt and Road' summit.
4 President Xi Jinping on television.
5 Frederick Starr and the Johns Hopkins University 'think tank'.

References and further reading

Brunner, H.-P. 2013. What is economic corridor development and what can it achieve in Asia's subregions? ADB Working Paper Series on Regional Economic Integration. ADB, Manila.

Dorian, J.P., Wigdortz, B. and Gladney, D. 1997. Central Asia and Xinjiang: Emerging Energy, Economic and Ethnic Relations. *Central Asian Survey* 16 (4): 469–71.

Ferguson, R.J. 2001. *China and the emerging Eurasian agenda: From special interests to strategic cooperation*. Centre for East-West Cultural and Economic Studies, Bond University, Robina, Australia.

Hagg, W. 2018. Water from the mountains of Greater Central Asia: a resource under threat. In this volume, pp. 237–48.

Hofman, I. and Ho, P. 2012. China's 'Developmental Outsourcing': A critical examination of Chinese global 'land grabs' discourse. *Journal of Peasant Studies* 39 (1): 1–48.

Maini, T. 2012. The 'New Silk Road': India's Pivotal Role. *Strategic Analysis* 36 (4): 651–6.

Melmet, Y. 1998 China's Political and Economic Relations with Kazakhstan and Kyrgyzstan. *Central Asian Survey* 17 (2): 236.

Perdue, P.C. 2010. *China Marches West: The Qing Conquest of Central Eurasia*. Belknap Press of Harvard University Press, Cambridge MA.

Puri, B.N. 1987. *Buddhism in Central Asia*. Motilal Banarsidass Publishers, Delhi, 259 pp.

Puri, M.M. 1997. Central Asian geopolitics: the Indian view. *Central Asian Survey* 16 (2): 237–68.

Razumov, Y. 2001. Central Asian States Seek to Bolster Transport Links with China. *Eurasia Insight*, 13 December.

Squires, V.R. and Lu, Q. 2018a. Greater Central Asia: its peoples and their history and geography. In this volume, pp. 3–22.

Squires, V.R. and Lu, Q. 2018b. Unifying perspectives on land, water, people, national development and an agenda for future social-ecological research. In this volume, pp. 283–305.

Starr, S.F. (ed.) 2004. *Xinjiang: China's Moslem borderland.* M.E. Sharpe, Armonk, NY.

Swanström, N. 2011. *China and Greater Central Asia: New Frontiers?* Central Asia-Caucasus Institute & Silk Road Studies Program and Johns Hopkins University-SAIS, Washington DC and Institute for Security and Development Policy, Stockholm.

Swanström, N. 2018. Greater Central Asia: China, Russia or multilateralism? In this volume, pp. 273–82.

Yang, Y., Pak Sum Low, Yang Liu and Jia Xiaoxia. 2018. Mitigation of desertification and land degradation impacts and multilateral cooperation in Greater Central Asia. In this volume, pp. 179–207.

13 Greater Central Asia

China, Russia or multilateralism?

Niklas Swanström

SILK ROAD INSTITUTE, NORWAY

Introduction

Greater Central Asia (GCA), defined here as encompassing the five Central Asian republics, Afghanistan, and western China (Xinjiang and Tibet), formed a coherent and interconnected sub-region for millennia, until the establishment of the Soviet Union and its borders destroyed the old trade and cultural routes. Indeed, the region was not only witness to impressive levels of trade and cultural exchanges, but also military and political competition that had ramifications far beyond the region. Today, something similar to the old GCA region is being reconstituted after a brief interlude (seen from a historical perspective) and it has once again become a locus for great power competition.

GCA has been growing in importance for many reasons, not only in terms of energy security and the struggle against terrorism and fundamentalism, which tends to dominate the news headlines when it comes to the region. These are all important, particularly for the bordering states, but, moreover, we have seen that GCA has increasingly positioned itself as a nexus for inter-regional trade between China, Iran, India, the EU, and Russia. This has increased the importance of the region in a positive way. More negatively, however, the region has become a transit route as well as a source of organized crime, particularly heroin production, weapons and human trafficking, as well as growing Islamic extremism and radicalization. This is further exacerbated by the presence of weak states and institutions across the region.

With Russia's preeminence in the region challenged after the collapse of the Soviet Union, GCA has also come onto the radar of other powers, including the United States, India, Turkey, Iran, the EU, and, most notably, China. The reemergence of GCA as a region of interest has begun to redefine the strategic landscape in GCA. The Chinese "March west" and its "Belt and Road" strategy have clearly positioned China as the single most important actor in the GCA region, in spite of important limitations and challenges to such a role.

This chapter starts by seeking to understand the reemergence of GCA as a regional concept and why it is relevant, despite the relatively low levels of regional integration. It then looks at the Sino-Russian "power transition" and how both actors have sought to exert influence in the region. Finally, it considers the present-day regional cooperation structures, notably the Shanghai Cooperation Organization (SCO), and their limitations, and what role China and Russia play in them.

Understanding the GCA region: commonalities and differences

The concept of Greater Central Asia is a contested one; and yet it is arguably better, from both a historical and a contemporary perspective, than others such as Eurasia and Central Asia. Indeed, the concept of Eurasia reflects ideas of Russian chauvinism and the notion of the area being in Russia's "backyard" and thus sphere of influence. In fact, the region has come under the control of several great empires but outlived them all. Moreover, the concept of Central Asia, if understood as comprising the five Central Asian republics, is too narrow to reflect the many interlinkages with Afghanistan and Xinjiang. It is also similarly inadequate to merely "integrate" the region into other regional entities, as this fails to take into account the extent to which the different parts of GCA in fact share many commonalties – i.e. historical, intellectual, religious, cultural, economic – that distinguish it from other regions (Starr, 2009). In sum, how to conceive of the region should be based on its intrinsic features and characteristics and not on externally imposed ideological or power-based definitions.

Two thousand years of free movement across the region, with intermarriage between the different groups, has created a regional identity that is hard to deny, even if local identities (i.e. clan-based etc.) are very important as social and economic security lies not with the state but with local communities. Many of the current distinctions between the different groups are a result of the lack of clear ethnic and cultural borders rather than the opposite. In fact, especially when it comes to Uzbekistan and Tajikistan, the distinction between Turkic and Persian is today a matter of preference rather than of genes. Starr has convincingly argued that the fact of centuries of economic and social interaction is more relevant than the factors that keep the states divided today (Starr, 2009). As will be examined, this is not to say that the states or governments in the region always define themselves as a region or even kin; but rather that there are more commonalities than divides between states and peoples.

A further commonality is that the different states (including Chinese Xinjiang) of GCA are landlocked. While this has traditionally been a problem in accessing markets, the prospect of a revived "Silk Road" that centrally posits the region between East Asia, South Asia, the Middle East, and Europe serves to further consolidate GCA as a geographically coherent region. Linn (2012) sees the transition from the maritime routes to a land-based route as an important factor both for regional integration and for economic development (Linn, 2012). It is difficult to envisage such a transition without a regional infrastructural network that ties the region together and connects it with other regions via land routes. This necessitates deeper cooperation and integration between the GCA states (Linn, 2013)

In spite of the many unifying factors that allow us to envisage GCA as a coherent and cohesive region, there are also many important differences between the states. Indeed, relations between the states and their leaders have been characterized by tensions and competition, notably between Uzbek President Islam Karimov and Tajik President Emomali Rahmon. The Soviet division of borders has left a lasting legacy with cooperation over once shared resources – such as water – having largely broken down. Furthermore, the civil war in Afghanistan and the intellectual and economic separation of Xinjiang from the rest of GCA during the Soviet era also decreased interaction within the region. In the case of Xinjiang, the Chinese fear of separatism has prompted Beijing to view with suspicion any greater identification with the rest of GCA, especially in terms of linkages between fundamentalist groups that challenge China's right to control Muslim Xinjiang.

In military and economic terms, furthermore, the states differ greatly. Due to its abundance of energy resources, Kazakhstan has become the region's economic engine, with a gross domestic product (GDP) far higher than the other states, while Uzbekistan is the military strongman with the largest population (excepting Afghanistan). The two poorest states, Tajikistan and Kyrgyzstan, lack energy resources and are wracked by instability and failing institutions. Even more so, Afghanistan is characterized by insecurity, poverty and terrorism, and, despite international intervention, the situation remains very insecure. Insecurity in the region is rampant and much of the investments that China and Russia have engaged in will be dependent on the region and their neighbors being able to increase security and stability.

Moreover, the strategic orientations of the GCA states differ. Tajikistan and Kyrgyzstan tend to be politically and culturally more aligned with Russia than with China compared with the other GCA states, with both states also harboring concerns over their territorial integrity, not least from their western neighbor Uzbekistan. They also rely on Russia for remittances and as a security guarantor. Kazakhstan, on the other hand, is increasingly concerned with how Russia will react over its own Russian minority in the north of the country and whether the Crimean situation could be repeated in Kazakhstan. Turkmenistan has adopted a policy of "permanent neutrality" which has seen it refrain from joining regional initiatives. Meanwhile, bilateral relations between Uzbekistan and Russia (as well as with the US) have vacillated. Xinjiang is different due to the fact that it is part of China. Accordingly, Beijing views the GCA region as a strategic opening to break the perceived US encirclement.

In spite of their differences, the Central Asian states have an interest in diversifying their strategic alternatives so as to avoid dependence on any one power, namely Russia or China. Indeed, the role of the two most influential powers in the region (excepting the United States in Afghanistan) is considered below.

The rise of China, Russia's decline and the impact on GCA

The collapse of the Soviet Union in 1991 and Russia's own pressing domestic problems saw a drastic reduction in Moscow's clout and influence. The establishment of newly independent states also saw the (re)emergence of other powers to partially fill the "void" left by Moscow and to exploit the new opportunities in diplomatic and economic relations. Notwithstanding, Russia has sought to partially regain its former military, economic, and political clout in what Russia's former foreign minister, Andrey Kozyrev, termed Russia's "near abroad" (*ближнее зарубежье*) – the implication of which is that Russia maintains privileged interests in regard to the region. However, while Russia remains an important player, especially in the domain of security, it has increasingly been eclipsed by China.

The economic arena is where China has made the biggest inroads into the region and which has been its primary interest. Indeed, China has emerged as the most important trading partner and investor for GCA as a whole, eclipsing Russia as of 2013 (in the cases of Tajikistan and Kyrgyzstan, China's share of trade is more than twice that of Russia's, at 36.3% and 47.3%, respectively). The bulk of Chinese trade with the states of GCA is in strategically valuable resources such as energy commodities and, to a lesser degree, other natural resources. In the cases of energy-poor Tajikistan and Kyrgyzstan, however, Russia provides energy assistance with the relationship increasingly characterized by dependency.

Additionally, the region has also become increasingly dependent on Chinese infrastructure investments and consumer goods. With the new energy deals signed by President Xi Jinping with his Central Asian counterparts in October 2013, valued at approximately USD 100 billion in total (the deal with Uzbekistan is worth USD 15 billion, Kazakhstan USD 30 billion, and Turkmenistan USD 50 billion),[1] China's dominant position will become even more consolidated and China is likely to remain the most important trading partner for all states. This is particularly true as these latest energy deals come on top of earlier investments in Central Asia and Afghanistan; a notable investment in regard to the latter was the USD 3.5 billion for the Aynak copper mine as well as USD 2 billion invested in infrastructure related to the mine.

There has been some concern in China that the Eurasian Customs Union could impact Chinese trade negatively. But, in fact, Chinese trade with the region has grown even faster than Russian trade (and international trade), not least in energy resources. Besides energy, of particular significance is the bazaar trade, which is estimated by the World Bank to be worth USD 7 billion, of which China controls 93–95% (Kaminski & Saumya, 2012). The economic interdependence between Xinjiang and the rest of Central Asia is also especially striking. Eighty-three percent of Xinjiang's trade is with the Central Asian states, with an increasing share of China's overall trade also transiting Xinjiang and GCA to the west. Russian trade with GCA is much more limited by comparison.

The above is not to say that Russia has been fully supplanted by China in GCA. Indeed, Russia still plays an instrumental role in some of the key industries, especially in the energy sector. As such, Russia has been eager to continue to control exports from the Central Asian states – not because Moscow needs the extra oil and gas imports but rather because it benefits from re-exporting the petroleum for profit. The Central Asian energy market has in fact sustained the Russian market with low-cost energy while sales to Europe are made at much higher prices. For example, Russia imports gas from Turkmenistan and Uzbekistan at prices as low as USD 100 per 1,000 cubic meters, but sells on the gas for as much as USD 250 per 1,000 cubic meters to European markets (Frickenstein, 2010). Accordingly, a primary motive for the Russian strategy in Central Asia has been to prevent any attempts to circumvent Russia's control over Central Asian energy sales, such as the Trans-Caspian, Nabucco, Turkmenistan–Afghanistan–Pakistan–India, and Trans-Anatolian pipelines. Moscow has applied significant pressure to prevent circumvention. As a result, Turkmenistan temporarily rejected the Southern Corridor Energy Projects even though China, Iran and Europe pushed for a more diversified energy network that would avoid Russia to a significant degree. Russia's strategies to control energy have decreased internal trade among the Central Asian states such that today there is very low intra-regional energy trade. Turkmenistan, Uzbekistan, and Kazakhstan are now primarily exporting their energy products out of the Commonwealth of Independent States (CIS) and European Economic Community (EEC) areas (apart from Russia).

In sum, Russia's stranglehold has benefited neither the producers nor the consumers. As a result, Central Asian governments have chafed at this policy, realizing that they could benefit more from avoiding Russia as a transit route. Increasingly, Chinese, European, and Central Asian governments are interested in developing alternatives to today's existing energy infrastructure and energy agreements controlled by Russia. As a developing economy hungry for energy, China now represents an attractive alternative as a customer for Central Asian oil and gas, but energy and trade routes are increasingly connecting South Asia – and then, most interestingly, India – to the overall infrastructural web through the proposed Southern Route that will be essential to connect the whole region in a functional network. In addition, China and South Asia are not only, or even primarily, interested in connecting GCA energy pipelines, but, more importantly, the energy networks from the Middle East and the trade network to Europe will be more significant over time and GCA is primarily a transit region, but a central and potentially lucrative transit region.

Energy and trade aside, it is clear that Russia still occupies a preeminent position when it comes to military and security cooperation in GCA, even if China has also increased cooperation in these domains. China has boosted military aid to all Central Asian states, even if in relative terms it remains modest. China has pledged large increases in military aid to Kyrgyzstan and Tajikistan, the two Central Asian states most affected by internal strife, and in September 2014 the Kyrgyz Armed Forces announced that China would give 100 million RMB in aid; this comes on top of newly constructed apartment buildings for officers that China has built. And, in April 2014, Defense Minister Chang Wanquan pledged "hundreds of millions of dollars" for new uniforms and training for the police in Tajikistan. Similar projects are ongoing in all Central Asian states, and the Chinese National Defense University, as well as other military organizations in China, have seen a great increase in visitors and students from the Central Asian states. China has also begun modest arms sales to the Central Asian region as well as military aid. Currently, this primarily involves light weapons, such as sniper rifles to Uzbekistan, and military equipment, such as flak jackets and night vision equipment and vehicles, but more substantial systems are being considered. Moreover, there has been an impressive number of military exercises between SCO member states (as well as involving China bilaterally with individual states). In August 2003, the first multilateral military exercise was held between China and the Central Asian states; this was preceded by a bilateral exercise with Kyrgyzstan in 2002. This has since grown to encompass more than 25 exercises over time, of which the exercises "2013 Peace Mission" and the "Naval Interaction 2013" in many ways were milestones of military cooperation in the region. However, all this is still only a fraction of the military cooperation and sales that Russia has with the GCA states.

This military cooperation with the GCA states is improving their ability to counter some of the threats that exists in the region, but insecurity is built into the system and the GCA states would need to reform themselves and strengthen state institutions, not least education, social security, policing and legal institutions, to be able to create long-term sustainable development. Unless this happens, much of the insecurity will remain in the system, and neither China nor Russia has been willing to do this and China has been much more reluctant than Russia to intervene in the internal affairs of the GCA governments. The reality is still that GCA will not manage the transition itself, and without external support there will be a continued situation of terrorism/political opposition, corruption, economic decline and political instability that will prevent many of the agendas that the outside world would like to see implemented in terms of economic development.

Russia can also exploit its soft power to a greater extent than China. While on the decline in some states due to the promotion of national languages, Russian nonetheless remains the *lingua franca* in the region, and leaders often speak better Russian than their local languages. Culturally, many, at all levels in society, are closely connected to Russia, not least in music, literature, and other forms of popular culture. Furthermore, while after the collapse of the Soviet Union many ethnic Russians left Central Asia for Russia, most countries still have small but significant pockets of ethnic Russians remaining. Most significant is Kazakhstan, which has a large Russian minority that dominates the northern part of the country; furthermore, ethnic Russians typically make up a large part of the Kazakh military's officer corps. While China is actively trying to become a more important actor in the region in terms of culture and language, such as by setting up Confucius Institutes, it is unlikely to supplant the Russian language and culture any time soon.

A further factor that accentuates Russia's role and leverage, especially over Tajikistan and Kyrgyzstan, is that the Russian labor market provides employment opportunities for migrant laborers from Central Asia that do not exist in the weak economies of their home countries. Indeed, Russia is host to some 3 million to 4 million Central Asian guest workers, many of whom live without the proper paperwork and under harsh conditions. Nevertheless, the remittances they send back have proven crucial. For instance, between 35% and 45% of Tajikistan's (legal) GDP was made up of remittances in 2011, with 1 million Tajik migrant workers in Russia. Without this outlet, a large unemployed (and youthful) labor force raises concerns of socio-economic instability in the weakest Central Asian states.

Russia has exploited this vulnerability by threatening to tighten regulations or even deporting such workers to keep the governments at heel with Moscow. At present, furthermore, China does not represent a viable alternative employment market as it is exporting its own laborers rather than seeking foreign workers. While predictions of Russia's demise are premature, the long-term dynamics work in favor of China and the Chinese influence is felt very strongly. The GCA states have become concerned with the transition from one strongman to another, and some effort has been put into diversifying their foreign policies, not least by engaging in multilateral cooperation and structures.

Multilateral structures and initiatives

There have been few truly regional initiatives involving the states of Central Asia. Notable examples are the Central Asian Union and the Central Asian Nuclear Free Zone, of which the former was merged into the Eurasian Economic Community (EEC or EurAsEC). However, regional integration and the success of such initiatives remain low and are marred by a lack of trust between the states. More significant has been the establishment of extra-regional structures involving Russia and China.

There have been a number of regional attempts initiated by Russia with some success that have aimed at strengthening Russian leverage over Central Asia, such as the Commonwealth of Independent States (CIS) in 1991, the EEC in 2002, and the Collective Security Treaty Organization (CSTO) in 2002. These organizations were designed to promote specifically Russian interests and have effectively circumvented any involvement from other significant actors such as China and the United States. However, Moscow lacks a clear and unified vision about the future of regional structures such as CSTO and CIS, which cripples the potential of these organizations. The Eurasian Union established by Russia is another direct attempt to decrease external influence (of China, the EU and others); and, while not favorably viewed by Beijing, even if its displeasure is not voiced, the limitations of the Union are apparent, especially if energy prices remain low and the Russian economy continues to falter. Nevertheless, military cooperation through the CIS and CSTO is still important in Central Asia. Military exercises and training in general are crucial for the Central Asian governments since these will both train them for possible domestic unrest and secure Russian support to counter such events. This was particularly noticeable after the Tsentr-2011 exercises, which involved 12,000 mostly Russian soldiers. Although publicly presented as an anti-terrorist exercise, it had all the characteristics (in terms of armament, size of troops, and strategy) of fighting a more traditional conflict, potentially aimed at securing pro-Moscow governments in the region.

The SCO has arguably emerged as the most interesting regional organization in Central Asia, with the potential to establish multilateral structures that could make deep inroads with regard to regional integration in terms of economic, security and political cooperation. Created in 2001 (it evolved from the Shanghai Five created in 1996), the organization includes both Russia and China together with the Central Asian states (excluding Turkmenistan) as well as five observers (Afghanistan, India, Iran, Mongolia and Pakistan), three dialogue partners (Belarus, Sri Lanka and Turkey) and three guests (ASEAN, CIS and Turkmenistan). The SCO can be seen as a weathervane both for Sino-Russian relations and for Chinese influence in the region, as China plays the leading role in the organization. In fact, the SCO has been portrayed as an anti-Western organization or even a counter-organization to NATO. Yet the reality is far from this, and even if it has interesting security aspects, Russia has viewed it as a structure to control China, while China views it more as a tool to increase influence in the region but without unduly provoking the concern of Russia or the regional states.

Thus, while China strives to increase security multilateralism (i.e. through multilateral military exercises) within the SCO, its success has been rather modest but growing since the June 2017 meeting in Kazakhstan. This is very much due to the inherent competition and lack of real reasons for cooperation between Russia and China; currently, the main reasons for cooperation are external – i.e. tensions with foreign powers – and there are few grounds for sustaining deeper cooperation should China's relations with the international community improve. Russia has thus become a partner out of necessity and not choice. GCA states are also reluctant to engage in any multilateral structure that limits their own sovereignty or increases the influence of their much larger neighbors. In fact, it would appear that the greatest value the SCO could have in terms of commonality of interest is as an anti-terrorist structure: each of the member states has major issues with alleged terrorist organizations and there is strong government-to-government support in combating terrorism. Although criticism has emerged that the means to combat terrorism are used equally in the Central Asian states to combat opposition, the development of the anti-terrorist center in Tashkent (RATS) has been relatively successful in coordinating national anti-terrorist activities and sharing information among the members, but it would not be wise to exaggerate its impact so far. However, RATS will only be as effective as its member states allow it to be, and currently there is very little trust between the member states, not least Russia and China, which will cripple any deeper engagement.

The problem is that the regional structures focus on Russia and China and to a lesser degree on the GCA states; this is very much due to the fact that the economic resources are concentrated in the larger nations. The view of multilateralism differs greatly between GCA, China and Russia, largely because China views it from a position of strength and Russia from a position of decline, while the GCA states are interested in keeping a balance not only between China and Russia but also with other external powers. China is more interested in stabilizing states, increasing economic development, and increasing international trade and interaction (even if not necessarily at the expense of China). Russia, on the other hand, has a more traditional position of political control and military cooperation per se and views China's return to GCA as a zero-sum game. From the perspective of the Central Asian states, they are not interested in exchanging one overlord for another, and both Russia and China are viewed with significant skepticism.

Despite this, attempts on the part of the Central Asian states to extend political and economic relations beyond Russia (and China) have been made difficult by problems of infrastructure, geographical location and history, as well as by failures on the part of external actors (including Europe and the United States) to act in Central Asia, despite engagement in Afghanistan by the US and EU. However, the recent competition between China and Russia has created some space for the Central Asian states to decrease one-sided reliance, something seen prominently in energy relations. In effect, despite the supposedly strong Sino-Russian relations, the failure of Russia and China to cooperate over a broad range of issues has opened up space for other actors in Central Asia, if there were any interest in doing so.

Concluding remarks: the future of GCA

The notion of GCA as a hub of economic, and cultural, interaction is nothing new but is founded in a 2000-year-old tradition that was temporarily interrupted by Soviet (and Russian) occupation and will again grow in importance. Therefore, GCA as an analytic concept is very useful in both a historical and a contemporary perspective, but the region is also characterized by internal tension. As has been noted, despite the fact that the region is a natural analytical unit and has the potential to become well integrated, it is currently fragmented and has very low levels of intra-regional trade. It will be necessary to build more cooperation in extra-regional structures, such as the SCO and the Asian Infrastructure Investment Bank (AIIB), or in Russian attempts that could create more stability and reduce competition and suspicion between the states. This sets up a new challenge as these structures are primarily designed to promote the interests of Russia and China than necessarily being of long-term benefit to the countries and populations of GCA.

Since the challenges in GCA are very much interconnected, not least in terms of economic and security terms, it is necessary to find solutions that connect the region in a more natural way. It is important to connect not only the capitals but also the underdeveloped countrysides in the overall development. The greatest challenge for successful development is the GCA states themselves – corruption, "totalitarian" systems, a failure to engage the region or states as a whole, grassroots dissatisfaction, the failure to provide public goods, etc. One of the major hurdles will be to strengthen the states and the state functions so that they can provide public good to all parts of the individual countries. Institutional development remains a constraint as few or no states, regionally or internationally, are ready to put in the political or financial capital to make these changes possible. Without an improved security situation, the transaction costs will be prohibitive and it will be impossible to implement the very ambitious projects that are in the pipeline.

The future economic development of the GCA countries, especially the five "stans" and Afghanistan, is very unclear as well, but the AIIB and other infrastructural investment projects will be essential for this. Without well-developed infrastructure, and strengthened institutions to manage this, the picture looks gloomy both for the individual states and for the region as a whole. The transit trade will potentially provide the much needed investments and economic development for the GCA region but the question remains of how to connect the whole region in an extensive and broad network. It will also be necessary to connect

South Asia to GCA through the Southern Route via Pakistan, with links to the Maritime Silk Road (Squires, 2018), as excluding this important route would cripple the whole project, but that will create new difficulties due to Indo-Pakistan relations. In many ways, the new infrastructural projects we see in GCA are, like the Coal and Steel Community in Europe, a peace project that will connect old adversaries together and increase economic prosperity in the region at large. The economic effects will undoubtedly be positive, but, if this project is successful, the real value will be increased security and lasting peace in the region.

China's rise, and China's economic focus on GCA, is essential for any engagement in GCA, since Chinese resources and political commitment will bring more potential for opening up and developing the region. Russia is declining very quickly and there is no apparent strategy from the US or the EU to increase their engagement in the GCA region after their departure from Afghanistan. Moreover, EU investments in infrastructure are uncoordinated and the US seems to be reluctant to engage in the region in any coherent way. The problem is that China views this solely as an economic project and has failed – or refuses – to see the greater benefits for the region. This is largely due to China's own inability to accept that it will have to engage in the region in a broader and more assertive way than it – and the rest of the world – is willing to do at this moment in time.

Note

1 At the June 2017 SCO summit in Astana, Kazakhstan, a further US$230 billion worth of investment projects were approved.

References and further reading

Bedeski, R. & N. Swanström (eds). 2012. *Eurasia's Ascent in Energy and Geopolitics: Rivalry or Partnership for China, Russia and Central Asia?* Routledge, Abingdon.

Frickenstein, S. 2010. The Resurgence of Russian Interests in Central Asia. *Air & Space Power Journal* (Spring): 67–74.

Kaminski, B. & M. Saumya. 2012. *Borderless Bazaars and Regional Integration in Central Asia.* World Bank, Washington DC.

Linn, J. 2012. *Central Asian Regional Integration and Cooperation: Reality or Mirage? (The Economics of the Post-Soviet and Eurasian Integration).* Washington D.C.: Brookings Institution. https://www.brookings.edu/research/central-asian-regional-integration-and-cooperation-reality-or-mirage/

Linn, J. 2013. Central Asian Regional Integration and Cooperation: Reality or Mirage? In: *EDB Eurasian Integration Yearbook 2012.* Eurasian Development Bank, Almaty, pp. 97–8.

Squires, V.R. 2018. Greater Central Asia as the new frontier in the twenty-first century. In this volume, pp. 251–72.

Starr, F. 2009. In Defense of Greater Central Asia. Policy Paper,. Institute for Security and Development Policy, Stockholm, 11 pp.

Swanström, N. 2007. China's Role in Central Asia: Soft and Hard Power. *Global Dialogue* 9 (1–2).

14 Unifying perspectives on land, water, people, national development and an agenda for future social-ecological research

Victor R. Squires and Lu Qi

INSTITUTE FOR DESERTIFICATION STUDIES, BEIJING

Introduction

Five aspects (economy, ecology, sociology, technology and policy) intersect and interact in the Greater Central Asia (GCA) region. This book examines and analyses the complex situation in this vast region using an integrated and regional approach to achieving sustainable land management (SLM). Internationally, SLM has been identified as the most important agenda for countries that are signatories to the UN Convention to Combat Desertification (UNCCD) as it strives for a Land Degradation Neutral World (LDNW), as defined as a Sustainable Development Goal by the Rio+20 conference (Wang et al., this volume). The goal is to secure the contribution of our planet's land and soil to sustainable development, including food security and poverty eradication. Responses by the governments of seven countries in GCA to the impending crisis brought on by past and ongoing mismanagement of the resource base (land, water and minerals) have been brought into sharp focus by increasing interest in GCA, especially by China but also by other countries as diverse as those in the Gulf Arab region, the EU and Russia (Squires, this volume; Swanström, this volume).

GCA encompasses a vast mosaic of diverse and contrasting landscapes, plant and animal species, and human populations (Orlovsky & Orlovsky, this volume) leading very different and unique lifestyles. However, when we look at the entire region, we tend to be struck by one significant characteristic: much of it is dry. GCA has some of the world's driest deserts and highest mountains. Much of it is covered by barren deserts, savannah, grassland, scrubland, woodlands and dry forests. Indeed, drylands comprise 67% of the GCA region, and drylands are home to a rapidly growing population that currently stands at about 230 million people. In spite of their environmental sensitivity and perceived fragility, and despite the prevailing negative perceptions of drylands in terms of economic and livelihood potentials, these ecosystems have supported human populations for centuries.

As the post-independence era has unfolded in the five "stans" and new political realities have set in, inter-state tensions and diverging priorities over the use of land, water and mineral resources have started to dominate the political, economic and environmental agenda in the region. The decade ending 2010 in particular has been characterized by an increase in disputes over water usage, particularly in countries dependent on agriculture (Squires & Lu, this volume; Orlovsky & Orlovsky, this volume; Krutov et al., 2014; Bekchanov et al., this volume). Tensions also exist over access to and exploitation of oil and gas and strategic minerals and the damage done by mining activities (including prospecting and surveys). A lack of political will and the absence of any effective mediation mechanisms have only exacerbated the problems. Recently, tensions between the highland countries of Kyrgyzstan and Tajikistan and the lowland countries of Uzbekistan, Turkmenistan and Kazakhstan have largely been generated by disparities in levels of prosperity and stability, energy accessibility and different priorities for water usage (Hagg, this volume). The emergence of China, India and Pakistan as dominant regional players and major water and energy consumers is also altering the political, economic and environmental landscape of GCA. Similarly, earlier agreements on water allocations during the Soviet period did not consider Afghanistan, whose interests in the basin have only recently begun to gain prominence, especially their claims for water from the Amu Darya river. Afghanistan, as a riparian state of the Amu Darya, must be fully integrated into future arrangements on water issues, along with Uzbekistan and Kazakhstan. Such coordination is essential if regional potentials in agriculture and the generation of hydroelectric power are to be achieved (Bekchanov et al., 2010; Conrad et al., 2007; Dukhovny & de Schutter, 2011; Krutov et al., 2014; SIC-ICWC, 2011; Varis, 2014).

In GCA countries, agriculture is the main source of employment. Social stability therefore requires a viable agricultural sector. Without improvements in farmers' incomes, progress in the war on narcotics is inconceivable. This calls for attention to the entire process of agricultural production and marketing and close coordination with other sectors, including banking and transport.

Issues that are addressed in this book are ecological constraints (including demographic pressures), modifying land tenure arrangements (including use rights) and the effects of international trade arrangements and the land management policies of other countries both in the region and elsewhere. Accounts of land use, areal extent and severity of land degradation in the five "stans", Mongolia and Western China are presented as a background to how governments have responded to the Rio+20 outcomes as set out in *The Future We Want* and initiatives such as Land Degradation Neutral World (Wang et al., this volume).

After a brief introduction to GCA landscapes and societies, the book considers determinants of and constraints to resource utilization. Current issues such as population, political instability, and misplaced government optimism, desertification, land degradation, and lost cultural identity are considered. We provide an overview of policy challenges, especially those arising from institutional and administrative issues. Resource rights are given consideration, especially the legal frameworks for land use (including land privatization), land reforms currently being considered, measures to empower minority groups, and their relationship to common-property systems. We also consider the opportunities to provide incentives for SLM, with special reference to recognition of the natural potential of GCA's vast land and water resources.

Outside of the cities, many dryland inhabitants are either pastoralists, sedentary or nomadic, or agro-pastoralists, combining livestock rearing and crop production where conditions allow. Intensive irrigated agriculture is an important contributor to gross domestic product (GDP) and to people's welfare. Over millennia, people have lived with variable rainfall and frequent droughts using a range of coping strategies, but changing circumstances (including the transition to the market economy) mean that these traditional methods must be capitalized upon and enhanced. Unsustainable practices are contributing to significant land degradation, and it is predicted that climate change will further compound the already tenuous situation, especially in irrigated areas where water supplies will become less certain (Li et al., 2015). Without significant efforts to address the impact of climate change and land degradation, the livelihoods of the GCA dryland populations will be in jeopardy (Kienzler et al., 2012).

There are many adaptation[1] options, which, if adequately designed and applied in response to specific local contexts and realities, can limit the negative effects of climate change and land degradation on drylands livelihoods in GCA. Some are based on natural resource management (NRM), which combines land conservation and productivity enhancement practices, such as improved rangeland restoration, herd management, improved animal nutrition, rain-water harvesting, and home gardens. Others are market-based, which aim to improve market access and increase incomes, thereby reducing vulnerability. Institutional options form a third category, focused on local-level structural change such as extension and education, micro-credit and migration. Many of these adaptation measures are already familiar to the inhabitants of the drylands, but their effectiveness now depends on careful selection and application, so that they are implemented in the right place and at the right time. An enabling policy environment is also critical for the adaptation options to be practiced in a sustainable manner. The key to realizing this potential is to recognize the continuing need for poverty alleviation support in the dry parts of GCA, and also to recognize the importance of carbon sequestration in drylands soils (Glenn et al., 1993; Squires, 1998) so that custodians of dryland ecosystems have incentives to manage them for long-term carbon sequestration compensation (Lal et al., 2007).

Inhabitants of drylands in GCA have learnt, over millennia, to cope with permanent water scarcity, variable inter- and intra-seasonal rainfall and the recurrent risks of weather-related shocks. However, as a result of high poverty rates, changing socio-economic and political circumstances and demographic growth, traditional coping strategies are increasingly becoming insufficient. Unsustainable land management practices, including overgrazing, over-cultivation, illegal and excessive fuelwood collection and poor irrigation technologies (among others) have become prevalent, often due to institutional or tenurial barriers. As a result, in the recent past, millions of hectares of the already fragile GCA drylands have been further degraded or desertified. This has been compounded by poorly conceived policies and ineffective governance (UNEP, 2006; Squires, 2012; Freedman & Neuzil, 2015).

Climate change: an impending calamity

Projected climate shifts over GCA will present substantial challenges for primary producers, land managers and governments over the next 50 years, with predicted declines in pastoral production and forage quality, increased livestock stress, more frequent droughts and high-intensity rainfall events, and increased land degradation and invasion by exotic pests and diseases. Falling incomes derived from dryland cropping, increasing reliance on government assistance, and lower economic sustainability will create substantial social problems for rural communities.

Vulnerability to climate change and other hazards constitutes a critical set of interactions between society and environment. Climate change adds another layer of risk to this precarious situation (Bekchanov et al., this volume; Hagg, this volume). Its impacts threaten to exacerbate the existing land degradation problem and add to the vulnerability of the drylands' inhabitants, unless significant measures are taken to improve management of the natural resource base. There is considerable variability and uncertainty in current climate change projections. Nevertheless, there is now reasonable agreement from a number of different models, including the IPCC's Fourth Assessment Report (AR4) on Climate Change, that the GCA region is at the highest risk from climate change, given the magnitude of existing stresses in the continent (IPCC, 2007; Lioubimtseva & Henebry, 2009; Pilifosova et al., 1997; Li et al., 2015). It is highly likely that in the coming years significant areas of the drylands in GCA will see changing rainfall patterns with more frequent and more intense extreme events such as droughts and floods. Observations indicate that, although average temperatures in most Central Asian countries showed almost no increases from 1997 to 2013, they have been in a state of high variability. Despite the lack of a clear increasing trend, this 15-year period is still the hottest in nearly half a century. Precipitation in the GCA countries remained relatively stable from 1960 to 1986 and then showed a sharp increase in 1987. Since the beginning of the 21st century, however, the increasing rate of precipitation has diminished. Dramatic changes in meteorological conditions could potentially have a strong impact on the region's natural ecosystems, as some significant changes have already occurred. Increased temperatures are expected to add to water problems by causing additional loss of moisture from the soil (Glantz, 2005; Bekchanov et al., this volume). The AR4 estimates that by 2020 between 75 million and 250 million people are likely to be exposed to increased water stress and that rainfed agricultural yields could be reduced by up to 50% in some GCA countries if production practices remain unchanged. These changes will undermine further the livelihoods of pastoralists, agro-pastoralists and irrigators in drylands. It is worth noting, though, that the impact of increased temperatures on low-input agriculture will be minimal, as other factors will remain the dominant constraints. Mountainous areas are especially vulnerable to climate change (Oxfam, 2009; Hagg, this volume).

Uncertainties

The uncertainties related to assessing climate impacts and consequences at the local, national and regional level stress the importance of implementing development activities that reduce the underlying vulnerability of a system in general, and enhancing the adaptive capacity as a strategy for integrating climate change adaptation in development. Fortunately, numerous climate change adaptation measures and development activities that foster climate change adaptation are associated with benefits that equal or exceed their costs to society – even without taking their climate change adaptation benefits into account. These opportunities are termed "no-regrets" or "net-negative costs" opportunities, which can and should be implemented even in the presence of uncertainty about future climatic conditions. Removing or decreasing maladaptation is one example of a no-regret opportunity. Still more development and climate change adaptation measures can be implemented at low cost without considering their climate change adaptation benefits, and are associated with net-negative costs if climate change adaptation benefits are included. These are frequently termed "low-regrets" opportunities (OECD, 2009).

Adaptation

GCA communities already have a long record of adaptation to climate variability. However, as previously mentioned, the impacts of climatic and other man-made stresses have been growing continuously at a rate that often exceeds human and ecosystem tolerance levels. Consequently, many traditional adaptive knowledge and livelihood strategies practiced in drylands for centuries no longer suffice or are inefficient. Efforts to reduce the vulnerability of drylands populations, therefore, must reinforce their risk management and coping capacities by augmenting existing adaptation mechanisms and supplementing them with new options that are tailored to the unique local contexts (Mizina et al., 1999). Several key livelihood challenges confront the dryland populations of GCA in the face of climate change. Potential adaptation options need to be evaluated and promoted as appropriate in the three main drylands agro-ecological zones: arid, semi-arid and dry sub-humid zones. Irrigated agriculture has been established in all three zones – some is large scale (artificial oases), while other agricultural practices are based on supplementary irrigation from wells or springs. Adaptation options are identified as: (1) natural resource management (NRM)-based; (2) market-based; and (3) institutional, and they could be implemented individually or simultaneously depending on the conditions on the ground. NRM-based options focus on the sustainable management of land, water, soil, plant and animal resources; market-based options are those that aim to improve market access and result in increased incomes, thereby reducing vulnerability. Table 14.1 sets out some adaptation options.

Drylands in GCA are under constant threat from multiple stresses and challenges, which occur as a result of a complex interplay of natural processes

Table 14.1 The grassroots actions that lead to adaptation

Adaptation options	Implementation mechanisms
Community ownership of cultural landscapes by traditional pastoral communities or newly emerged herders' groups	Reform law on pastureland access and users' rights, providing security of tenure (e.g. Kyrgyzstan)
Reduction of livestock inventories in excess of carrying capacity	To increase turn-off rates – reduce fodder requirements in winter and improve conception rates in breeding females
Increase in hay and fodder production from hay fields that are reserved (grazing bans)	Government support, seed, fertilizer, machinery hire
Perception of water resources (springs, streams, etc.) and riparian ecosystems as natural "green walls"	To decrease over-exploitation of water and riparian ecosystems
Diversification of income through tourism, handicrafts, value adding to natural products such as wool, cashmere, hides, etc.	Co-management
Enlargement of current administrative-territorial units to restore cultural landscapes – to give access to traditional cultural landscapes (e.g. to summer pastures)	Local initiatives and government decisions to regulate access to seasonal pastures/water (e.g. pasture user groups)

(such as weather variability, recurrent and unpredictable droughts and the concomitant floods caused by the typically short and heavy intervening rains) and human-induced processes (including land degradation and desertification caused by unsustainable and inadequate land use practices on a fragile resource base of low fertility). These processes are fueled by local forces, such as demographic pressure, poverty, high dependence on subsistence rain-fed agriculture, prevalence of infectious and chronic diseases, and periodic outbreaks of civil conflict. They are also often driven by external forces, including inadequate governance mechanisms, ineffective land tenure systems and poorly conceived national policies, protectionism measures by developed countries, import restrictions and fluctuations in the world economy (Millennium Assessment, 2005).

Degradation is the result of a number of interrelated factors that can result in a downward spiral, ending in land that is chemically or physically too degraded for productive use or environmental service, and often also results in degraded visual amenity. The existing driving forces produce pressures that result in the current state of land and water resources, with a negative impact on society and the environment. Human activity may directly or indirectly influence the degradation or rehabilitation process at every stage (Squires et al., 2009).

The driving forces fueling land degradation may result from increased human population (a particular problem in some GCA countries such as Afghanistan and Uzbekistan, where the population growth rate is over 3% per annum – a doubling of the population in about 20 years), urban expansion, war, tourism, agricultural production demands, transport, infrastructure, industrial or extractive activity and natural events such as climate change. These forces introduce pressures into the environment through noxious emissions to the air, water and land; nutrient mining; deforestation and forest fires; and rangeland overgrazing leading to denudation of vegetation cover. We analyze land degradation against the background of the immediate past history and identify the pressures that have brought about the current state. Land degradation includes biological or physical deterioration, eutrophication or nutrient depletion, salinization, acidification or chemical contamination (Squires, 2015, 2016). These changes in the state of land resources may have significant impacts on food supply, rural incomes and environmental services normally provided by land resources such as habitat for diversity and hydrological functions. Chapters 7 and 8 (Aralova et al., this volume; Feng et al., this volume) describe tools for monitoring and assessment. These methods have revolutionized the study of land degradation. It is now widely recognized that land degradation affects all facets of life. Many issues that confront those working in the domain of land resource management include technologies to reduce land degradation and also techniques to assess and monitor it.

A number of questions remain unanswered. These include:

- Is land degradation (LD) inevitable?
- Are there adequate early warning indicators of LD?
- Does the absence of property rights and use rights result in poor land stewardship (Squires, 2012) and how can this be resolved?
- Declining productivity resulting largely from human-induced degradation triggers social unrest – what is the societal responsibility of scientists and policymakers?
- Local actions have a global impact via such matters as dust and sandstorms, carbon and methane emissions – what are the areas for international collaboration?
- Who pays and who wins in the economics of LD?
- Degradation of resources such as water and land results in the loss of intergenerational equity and diminished value of bequests – how do we quantify and raise awareness of this?
- Is there a link between LD and the health of humans and their livestock?
- LD often destroys or reduces the beauty of natural landscapes – how might the aesthetic value be quantified?
- How can greater awareness of the perils of LD be created in the broader society and among the political leadership?

We also attempt to describe the responses people have made to the problems generated by land and water degradation. Sustainable traditional practices of resource management and utilization should be nurtured and replicated as a way to combat desertification and land degradation processes. Traditional knowledge, beliefs and values comprise a wide range of accumulated experience to manage natural resources in farming, grazing and landscape restoration. But they have relevance to institutional and organizational arrangements as well. Traditional knowledge and local technology are part of complex social systems and represent far more than a simple list of technical solutions. Traditional knowledge can be an elaborate and often multipurpose system that is part of an integrated approach between society, culture and economy. Indigenous knowledge systems of nomads are cumulative, representing generations of experience herding livestock, careful observations, and trial-and-error experiments. This knowledge enabled nomads on the Tibetan Plateau, for example, to develop sophisticated range-livestock management practices in an environment that posed considerable risks. Indigenous knowledge systems are also dynamic as new information is constantly being added. Doubts have been raised as to whether the long period of dominance of the top-down 'command-and-control' governance system has extinguished traditional knowledge in GCA, but sociologists and anthropologists working with rural communities in the more remote areas of GCA report that local ecological knowledge (LEK) is still strong and that ecological restoration efforts would be enhanced if more LEK were applied. Regrettably, a disproportionate amount of rangeland research is directed towards livestock production rather than understanding how livestock fit into the wider ecological system and how to optimize production in an environmentally and socially sustainable way. The social dimension of rangeland ecosystems should be an important aspect of research and development in pastoral areas, but, unfortunately, it is not.

Barriers to implementation of SLM in GCA

The GCA region has seen a change in both the concept and approaches to sustainable development. In the recent past the region has seen an impressive increase in awareness of the concept of sustainable development. Civil society, NGOs, policymakers and the private sector have an increasing understanding of the need to integrate economic development, social development and environmental protection and have developed models that can be built upon and adapted. There are now sustainable development strategies and regulations in almost all countries in the region. The institutional and regulatory framework for sustainable development is in place but the region has need of stronger political will for effective sectoral policy integration as well as for internal financial and technical resource allocation to implement them properly. Environmental problems and land degradation issues are extremely serious in the region. Regrettably, domestic resources that are allocated to sustainable development activities often do not reach the beneficiaries in the most efficient manner (Akramkhanov et al., this volume).

Progress towards sustainable development in this region has not taken a linear path. This has been especially so since 1992 for the five "stans" and Mongolia.

Sustainable development efforts have been slowed and in some cases interrupted by financial crises, environmental and natural disasters, armed conflicts and some of the negative consequences of globalization. The implementation of SLM in the GCA region faces a number of impediments:

- The governments of all seven countries face the enormous task of satisfying the basic human needs of growing populations. Poverty and marginalization of the rural poor, in particular in remote areas, due to a lack of capability to participate in national economic development processes compounded by an exclusion from reaping the benefits of the globalization process, have in some cases led to severe hardships, poverty, social conflicts, resource degradation and further development of desertification.
- In mountain areas, along with some arid and inland sea coastal areas in the region, e.g. in the Aral Sea Basin and around the Caspian Sea, there are particularly fragile ecosystems. Arable land, and water to irrigate it, is particularly scarce and the population often direly poor and lacking access to basic social services (Bekchanov et al., this volume; Hagg, this volume). Unfortunately, these are often areas where conflicts and unrest are common. Addressing the underlying economic, social and environmental causes of conflicts is the best possible way of preventing them. Thus, development and security are inalienable concepts.
- The natural resources to be maintained and protected are variable in space and time and. though important, are thinly spread. A catalogue of environmental problems will include: (1) degraded and polluted soils; (2) desertification; (3) silting and salinization; and (4) compromised quality of the region's inland rivers, seas and wetlands (Orlovsky & Orlovsky, this volume). Intense irrigation and deforestation of the watershed areas have affected freshwater supplies. The ability of the region to maintain sustainable agricultural production stands in question if the quality and carrying capacity of the natural resource base is not addressed. The region also contains some unique ecosystems that need to be protected and sustainably developed, such as alpine pastures, mountain forests and deserts, which also harbor populations with traditional lifestyles and unique cultures as well as large centers of global biodiversity.
- Natural disasters (earthquakes, landslides, floods and severe dust and sandstorms) are also negative factors. Calamities and natural disasters are often irreparable due to a lack of fundamental economic and social safety nets.

Some countries in the region have seen improvements in such indicators as economic growth rates, better literacy rates and levels of education, and slower population growth rates, and there is an increasing awareness of the environment and development interrelationships. However, improvements are often undone by the negative consequences of environmental degradation and natural resource depletion. There are three major causes for concern.

Desertification

Accelerated land and water degradation is happening in more and more places, often exacerbated by major infrastructure developments, mining activities (both small-scale artisanal mining and large-scale) and urban expansion onto productive arable land (Aralova et al., this volume). The nature and extent of these changes is being monitored by remote sensing (Aralova et al., this volume; Feng et al., this volume). Land surface dynamics is one of the key drivers of global environmental change, and earth observation is the most important tool for its monitoring. Remote sensing-based land dynamics monitoring relies on a wide range of change detection methods. One set of such techniques utilizes image classification to derive multi-temporal land use/land cover (LULC) maps. These maps are then compared to identify changes in mapped classes. The trend analysis/time series methodology itself also has some limitations for the detection of vegetation productivity decline (Aralova et al., this volume). Different methods applied to the same data may lead to contradictory results (Feng et al., this volume). The temporal aspect of a particular productivity decline is also significant. The decline occurring at the beginning or end of the time series was often not detected by trend analysis, while a decline beginning in the middle of the time series typically was usually detected. Furthermore, the rate of decline affects its detectability (Acreman et al., 2009; Wang et al., 2012). Normalized difference vegetation index (NDVI) trends have been used for numerous purposes, including the assessment of ecological responses to global warming, crop status (Tottrup & Rasmussen, 2004), and desertification (de Jong et al., 2011).

The monitoring of terrestrial vegetation dynamics thus underpins efforts to better understand the relational feedback between vegetation and the atmosphere (Bounoua et al., 2010; Angelini et al., 2011). The NDVI of natural vegetation in Central Asia during 1982–2013 exhibited an increasing trend at a rate of 0.004 per decade prior to 1998, after which the trends reversed, and the NDVI decreased at a rate of 0.003 per decade. Moreover, shrub cover and patch size exhibited a significant increase in 2000–2013 compared with the 1980s–1990s, including shrub encroachment on grasslands. Over the period to 2010, 8% of grassland has converted to shrubland. Precipitation increased in the 1990s, providing favorable conditions for vegetation growth, but precipitation slightly reduced at the end of the 2000s. Meanwhile, warming intensified 0.93°C since 1997 compared with the average value in 1960–1997, causing less moisture to be available for vegetation growth in Central Asia.

Vegetation activity is regarded as one of the most important indicators for evaluating interactions between climate and terrestrial ecosystems. Vegetation dynamics are highly sensitive to climate change, especially in arid and semi-arid regions (Sitch et al., 2003).

People living in the drylands often suffer chronic food insecurity, malnutrition and other health problems as household incomes fall, and the demands for cash, as result of the growth of market-based economies, are becoming almost universal

Mountains

For GCA, a comprehensive framework to assist the mountain poor and protect and respect the highlanders' way of life is being elaborated in cooperation with international agencies (Yang et al., this volume). A regional strategy for sustainable development of mountains, recently agreed by several Central Asian countries, could be further elaborated for consideration as a global action plan (Anon, 2012). The outcomes of the 2002 International Year of Mountains provided an opportunity to raise public awareness and support for sustainable mountain development (see also Kerven et al., 2011).

The problems constraining sustainable mountain development are formidable, among them: environmental challenges and poor natural resource management; limited infrastructure and local development opportunities; poor economic performance and governance inefficiencies; poverty; and the erosion of education. The demand and the will to tackle these problems are on the rise and are genuine.

Some demands have roots in the previous high standards and levels of education, security, energy and food sufficiency. In the 20 years ending 2014, the mountains of GCA have benefited from numerous sustainable development projects and initiatives. The sponsors and participants have included governments, international organizations, NGOs – both global and local – and educational and scientific institutions. The literature on sustainable mountain development is rich (Breu & Hurni, 2003, 2005; Anon, 2012), with advice on practices that have proven effective over time and across space. Extensive research and field experience have led to a broad agreement on the important considerations for successful sustainable development. Professionals in the field are likely to advocate for:

- a decentralized approach that provides local participants with a share in decision making;
- a capacity-building function that assists participants to acquire the tools and knowledge necessary to succeed;
- the broad participation of civil society, NGOs and decision makers at all levels;
- a strong and effective process for incorporating the views of stakeholders;
- the inclusion of all relevant sectors;
- a process that honors traditional knowledge;
- a multidisciplinary and geographically focused approach; and
- a balance among the three components (environmental, economic and social) of sustainable development.

Conforming to this guidance may not guarantee a project's success, but failure to conform may increase the likelihood of failure. Where there is no tradition of local participation in civic affairs, adherence to the best sustainable development practices may be more difficult, but the experience of the Central Asian mountain projects suggests that the effort to overcome the barriers to broad participation is rewarded by the success of the projects. The capacity-building component of sustainable development can include a wide range of activities – from workshops on the processes to be followed, to training on the specific tasks necessary to implement a project, to institution building.

The analysis of mountain development in Central Asia beyond the case studies (Anon, 2012) also shows the following:

- Political stability and conflict avoidance are the key factors for sustainable mountain development.
- Personal safety, food and energy security, decent jobs, health and education, and poverty alleviation are the key priorities for people in the Central Asian mountains. If these basic necessities are not addressed and balanced, sustainable mountain development and environmental protection cannot be ensured.
- Good governance, corruption prevention, transparency and participation in decision making in the main economic and social sectors are paramount for the success of development projects in the mountains.
- Communication of easily understandable, reliable information is crucial for public understanding, support and motivation to act responsibly.
- The absence of well-defined property and management rights and responsibilities puts constraints on, and adds uncertainties to, sustainable mountain development.
- Heavy reliance on subsidies (as in the Soviet period), natural resource extraction and use without benefit sharing (as in the energy and mining sectors) and continuing reliance on substantial external donor inputs may lead to unsustainable mountain development patterns that could hit hard in times of abrupt change.
- Affordable microfinance, successful demonstration projects and new knowledge often lead to self-reliance.
- The valuation of mountain ecosystem services and the provision for ecosystem carrying capacity, including the regulation and mitigation of man-made pressures, are essential to mountain development and benefit sharing.
- Legislation and programs on mountain development are essential, and need to be supported by efficient institutions and resources.
- The lack of willingness to cooperate and the tensions between upstream and downstream countries (mainly on the region's delicate and politicized water issues) impede regional cooperation (see Hagg, this volume).

Sustainable agriculture and food security

As Bobojonov et al. (2013) indicate, the matter of food security looms large in some GCA countries. The relationships between population size and the basic resources for food production, soil and water require careful attention to food supply and contingency planning to ensure the adequate storage of food grains. With the projected increase in the urbanized population and change in food habits, the future food demand for cereals and meat products will increase drastically. Therefore, among the principal concerns of the 21st century are: (1) food security due to a rapid increase in the urban population; (2) soil degradation through land misuse and soil mismanagement; and (3) anthropogenic increases in atmospheric greenhouse gases. All of these issues are linked to the issues surrounding food security. In the five "stans", agriculture and water management have ranked as the two most important economic activities in this arid environment. These activities gained even more prominence during the Soviet era as planners expanded irrigation into previously marginal land, bolstering their vision that the best land be allocated exclusively for cotton production. In the wake of the fall of the Soviet Union, laws were enacted to expand and clarify land use categories to meet the dual targets of expanding food production while maintaining as much land as possible in cotton production – their economic mainstay (Rowe, 2010).

The agricultural system as a whole will have difficulty supplying adequate quantities of food to maintain constant real prices. And the challenges extend further: to national governments, to provide the supporting policy and infrastructure environment; and to the global trading regime, to ensure that changes in comparative advantage translate into unimpeded trade flows to balance world supply and demand. Farmers everywhere will need to adapt to climate change. For a few, the changes might ultimately be beneficial, but for many farmers there are major challenges to productivity and more difficulties in managing risk. But neither food security nor climate change should be viewed in isolation. Agricultural productivity is strongly determined by both temperature and precipitation. Nelson et al. (2010) identify the following key points:

- **Broad-based economic development is central to improvements in human well-being, including sustainable food security and resilience to climate change**. Rural families with more resources at their disposal are better able to cope with whatever uncertainties mother nature or human activities cause. Lifting household incomes, making rural credit available, improving land tenure arrangements and rehabilitating degraded land and water resources will all contribute to better outcomes. Unsustainable land management practices, including overgrazing, over-cultivation, illegal and excessive fuel wood collection and poor irrigation technologies, among others, have become prevalent, often due to institutional or tenurial barriers. Tenure systems define and regulate how people, communities and others gain access to natural resources, whether through formal law or informal arrangements. The rules of tenure determine who can use which resources, for how long, and under what conditions. They may be based on written policies and laws, as well as on unwritten customs and practices. The responsible governance of tenure ensures that access to land, fisheries and forests are equitably shared. It protects economically and socially marginalized people from alienation from the resources they need to live. Weak governance of tenure is also associated with the overexploitation of natural resources and consequential environmental degradation.
- **Climate change offsets some of the benefits of income growth**. International trade flows provide a balancing mechanism for world agricultural markets. Trade flows can partially offset local climate change productivity effects,

allowing regions of the world with positive (or less negative) effects to supply those with more negative effects. Countries with a comparative advantage in a crop can produce it relatively more efficiently and exchange it for other goods with other countries, whose comparative advantage lies elsewhere. But comparative advantage is not fixed. Climate change alters comparative advantage, as do changing consumer preferences. Economic development itself changes the mix of goods demanded by consumers.

Options and opportunities for climate change adaptation and mitigation

Analysis by Nelson et al. (2010) for the International Food Policy Research Institute (IFPRI) suggests that, up to 2050, the challenges from climate change are "manageable" in the sense that well-designed investments in land and water productivity enhancements might conceivably substantially offset the negative effects from climate change. Introducing the effects of climate change scenarios into overall food and agriculture scenarios presents a particular challenge, to take into account the range of plausible pathways for greenhouse gas emissions. Moreover, the general circulation models translate those emission scenarios into varying temperature and precipitation outcomes.

Geopolitical considerations

A thread that runs through this book from Chapter 1 (Squires & Lu, this volume) onwards is the significance of GCA from a strategic and a geopolitical point of view. The ongoing and planned investments in infrastructure (pipelines, highways, railways and ports) but also in mines will exacerbate some land and water degradation through disturbance they create, but will have a more permanent impact in terms of population increase, urban expansion and (hopefully) higher GDP for GCA states. As populations rise, urbanization expands and disposable incomes rise, there will be more pressure to produce food, especially "organic" red meat to meet the needs of a burgeoning population of more discriminating consumers.

Three issues – oil, water and food security – are driving China toward a short-term assertion of its national interests (Squires & Lu, this volume). Central Asia is seen as an arena that is critical to larger geopolitical competitions and realignments. The Indian strategists who are looking beyond India's borders see a number of threats and opportunities in Central Asia. First, Indians are thinking about how to contain China in Central Asia. They see the Chinese threat encircling India, as China strengthens its influence in Pakistan, Myanmar, Bangladesh and Central Asia, and some Indians believe that China's growing economic and political penetration in Central Asia, especially in Kazakhstan, must be countered before it leads to strategic realignments, possibly alliances, that might threaten India's interests. In this context, Uzbekistan was referred to as the "key to the region", and India has taken steps to cultivate close ties to Uzbekistan.

Second, India and the Central Asian states share an interest in controlling the highly unstable and unpredictable situation in and around Afghanistan and

Pakistan, from which instability could cascade in all directions. India is looking to Iran and to Central Asia – Uzbekistan in particular – to contain this chaos in the center. Third, Central Asia is considered to be the northern border of an emerging east–west economic corridor that connects Israel, Turkey, Iran, India and East Asia.

Finally, Central Asia is seen as a conduit to the United States. It is a region where Indian and US interests have the potential to intersect in areas such as countering Chinese influence, containing terrorism and the drug trade, and promoting stability. But US and Indian interests could also come into conflict, particularly in the area of Central Asian energy development and transport for South Asian markets, in which, for India, Iran almost certainly has to play a central role.

A further aspect of South Asian interest in Central Asia is economic. Both India and Pakistan seek to develop stronger trade relations with the Central Asian states. India's pharmaceutical companies and hotels have built a strong presence in the region. Pakistani traders are increasingly active across the region. Pakistan also sees itself as a potential trade route for Central Asian states by offering them access to the Arabian Sea. China is already committed to developing port facilities in Pakistan to allow the transport of goods overland from Kashgar in Xinjiang to Pakistan (Squires, this volume).

Energy will also be a major driver of relationships between Central Asia and South Asia. India and Pakistan – with their burgeoning demand for gas – are natural markets for Central Asian gas, and the governments in both states have high expectations of obtaining Central Asian energy, from Turkmenistan in particular, but also from Kazakhstan and Uzbekistan.

India in particular will be importing energy from multiple directions. Central Asia is considered a key component of India's energy strategy to mitigate its dependence on the Middle East over the long term. On the prospects for pipelines there are several possible routes. Iran offers more promise as a transit state for Turkmen gas. Some Indians suggest that the Afghan pipeline be shifted a few hundred kilometers west, so it can link into Iran's existing north–south infrastructure. Indians have had a late start, but they are seeking to develop a presence in Central Asia's energy sector: for example, several years ago they teamed with the Chinese Natural Gas Exploration and Development Corporation to develop a small field in Kazakhstan. The Indians are also exploring the possibility of importing electricity from Kyrgyzstan, which has an excess of hydropower. This scheme would probably involve transmitting the power across Chinese territory. For all of these reasons – South Asian competition, larger geopolitical considerations, economic interests – energy-poor South Asia, particularly India, is likely to be an important player in Central Asia in the next decades. India's relationships with the Central Asian states will be multifaceted, based on shared strategic interests, and part of a broader Indian strategy in Eurasia. Pakistan will also be involved, although at the moment it is hard to see the outlines of a coherent Pakistani strategy. Currently, Pakistan is viewed more as a threat to GCA stability than as a constructive player, although this could change as China's partnership with Pakistan in the "Belt and Road" initiative CPEC project gains traction (Squires, this volume).

GCA is important, however, not only because of India's interest in the region's energy, but also because it is linked to how India thinks about other major actors such as Iran, Pakistan and China. For example, the Indians worry about growing Chinese influence in Kazakhstan. The GCA region is at the center or on the near periphery of the national security interests of all the states in the region. In the view of nearby states, GCA is not just a small, isolated group of former Soviet colonies that is of interest only because of historical quaintness. Most countries in the region have multiple interests that converge in GCA. India, which seems far away, is a case in point. It views Soviet Central Asia as its "extended strategic neighborhood" in the competition with Pakistan, and it has a direct interest in controlling the political chaos in Pakistan and Afghanistan; its interest in Central Asian energy is paramount; and it sees Central Asia as a key piece of its effort to contain Chinese expansion and prevent "encirclement" by China. In this sense, Central Asia is a critical strategic juncture point in its national security planning for China, the Middle East, and South Asia. Similarly, for nearby states such as Turkey, Iran, Pakistan, China and, of course, Russia, as well as for some more distant states such as Israel, Central Asia and the South Caucasus are no less important in their strategic calculations.

The main factors contributing to economic growth in Central Asian countries have been foreign trade, foreign investment, and foreign loans and credits. The economies of the countries of the region are critically dependent on foreign trade. The foreign trade turnover in these countries represents 60% to 70% of the GDP. In all countries, imports exceed exports. So, the dynamics of most GCA economies are totally determined by the conditions prevailing on world raw material markets. At any moment, a sudden decline in world prices for oil and gas and for cotton could threaten prosperity. It is increasingly apparent that the model of development chosen by Central Asian governments is in need of serious correction. To judge from the available evidence, the opportunities for extensive expansion of exports are diminishing. The exception is Kazakhstan's oil: despite some delays, the Caspian Pipeline Consortium enables the shipping of oil through the Russian port of Novorossiisk. Other pipelines to China are also important. Kazakhstan can export crude oil and gas to a profitable market. Achievement of macroeconomic stabilization, together with political stability, make it possible to begin large-scale structural changes and to give greater attention to the domestic market. The agrarian sector should become the main priority for development, at least in Uzbekistan and Kyrgyzstan. In both countries, agriculture provides employment for about 45% of the total labor force, produces between one-third and one-half of the GDP, and accounts for a significant part of exports that earn hard currency. In both countries, economic policy actually discriminates against the agrarian sector, which has been transformed into a source of reserves for import substitution in industry (Uzbekistan), or which is used in the interests of commercial intermediaries (Kyrgyzstan). The proportion of budgetary, credit and investment resources for the agrarian sector does not correspond to its role and significance in the economies of the Central Asian states. By functioning as a "donor" for the other sectors of the economy, the potential is rapidly increasing for a crisis to beset the agriculture sector itself. This seriously undermines efforts to implement more sustainable land management (Akramkhanov et al., this volume). To a significant degree, this discrimination against the agrarian sector explains the depressing social and economic situation in these countries. The eradication of poverty and indigence is not possible without a change in economic policy addressing the needs of the agrarian sector.

Top priority in development policy must also be given to the expansion of light industry and the processing of agricultural commodities (value adding). It is precisely these branches of industry that have been subjected to the greatest destruction during the processes of transformation since the 1990s and that are now situated on the periphery of attention in official economic policy. Without a reorientation of investment resources to these branches, most GCA states are doomed to remain exporters of agricultural commodities and products with a low level of processing – as is seen at the current time.

As mentioned earlier, China, Russia, India, Pakistan, Turkey, Iran and even some Arab Gulf states are seeking to develop partnerships with GCA countries. Linn (2013) has observed that the transition from maritime routes to a land-based route is an important factor both for regional integration and for economic development (see also Annex 1 in Wang et al., this volume).

As Niklas Swanström (this volume) says:

China's rise, and China's economic focus on GCA, is essential for any engagement in GCA, since Chinese resources and political commitment will bring more potential for opening up and developing the region. Russia is declining very quickly and there is no apparent strategy from the US or the EU to increase their engagement in the GCA region after their departure from Afghanistan. Moreover, the EU investments in infrastructure are uncoordinated and the USA seems to be reluctant to engage in the region in any coherent way. The problem is that [up until recently] China views this solely as an economic project and has failed – or refuses – to see the greater benefits for the region. This is largely due to China's own inability to accept that it will have to engage in the region in a broader and more assertive way than it – and the rest of the world – is willing to do at this moment in time.

As Chapter 12 (Squires, this volume) shows, there is a view in China now that much can be gained in cultural, scientific and technological cooperation with GCA countries and that better relations there will help with China's cross-border problems, drug trade, human trafficking, spread of radical Islam, tension over trans-boundary rivers, disputed territorial claims, etc. From another angle, there is much interest in, and opportunity for, the export of Chinese expertise and personnel to tackle problems of land and water degradation in neighboring countries (Yang et al., this volume). The five "stans" and Mongolia are at different stages of economic development, and possess varying levels and types of technical expertise, experience and assets. They have common as well as unique problems, even though some of them have already experienced and devised resolutions for some of those problems. Policy dialogue, pilot demonstration, capacity building and information sharing would be efficient and economical means of solving common issues and enhancing the sharing of experiences and transfer of technology. El-Beltagy and Madkour (2012) argue that there is a need for a new paradigm for agricultural research and technology transfer. A key aspect of their submission is that such a new paradigm will require much more investment by international agencies and national governments alike, for supporting research and sustainable development efforts that include the full participation of who they refer to as the "target communities". Safriel and Adeel (2008) emphasize the advantages of shifting from a focus on the lands themselves and on their potentiality (or otherwise) as productive resources, to the sustainability of alternative livelihoods and well-being of the ever-growing populations in the arid zones (Adeel & Safriel, 2008). They also highlight the need for inclusion of all key stakeholders in the process of development as well as the integration of the bio-physical with the socio-cultural.

Summary and conclusions

In this book we have attempted a synthesis that includes a search for new policies that should be based on better understanding of existing management systems and land capabilities, including a full assessment of existing cultural and ecological constraints to development and of all the costs and benefits of the proposed changes. Policy opportunities include: strengthening social technology at the local level; formally documenting and registering traditional rights; adopting administrative arrangements which are ecologically sensitive to the episodic, variable nature of land use systems; devolving responsibilities to local communities; and emphasizing community needs for regional rather than sectoral development. Land systems include alpine regions that are cold and arid; deserts that are hot and dry; lakes and rivers; and vast areas of agricultural land on which some livelihoods depend.

The objective of the technical assistance being provided by bilateral and multilateral donors (Yang et al., this volume) is to contribute to policies that promote environmentally sustainable and profitable pastoral livestock industries in GCA countries and improve the capacity of Central Asian research institutes to conduct similar work in the future. To achieve this goal, remote-sensed data have been used to identify desertification patterns at selected study sites (Aralova et al., this volume; Feng et al., this volume). Satellite-based environmental assessments should be combined with interdisciplinary field studies that show why land use systems are changing and how these changes affect land degradation processes. The environmental impact of alternative policies should be carefully assessed. Results will be disseminated to national policymakers, international donor agencies, and representatives of local producer groups, with a view to improving productivity and lifting household incomes.

Drylands livelihoods represent a complex form of natural resource management (NRM), involving a continuous ecological balance between pastures, livestock, crops and people. The people living in the drylands of GCA are heavily dependent upon ecosystem services, directly or indirectly, for their livelihoods. But those services – from nutrient cycling, flood regulation and biodiversity to water, food and fiber – are under threat from a variety of sources such as urban expansion, mining and unsustainable land uses. As a result, these fragile soils are becoming increasingly degraded and unproductive. Climate change is now aggravating these challenges. However, combating climate change and adapting communities to its impacts represents an opportunity for new and more sustainable investments and management choices that can also contribute to improved livelihoods and fighting poverty among dryland communities. An ecosystem approach which includes restoration and renovation is one important direction that needs to be urgently undertaken.

Unsustainable practices are contributing to significant land degradation, and it is predicted that climate change will further compound the already tenuous situation, especially in irrigated areas. Without significant efforts to address the impact of climate change and land degradation, the livelihoods of the GCA drylands populations will be in jeopardy.

Note

1 Adaptive strategies are longer-term (beyond a single season) strategies that allow people to respond to a new set of evolving conditions (biophysical, social and economic) that they have not previously experienced. The extent to which communities are able to respond successfully to a new set of circumstances will depend on their adaptive capacity, i.e. their ability to take advantage of opportunities, or to cope with the consequences of previously unencountered circumstances.

References and further reading

Acreman, M.C., Aldrick, J., Binnie, C., Black, A., Cowx, I., Dawson, H., Dunbar, M., Extence, C., Hannaford, J. and Harby, A. 2009. Environmental flows from dams: the water framework directive. *Proceedings of the ICE – Engineering Sustainability* 162 (1): 13–22. doi:10.1680/ensu.2009.162.1

Adeel, Z. and Safriel, U. 2008. Achieving Sustainability by Introducing Alternative Livelihoods. *Sustainability Science* 3 (1): 125–33.

Akiner, S. 2000. Central Asia: A Survey of the Region and the Five Republics. UNHCR Centre for Documentation and Research, WRITENET Paper No. 22/1999. United Nations High Commissioner for Refugees, Geneva, 50 pp.

Akramkhanov, A., Djanibekov, U., Nishanov, N., Djanibekov, N. and Kassam, S. 2018. Barriers to sustainable land management in Greater Central Asia: with special reference to the five former Soviet republics. In this volume, pp. 113–30.

André, K., Kraudzun, T. and Samimi, C. 2012. Land Stewardship in Practice: An Example from the Eastern Pamirs of Tajikistan. In: V.R. Squires (ed.) *Rangeland Stewardship in Central Asia: Balancing Improved Livelihoods, Biodiversity Conservation and Land Protection.* Springer, Dordrecht, pp. 71–90.

Angelini, I.M., Garstang, M., Davis, R., Hayden, B., Fitzjarrald, D.R., Legates, D.R., Greco, S., Macko, S. and Connors, V. 2010. On the Coupling Between Vegetation and the Atmosphere. *Theor. Appl. Climatol.* 78: 47–59.

Anon. 2012. *Sustainable Mountain Development. From Rio 1992 to 2012 and beyond. Central Asia Mountains.* University of Central Asia, Zoï Environment Network, Mountain Partnership, GRID-Arendal.

Aralova, D., Kariyeva, J., Menzel, L., Khujanazarov, T., Toderich, K., Halik, U. and Gofurov, D. 2018. Assessment of land degradation processes and identification of long-term trends in vegetation dynamics in the drylands of Greater Central Asia. In this volume, pp. 131–54.

Bekchanov, M., Djanibekov, N. and Lamers, J.P.A. 2018. Water in Central Asia: a cross-cutting management issue. In this volume, pp. 211–36.

Bekchanov, M., Lamers, J.P.A. and Martius, C. 2010. Pros and Cons of Adopting Water-Wise Approaches in the Lower Reaches of the Amu Darya: A Socio-Economic View. *Water* 2: 200–16.

Bobojonov, I., Lamers, J.P.A., Bekchanov, M., Djanibekov, N., Franz-Vasdeki, J., Ruzimov, J. and Martius, C. 2013. Options and constraints for crop diversification: A case study in sustainable agriculture in Uzbekistan. *Agroecology and Sustainable Food Systems* 37 (7): 788–811.

Bounoua, L., Hall, F.G., Sellers, P.J.A., Kumar, A.G.J., Collatz, G.J., Tucker, C.J. and Imhoff, M.L. 2010. Quantifying the negative feedback of vegetation to greenhouse warming: A modeling approach. *Geophysical Research Letters* 37: L23701.

Breu, T. and Hurni, H. 2003. *The Tajik Pamirs: Challenges of Sustainable Development in an Isolated Mountain Region.* Geographica Bernensia for Centre for Development and Environment, Bern, Switzerland.

Breu, T. and Hurni, H. (eds). 2005. *Baseline Survey on Sustainable Land Management in the Pamir-Alai Mountains: Synthesis Report.* Centre for Development and Environment, University of Bern, Switzerland.

Conrad, C., Dech, S.W., Hafeez, M., Lamers, J.P.A., Martius, C. and Strunz, G. 2007. Mapping and assessing water use in a Central Asian irrigation system by utilizing MODIS remote sensing products. *Irrig. Drain. Syst.* 21: 197–218.

Cowan, P.J. 2007. Geographic usage of the terms Middle Asia and Central Asia. *Journal of Arid Environments* 62 (2): 359–63.

de Beurs, K.M. and Henebry, G.M. 2004. Land surface phenology, climatic variation, and institutional change: analyzing agricultural land cover change in Kazakhstan. *Remote Sensing of Environment* 89: 497–509.

de Jong, R., de Bruin, S., de Wit, A., Schaepman, M.E. and Dent, D.L. 2011. Analysis of monotonic greening and browning trends from global NDVI time-series. *Remote Sensing of Environment* 115 (2): 692–702.

Dukhovny, V.A. and de Schutter, J.L.G. 2011. *Water in Central Asia: Past, Present, Future.* Taylor & Francis, London.

El-Beltagy, A. and Madkour, M. 2012. Impact of Climate Change on Arid Lands Agriculture. *Agriculture and Food Security* 1: 3–12.

Feng, Y., Yan, F. and Cao, X. 2018. Land degradation indicators: development and implementation by remote sensing techniques. In this volume, pp. 155–78.

Freedman, E. and Neuzil, M. (eds). 2015. *Environmental Crises in Central Asia. From Steppes to Seas, from Deserts to Glaciers.* Taylor & Francis, London.

Glantz, M.H. 2005. Water, climate, and development issues in the Amu Darya basin. *Mitigation and Adaptation Strategies for Global Change* 10 (1): 23–50.

Glenn, E.P., Olson, M. and Frye, R. 1993. Potential for carbon sequestration in the drylands. *Water, Air and Soil Pollution* 70: 341–55.

Hagg, W. 2018. Water from the mountains of Greater Central Asia: a resource under threat. In this volume, pp. 237–48.

IPCC. 2007. *Fourth Assessment Report (AR4) Contribution of Working Groups I, II and III to the Fourth Assessment Report of the Intergovernmental Panel on Climate Change.* Cambridge University Press, Cambridge.

Jansky, L. and Pachova, N.I. 2006. Towards Sustainable Land Management in Mountain Areas in Central Asia. *Global Environmental Research* 10 (1): 99–115.

Kerven, C. (ed.) 2003. *Prospects for pastoralism in Kazakhstan and Turkmenistan: From state farm to private flocks.* Routledge Curzon, London.

Kerven, C., Steimann, B., Ashley, L., Dear, C. and ur Rahim, I. 2011. Pastoralism and Farming in Central Asia's Mountains: A Research Review. Bishkek.

Kienzler, K., Lamers, J.P.A., McDonald, A., Mirzabaev, A., Ibragimov, N., Egamberdiev, O., Ruzibaev, E. and Akramkhanov, A. 2012. Conservation agriculture in Central Asia – What do we know and where do we go from here? *Field Crops Research* 132: 95–105.

Krutov, A., Rahimov, S. and Kamolidinov, A. 2014. Republic of Tajikistan: Its Role in the Management of Water Resources in the Aral Sea Basin. In: V.R. Squires, H.M. Milner and K.A. Daniell (eds) *River Basin Management in the Twenty-first Century: Understanding People and Place.* CRC Press, Boca Raton, pp. 325–44.

Lal, R., Suleimenov, M., Stewart, B.A., Hansen, D.O. and Doraiswami, P. (eds). 2007. *Climate Change and Terrestrial Carbon Sequestration in Central Asia.* Taylor & Francis, London.

Li, B., Gasser, T., Ciais, P., Piao, S., Tao, S., Balkanski, Y., Hauglustaine, D., Boiisier, J.-P., Chen, C., Huang, M., Li, L.Z., Li, Y., Liu, H., Liu, J., Peng, S., Shen, Z., Sun, Z., Wang, R., Wang, T., Yin, G., Yin, Y., Zeng, H. and Feng, Z. 2016. The contribution of China's emissions to global climate forcing. *Nature* 531: 357–61. doi:10.1038/nature17165

Linn, J. 2013. *Central Asian Regional Integration and Cooperation: Reality or Mirage? EDB Eurasian Integration Yearbook 2012.* Eurasian Development Bank, Almaty, pp. 97–8.

Lioubimtseva, E. and Henebry, G.M. 2009. Climate and environmental change in arid Central Asia: Impacts, vulnerability, and adaptations. *J Arid Environ.* 73: 963–77.

Millennium Assessment. 2005. *Ecosystems and Human Well-being: Synthesis.* Island Press, Washington DC, 160 pp.

Mizina, S.V., Smith, J.B., Gossen, E., Spiecker, K.F. and Witkowski, S.L. 1999. An evaluation of adaptation options for climate change impacts on agriculture in Kazakhstan. *Mitigation and Adaptation Strategies for Global Change* 4: 25–41.

Nelson, G.C., Rosegrant, M.W., Palazzo, A., Gray, I., Ingersoll, C., Robertson, R., Tokgoz, S., Zhu, T., Sulser, T.B., Ringler, C., Siwa, M. and You, L. 2010. *Food Security, Farming and Climate Change to 2050: scenarios, results, options.* IFPRI Research Monograph, Washington DC, 132 pp.

Nkonya, E., Mirzabaev, A. and von Braun, J. 2015. *Economics of land degradation and improvement – A global assessment for sustainable development.* Springer International Publishing.

OECD. 2009. *Economic Policy Reforms: Going for Growth.* OECD, Geneva.

Oldeman, L.R. 1994. The Global Extent of Soil Degradation. In: D.J. Greenland and T. Szabolcs (eds) *Soil Resilience and Sustainable Land Use.* Commonwealth Agricultural Bureau International, Wallingford.

Orlovsky, L. and Orlovsky, N. 2018. Biogeography and natural resources of Greater Central Asia: an overview. In this volume, pp. 23–47.

Oxfam. 2009. *Reaching tipping point: climate change and poverty in Tajikistan*. Oxfam International, Dushanbe, 22 pp.

Pender, J., Mirzabaev, A. and Kato, E. 2009. *Economic analysis of sustainable land management options in Central Asia: Final report submitted to ADB*. Washington DC.

Pilifosova, O.V., Eserkepova, I.B. and Dolgih, S.A. 1997. Regional climate change scenarios under global warming in Kazakhstan. *Climatic Change* 36: 23–40.

Rowe, W.C. 2010. Agrarian adaptations in Tajikistan: land reform, water and law. *Central Asian Survey* 29 (2): 189–204.

Rumer, B. and Zhukov, S. (eds). 1998. *Central Asia: The Challenges of Independence*. M.E. Sharpe, New York.

Safriel, U. and Adeel, Z. 2008. Development Paths of Drylands: thresholds and sustainability: *Sustainability Science* 3: 117–23.

Saiko, T.A. and Zonn, I.S. 2000. Irrigation expansion and dynamics of desertification in Circum-Aral region of Central Asia. *Applied Geography* 20: 349–67.

SIC-ICWC. 2011. CAREWIB (Central Asian Regional Water Information Base). Available online at: http://www.cawater-info.net (accessed on 27.01.2012).

Sievers, E.W. 2003. *The Post-Soviet Decline of Central Asia: Sustainable Development and Comprehensive Capital*. Taylor & Francis, London and New York, 264 pp.

Sitch, S., Smith, B., Prentice, I.C., Arneth, A., Bondeau, A., Cramer, W., Kaplan, J.O., Levis, S.W., Lucht, W., Sykes, M.T., Thonicke, K.S. and Venevsky, S. 2003. Evaluation of ecosystem dynamics, plant geography and terrestrial carbon cycling in the LPJ dynamic global vegetation model. *Global Change Biology* 9 (2): 161–85.

Squires, V.R. 1998. Dryland Soils: their potential as a sink for carbon and as an agent in mitigating climate change. *Advances in GeoEcology* 31: 209–15.

Squires, V. 2012. *Rangeland Stewardship in Central Asia: Balancing Livelihoods, Biodiversity Conservation, and Land Protection*. Springer, Dordrecht, 458 pp.

Squires, V.R. 2015. *Rangeland Ecology and Management and Conservation Benefits*. Nova Science Publishers, New York.

Squires, V.R. 2016. *Ecological Restoration: Global Challenges, Social Aspects and Environmental Benefits*. Nova Science Publishers, New York.

Squires, V. 2018. Greater Central Asia as the new frontier in the twenty-first century. In this volume, pp. 251–72.

Squires, V.R. and Lu, Q. 2018. Greater Central Asia: its peoples and their history and geography. In this volume, pp. 283–305.

Squires, V., Lu, X., Lu, Q., Wang, T. and Yang, Y. 2009. *Rangeland Degradation and Recovery in China's Pastoral Lands*. CABI, Wallingford, 264 pp.

Suleimenov, M., Iniguez, L. and Mursayeva, M. 2006. Policy reforms and livestock development in Central Asia. In: S.C. Babu and S. Djalalov (eds) *Policy Reforms and Agriculture Development in Central Asia*. Springer, New York, pp. 277–310.

Swanström, N. 2018. Greater Central Asia: China, Russia or multilateralism? In this volume, pp. 273–82.

Tottrup, C. and Rasmussen, M. S. 2004. Mapping long-term changes in savannah crop productivity in Senegal through trend analysis of time series of remote sensing data. *Agriculture, Ecosystems & Environments* 103: 545–60.

UNCCD. 2003. Discussion Paper presented at the Sub-regional Partnership Building Forum for the Central Asian Republics: Confronting Land Degradation and Poverty through Enhanced UNCCD Implementation, Tashkent, Uzbekistan, 30 June–4 July.

UNEP. 2006. *UNEP Finance Initiative: Innovative financing for sustainability*. UNEP, Geneva.

UNU. 2006. *Sustainable Land Management in the High Pamir and Pamir-Alai Mountains.* United Nations University, Tokyo. http://palm.unu.edu

Varis, O. 2014. Curb vast water use in central Asia. *Nature* 514 (7520): 27–9.

Wang, D., Morton, D., Masek, J., Wu, A., Nagol, J., Xiong, X., Levy, R., Vermote, E. and Wolfe, R. 2012. Impact of sensor degradation on the MODIS NDVI time series. *Remote Sensing of Environment* 1 (16): 55–61.

Wang, F., Squires, V.R. and Lu, Q. 2018. The future we want: putting aspirations for a land degradation neutral world into practice in the GCA region. In this volume, pp. 99–112.

Yang, Y., Low, P.S., Yang, L. and Jia, X. 2018. Mitigation of desertification and land degradation impacts and multilateral cooperation in Greater Central Asia. In this volume, pp. 179–207.

Index

T - #0115 - 111024 - C336 - 234/156/16 - PB - 9780367872663 - Gloss Lamination